new lives
for old

new lives
for old

cultural transformation—

Manus, 1928–1953

m a r g a r e t   m e a d

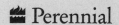 Perennial

An Imprint of HarperCollins*Publishers*

HarperCollins books may be purchased for educational, business, or sales promotional use. For information please write: Special Markets Department, HarperCollins Publishers Inc., 10 East 53rd Street, New York, NY 10022.

First Morrow paperback edition published 1975.
First Perennial edition published 2001.

Designed by Nancy Singer

Library of Congress Cataloging-in-Publication Data

Mead, Margaret.
    New lives for old : cultural transformation—Manus, 1928–1953 / by Margaret Mead
        p. cm.
    Includes bibliographical references and index.
    ISBN 0-06-095806-5
    1. Manus (Papua New Guinea people). 2. Acculturation—Papua New Guinea—Admiralty Islands—Case studies. I. Title.

DU520 .M4 2001
305.89'91209581—dc21

2001032184

01 02 03 04 05 WB/RRD 10 9 8 7 6 5 4 3 2 1

Dedicated in 1977 to
His Honour Judge Phillips
Chief Justice of the Territory
of Papua and New Guinea

contents

plates

words for a new century

by Mary Catherine Bateson

When my mother, Margaret Mead, was ready to seek a publisher for her first book, *Coming of Age in Samoa*, she found her way to William Morrow, the head of a new publishing company, and he gave her a key suggestion for the rest of her career, that she add "more about what all this means to Americans." This set a course she followed throughout her life, establishing not only the appeal of anthropology as a depiction of the exotic but as a source of self-knowledge for Western civilization. The last chapter of *Coming of Age* laid out a theme for the years ahead: "Education for Choice."

Even before World War II, still using the terminology of her time that now seems so outmoded and speaking of "primitives" or even of "savages," she believed that Americans should learn not only about the peoples of the Pacific, but from them. And after almost every field trip she went back to William Morrow, now HarperCollins, where many of her books have remained in print ever since, offering new meanings to new generations of Americans. A century after her birth, they are offered once again, now for a new millennium, and today they still have much to offer on how individuals mature in their social settings and how human communities can adapt to change.

Several of Mead's field trips focused on childhood. Writers have been telling parents how to raise their children for centuries; however, the systematic observation of child development was then just beginning, and she was among the first to study it cross-culturally. She was one of those feminists who have combined an assertion of the need to make women full and equal participants in society with a continuing fascination with children and a concern for meeting their needs. A culture that repudiated children "could not be a good culture," she believed. [*Blackberry Winter: My Earlier Years*, New York: William Morrow and Company, 1972, p. 206.]

After studying adolescents in Samoa, she studied earlier childhood in Manus (*Growing Up in New Guinea*) and the care of infants and toddlers in Bali; everywhere she went, she included women and children, who had been largely invisible to earlier researchers. Her work continues to affect the way parents, teachers, and policy makers look at children. I, for one, am grateful that what she learned from the sophisticated and sensitive patterns of childcare she observed in other cultures resonated in my own childhood. Similarly, I have been liberated by the way her interest in women as mothers expanded into her work on gender (*Sex and Temperament* and *Male and Female*).

In addition to this growing understanding of the choices in gender roles and childrearing, the other theme that emerged from her fieldwork was change. The first postwar account of fieldwork that she brought to her longtime publisher described her 1953 return to the Manus people of New Guinea, *New Lives for Old*. This was not a book about how traditional cultures are eroded and damaged by change but about the possibility of a society choosing change and giving a direction to their own futures. Mead is sometimes labeled a "cultural determinist" (so obsessed are we with reducing every thinker to a single label). The term does reflect her belief that the differences in expected behavior and character between societies (for instance, between the Samoans and the Manus) are largely learned in childhood, shaped by cultural patterns passed on through the generations that channel the biological potentials of every child, rather than by genetics. Because culture is a human artifact that can be reshaped, rather than an inborn destiny, she was not a simple determinist, and her convictions about social policy always included a faith in the human capacity to learn. After the 1950s, Mead wrote constantly about change, how it occurs, and how human communities can maintain the necessary threads of connection across the generations and still make choices. In that sense, hers was an anthropology of human freedom.

Eventually, Mead wrote for Morrow the story of her own earlier years, *Blackberry Winter*, out of the conviction that her upbringing by highly progressive and intellectual parents had made her "ahead of her time," so that looking at her experience would serve those born generations later. She never wrote in full of her later years, but she did publish a series of letters, written to friends, family, and colleagues over the course of fifty years of fieldwork, that bring the encounter with unfamiliar cultures closer to our own musings. Although *Letters from the Field* was published elsewhere, by Harper & Row, corporate metamorphoses have for once been serendipitous and made it possible to include *Letters from the Field* in this HarperCollins series, where it belongs. Mead often wrote for other publishers, but this particular set of books was linked by that early desire to spell out what her personal and professional experience could and should mean to Americans. That desire led her to write for *Redbook* and to appear repeatedly on television, speaking optimistically and urgently about our ability to make the right choices. Unlike many intellectuals, she was convinced of the intelligence of general readers, just as she was convinced of the essential goodness of democratic institutions. Addressing the public with respect and affection, she became a household name.

Margaret Mead's work has gone through many editions, and the details of her observations and interpretations have been repeatedly critiqued and amended, as all pioneering scientific work must be. In spite of occasional opportunistic attacks, her colleagues continue to value her visionary and groundbreaking work. But in preparing this series, we felt it was important to seek introductions outside of ethnography that would focus on the themes of the books as seen from the point of view of Americans today who are concerned about how we educate our children, how we provide for the full participation of all members of society, and how we plan for the future. Times change, but comparison is always illuminating and always suggests the possibility of choice. Teenage girls in Samoa in the 1920s provided an illuminating comparison with American teenagers of that era, who were still living in the shadow of the Victorian age, and they provide an equally illuminating comparison with girls today, who are under early pressure from demands on their sexuality and their gender. Preteen boys in Manus allow us to examine alternative emphases on physical skills and on imagination in childhood—and do so across fifty years of debate about how to offer our children both. Gender roles that were being challenged when Mead was growing up reverted during the postwar resurgence of domesticity and have once again opened up—but the most important fact to remember about gender is that it is culturally constructed and that human beings can play with the biology of sex in many different ways. So we read these books with their echoes not only of distant climes but also of different moments in American history, in order to learn from the many ways of being human how to make better choices for the future.

introduction to the Perennial edition

Margaret Mead was short and female. To manage that in a world of science dominated by tall males, she spoke firmly *and* carried a big stick—a long forked staff, set off by the dramatic cape she wore in public. Yet she was so compassionate a field anthropologist that informants named their children after her. Decades before Carl Sagan and other scientists learned to write directly to the public, Mead was writing bestsellers like *Coming of Age in Samoa* and had a regular column in the women's magazine, *Redbook*. That cost her with her peers.

She was sharp, in both senses. She could anticipate the argument of a scientist or political authority and cut it off with a pronouncement. She was queen bee at New York's American Museum of Natural History. She waded effectively and tirelessly into public policy in the mold of an Eleanor Roosevelt, yet as a liberal she was too intellectually honest to be predictable or "reliable." She never claimed feminism nor would feminists claim her, because, again, decades before it became a movement in America, she was advancing the study of child rearing, education, and the role of females and was herself a model of how to shatter the glass ceiling. She found many of the 1970s feminists to be too narrow-minded and narrow-agendaed for her taste and offered criticism which was as uninvited as it was astute.

In 1976, I spent one day with Margaret Mead, conducting a joint interview with her and her former husband, Gregory Bateson. Her insight and personality were so memorable that I find I still converse regularly with her shade, who is as impatient, amused, and helpful as ever. What a treat, then, to engage the substantial Margaret in this remarkable book.

It is a book about change, about the most difficult kind of transition for humans: cultural change. It explores important questions about how profound systemic change can occur successfully. Mead's island tribe, the Manus, underwent sudden, comprehensive cultural change—something usually thought to be impossible or so destructive as to be not worth attempting. Sudden comprehensive change, in the French Revolution of 1789, the Russian Revolution of 1917, the Chinese Communist Revolution of 1949, led to grotesque failure, yet the self-conversion of the "stone-age" Manus to a "modern" society in a single generation is a success story. What happened?

No one was better positioned to answer that question than Dr. Mead. She had the "before" data: she had studied the Manus in the depth in the late 1920s. Returning in the 1950s, she was talking to the same informants and studying the same families, families who were now living in a new world of their own making. She could document which practices were utterly transformed and which remained constant. She could try to tease out the details of traditional Manus character and behavior that might have allowed an anthropologist to predict the successful transition. Having been part of the founding of the science of cybernetics, she had a depth of understanding of dynamic systems that is still rare half a century later.

Margaret Mead lets you watch her learn. Puzzled by the emptiness of a government-mandated council meeting, she realizes, "While taxation without representation is tyranny. . . . *government without taxation is degradation.*" The Manus are handed governance instead of being required to invest in it; it becomes something done *to* them instead of *with* them. You can see Mead taking extra glee from insights that go against her own liberal instincts. Her love became a form of tough love, based in matter-of-fact common sense informed by constant inquiry. She lived for surprise, for data that would change her opinions, revise her theories. As a result, her theories have proved more durable than most, and the subjects she took on remain important.

New Lives for Old is a book about civilization. For the Manus, civilization is a choice, not a given. Why do they choose it? What elements of civilization are most important to them? What details from the American military and Australian governmental apparatus are seized on? How do those details work when thrown into a wholly different context? I'm thrilled by Manus statements such as, "Before, we couldn't count back, we only knew our fathers and our father's fathers." Calendric time, a fundamental tool of civilization, is a revelation to them; likewise, machines: "The Americans believe in having work done by machines so that men can live to old age instead of dying worn out while they are still young." And the Manus, it turns out, are highly adept at understanding and fixing machines. Mead explains how their traditional harsh seafishing life and child-training techniques made them naturally machine-friendly.

She theorizes that much of the success of the Manus transition can be attributed to its selective use of existing models, as opposed to trying to invent from scratch. Old models made new are refreshing to contemplate. Thus, the familiar idea of equality before the law is stated in a Manus court case: "All of us are human, all of us are weak, you do wrong and come before the court, I also do wrong and if I do, I must come before the court, none of us is without blame, this is the fashion of humankind." That's worth etching on the walls of anybody's Supreme Court. For the Manus it meant the replacement of constant harassment and quarreling as the social norm with a prevalence of laughter and singing around the village.

New Lives for Old was published in 1956. While America was in the midst of a postwar boom, the intelligentsia of the time was critical of everything American—America was deemed shallow in contrast to European depth; it was seen as being in a decline similar to the fall of the Roman Empire. In that environment, Mead's report was unwelcome good news. Its optimism went against the grain. Its implied praise of America was embarrassing. Neither for the first nor last time, she was criticized. Yet, she criticized right back and went on studying how to make things work, from the family to education to civilization.

Margaret Mead's anthropology stood tall, and female. The book wears well. It is welcome back.

—*Stewart Brand*

preface to the 1975 edition

This book is the record of one of the most astonishing and rewarding experiences that has fallen to the lot of man or woman in this century. In 1928 I made a study of the Manus, a small group of primitive people, on a little known archipelago, the Admiralty Islands, in a relatively unknown and very primitive part of the world called the Trust Territory of New Guinea. I finished my research, returned to America, wrote a book about them called *Growing up in New Guinea*, and some technical papers, and then went on to other problems in other places. I never expected to revisit them, nor did I have much hope that I would ever know their fate, although I knew there would be news, no doubt, of the progressive course of the loss of their aboriginal culture, as mission, trader and government moved in. I thought the children with whom I spent so many vivid months would remain forever in my mind as I had known them, two and four and six, bright eyed or sulky, never to be known as grown-up people because I would have no knowledge of their future. For the people of the small village of Peri, in Manus, my husband and I were passing into the realm of the unknown, never to return.

And then through an extraordinary series of circumstances attending World War II, the people of Manus Island moved to the center of international concern. They accomplished a nonviolent transformation of their society which was faster, more complete and more startling than anything recorded before. And in 1952, my Australian anthropologist friends insisted that I go back to study what had happened. I was in a unique and unusual position to do this because, unlike most anthropologists who, attempting to catch the details of a culture on the edge of change, work with the oldest members of a tribe, I had studied young children.

Most anthropologists returning after twenty-five years would have found a village of strangers. I found the small boys I had known so well, entrenched in positions of leadership, crossing the widest distance between cultural levels that had ever been known in human history—from a recent past in the stone age to the electronic age. It is quite possible that no such distance will ever be recorded after this century. Exploration from the industrialized world has become so pervasive and comprehensive that no people on this planet can hide for long within even the deepest jungle fastnesses without some mapping satellite noting the smoke from their fires.

This opportunity then was indeed unique. What I found changed my ideas about the possibilities of rapid and complete cultural change as compared with slow piecemeal change, where each change sets up a compensatory process which becomes in turn a drag on constructive progress. It was not that I believed that every primitive people could change so rapidly and constructively as the Manus had done, but rather that one instance of such rapid change could alter our whole notion of how change occurred. A careful investigation of the process by which a preliterate people with only ceremonial economic ties to bind them together could build themselves a society, and face the exacting modern world, gave us new ideas about the conditions necessary for such changes.

Among the Manus I found three conditions: the fact that they made the changes under their own steam; the sweeping and inclusive character of the changes they made; and the presence of a rarely gifted leader. These all relate to issues which were and still are agitating people all over the world: how much should change spring from the people themselves; how much can or should it be helped from outside? How many things have to be changed at once if any change is going to stick? And do we need leadership, charismatic leadership, for the people of a nation-state, or even the people of the smallest village, to change their way of life?

The new light that this return to Manus shed on these problems in 1955 is just as relevant in 1975. All over the world there are experiments in centralization and experiments in decentralization, from introducing nationwide programs of population control in Eastern Europe to the attempts to establish self-sufficient communes in the People's Republic of China, from tractor factories in one country making tractors for the world to attempts to develop middle technologies—little technologies using local materials. The question of how much dependence should be placed on local initiative bedevils the technical assistance planning of governments all over the world, from the poverty programs in the United States in the 1960s to the efforts of industrial missions, special cadres of highly trained workers, Peace Corps, International Volunteer corps, trained in one country and sent to inspire and implement the aspirations of the rural people of another.

Perhaps the most vexed question of all is the question of leadership. How essential is a leader? Is the present plight of the world—as the leaders, remnants of World War II, die off and are replaced by men who don't have their predecessors' apparent capacity to command a following—due to our having constructed political conditions which jeopardize rather than promote personal leadership? In Africa old patterns of rivalry among the sons of chiefs reassert themselves in rivalry among nominally elected leaders. In the Middle East, Asian styles of resolving succession survive. In the Euro-American world, assassination and kidnapping are new threats to those who accept the leadership role. How much does the actual personality of the leader matter and how much depends on surrounding conditions? Has TV made it harder for any human being to achieve and retain charisma? Can a team approach, more appropriate to the complexities of the modern, overspecialized and interdependent world, really work? These are all live questions today, as they were live questions in 1953 when this study was made and when this book was first published in 1955.

What the anthropologist is able to do is to bring the understanding of a microcosm, a small native village, to bear on the macrocosmic problems of a larger nation, or today, a globally interdependent community of peoples. Because this restudy of Manus was made with a lively appreciation of what our worldwide problems were going to be, it still presents matter for thought and analysis.

Furthermore, what concerned me then and concerns me now—the ability of Americans to provide an infusion of vital belief in the importance of human well-being into international contacts—is even more pressing today. One example of the possibilities of this contact was demonstrated by the way the people of Manus understood our institutions, fastened on our ideals and neglected many of our most conspicuous defects. They interpreted the tremendous effort put into the care of each individual serviceman in the American Armed Forces during World War II as primarily a demonstration of how important Americans thought each human life to be—not as a cold-blooded calculation of how much it would cost to train a substitute for an unrehabilitated serviceman. Both were true, but they saw the emphasis on human beings as paramount. They interpreted American GIs' willingness to give Uncle Sam's property away to them as generosity, which it was, because the Americans enjoyed the Manus and were glad to give them tools. They were in fact a kind of living tribute to how much good could be extracted from the basic ideals of American culture. They saw that all servicemen—black as well as white—wore the same clothes, ate the same food and drew the same pay—an enormous guarantee that they who had been treated as incapable of movement into the modern world could in fact do anything the white man could do. And they proceeded to try—with great success.

When this book was published, it was greeted by a torrent of pessimism from the perennial Jonahs of the American intellectual establishment who knew that no good could ever come from any contact between Americans and the peoples of other parts of the world. This was the age of The Ugly American, The Pretty American, and a mass of criticism of the way we were conducting our relationships abroad. Most of the criticism of our general policies was deserved, and the years of the disastrous Vietnam War have only served to confirm the most melancholy of these prophecies. And yet, this has not been so in New Guinea and in many parts of Oceania. The fact that the Manus could use an American model to inspire them was important and remains a ray of hope in the present gloomy picture of a world almost totally disenchanted with the results of technical change. It remains a promise that it is possible to have a kind response to inevitable modernization which is not disastrous.

In the years after this book was published, and in the course of lecturing about what the example of the Manus might demonstrate, I came also to understand much better the difference between building a society on an existing model—however idealized or misunderstood—and having to create a new kind of society which never existed before. At every step of their extraordinary, intelligent making-over of their small community, the Manus were able to point to the example of what the prestigious Americans had done or what they believed they had done, like letting small children vote, or combining ordinary schooling with the continued emphasis on a mastery of subsistence technology. We, on the other hand, have no such models, however much lazy revolutionaries have wanted to let the revolutionary experiments of other ages or other countries set models for us. When the Manus looked at a tape recorder and called it by a numeral used for living things, while they counted typewriters with a numeral for inert objects, they were recognizing how the electronic age differed from the pre-electronic age of steam and steel. But we who are trying to think out plans for utility corridors for a continent which can contain lasers or holographic communications have to pursue a completely uncharted course in trying to work out new relationships between our technical potential and the institutions that are needed to make it manifest.

New Lives for Old is an example of what one people, inspired by the institutional forms of another people, can accomplish. It should illuminate the difficult processes of modernization going on throughout the developing world. But only by the contrast with which it provides us does it help us to solve the new problems that confront us today—the dangers to the environment, the uses of computers, the consequences of instant communication, the frailties of an over-interdependent world with its lack of a sense of direction.

In a strange way this second visit to Manus, twenty-five years after the first, echoed the first, for when I said good-bye in 1953, I again did not expect to return. It was then that Pokanau, an old man who should long since have been dead, said to me: "Now, like an old turtle, you are going out into the sea to die and we will never see you again." Again I thought my task, as envisioned within the problems of the period, was finished. There was a sense of finality when I and my two young colleagues, Theodore Schwartz and Lenora Shargo (now Lenora Foerstel), shared the farewell feast for my departure and celebrated the attainment of the council status in the Territory of Papua-New Guinea for which the people had been working so long.

And as in 1953, I felt it important to return to see what had happened because of my long knowledge of the past, so Theodore Schwartz and I decided in the 1960s to mount another expedition to examine how the Manus had fared in the world that was changing more rapidly than even we had been able to imagine in 1953. We have made several return trips to Manus, watched a new village built and that village have to be moved again because of overcrowding and pollution. We watched Paliau rise to the position of member of the Assembly and exert important leadership in the newly established political structure of Papua-New Guinea. We watched first an intensification of the rules of the new society established in 1946, and then, on our last field trips in the early 1970s, we saw the strictness of the regime relaxed, as children again learned to dance the old dances, and houses made of thatch, but in a new design, were more prestigious than the tin roofs of the intermediate years. In 1973 there were five hundred young people from the Admiralty Islands (of which the Manus are simply the largest group of people who speak one language) in Port Moresby engaged in some form of higher education—students of medicine, law, economics, engineering and constitutional development.

The incredible changes go on. Papua-New Guinea is now a territory on its way to full nationhood with independence scheduled for 1975. In 1970 John Kilepak, who looms so large in these pages, came to California to visit my colleague, Barbara Heath, who started working with us in the 1960s. He introduced me before I spoke at commencement luncheon, as he had known me longer than anyone else there. The tremendous leap ahead that the Manus people took is continuing, beset, as all such great transformations are, by many difficulties.

This summer I am going back again.

Margaret Mead

The American Museum of Natural History,
New York
April, 1975

REFERENCES

Fortune, R. F. (1935) *Manus Religion*. Lincoln: University of Nebraska Press, 1965.

Mead, Margaret. (1928) *Coming of Age in Samoa*. Morrow Paperback Editions, New York, N.Y.

————. (1964) *Continuities in Cultural Evolution*. New Haven: Yale University Press, 1968.

————. "Kinship in the Admiralty Islands," *Anthropological Papers of The American Museum of Natural History*, 34, Pt. 2 (1934), pp. 183–358.

————. "Letter from Peri Village, New Guinea," *ATI-ACE Newsletter*, 1, No. 9 (March, 1966), pp. 6–7.

————. *Margaret Mead's New Guinea Journal*. Produced by Craig Gilbert for National Educational Television. 90 minutes, 16 mm, sound, color. Distributed by National Educational Television Film Service, Indiana University Audio-Visual Center, Bloomington, Indiana. 1968.

————. *New Lives for Old*. New York: William Morrow and Company, 1956, 1966.

————. *New Lives for Old*. Horizons of Science Series, vol. 1, no. 6. Educational Testing Service, Princeton, New Jersey. 20 minutes, 16 mm, sound, color. Distributed by Indiana University, Audio-Visual Center, Bloomington, Indiana. 1960.

Mead, Margaret, and Schwartz, Theodore. "The Cult as a Condensed Social Process." In *Group Processes*, Transactions of the Fifth Conference 1958, ed. Bertram Schaffner. New York: Josiah Macy, Jr. Foundation, 1960, pp. 85–187.

Schwartz, Theodore. "The Paliau Movement in the Admiralty Islands, 1946–1954," *Anthropological Papers of The American Museum of Natural History*, 49, Pt. 2 (1962), pp. 207–422.

Schwartz, Theodore, and Mead, Margaret. "Micro- and Macrocultural Models for Cultural Evolution," *Anthropological Linguistics*, 3, no. 1 (1961) pp. 1–7.

Manus revisited
—preface 1965

When the people of Peri beat the death drums as our canoe pulled away from the village in 1929, neither they nor I expected that I would ever return. I was sailing away toward a world of which they had no clear conception, taking with me all I needed to know about their strange and savage life. It seemed clear to me that within a few years this way of life would be fragmented and changed. Mission and government would bring them a new faith and new sanctions to replace the rule of the old ghosts. As far as one could tell then, their future was already marked out in the experiences of many other Pacific peoples. My work in Manus was done and I went on to study other peoples—an American Indian tribe, four tribes on the New Guinea mainland, and the people of Bali.

But, after all, I did make another field trip to Manus. This book is my account of that return. In 1953, twenty-five years after the first field work in Peri Village, I decided to go back in response to questions no one had answered about the incredible changes that had taken place in Manus and to find answers to new problems of the postwar world. As on that earlier occasion, when I left, no one expected me ever to come back. Pokanau, who had been a principal informant in 1928 and who was now the grand old man of the village, spoke at the farewell feast: "Now, like an old turtle, you are going out into the sea to die and we will never see you again."

In my response I reminded them that this time they could write letters telling me what was happening. They were now part of the world and we would no longer be separated by the great gulf that divides the totally illiterate from those who leave the confines of the village. Privately, half jokingly, I also said: "If you make as much progress in the next twenty-five years as you did in the last, I'll come back to see the changes." But no one really believed this, least of all Pokanau who had already lived far beyond his time while much younger men had aged and died.

In fact, I did make a third trip to Manus. In the autumn of 1964, I went back to Peri as part of a much more inclusive plan for the anthropological study of the whole Admiralty Island archipelago that Dr. Theodore Schwartz—who had been a student colleague in 1953—and I were undertaking under a grant from the National Institute of Health* We set up our expedition headquarters in Peri Village so that I could see my friends again, now eleven years older. I knew that my small boys of 1928 would be on the verge of old age, but I thought that the interval between field trips had been too brief for me to find changes that were in any sense as dramatic as those I had found in 1953. I was mistaken. This time the whole Territory of Papua-New Guinea had taken a great leap ahead and the people of Peri, no longer isolated within their small island world, also had moved forward.

What we were witnessing was the local response to a new climate of world opinion. Australia was bringing modern political institutions to the Territory, where a population of primitive peoples, speaking more than 500 languages, had so recently come

---

*A Field Study in Cultural Systematics, New Guinea, National Institute of Mental Health, Grant No. 730500, sponsored by The American Museum of Natural History, New York.

into contact with the larger world that not even one individual among them had completed a university education. Nevertheless, a territory-wide representative House of Assembly had been instituted and an electoral roll had been built, laboriously spelled out by young officers whose ears were untrained to catch the strange-sounding names of men and women, very few of whom could as yet read or write. An election had been held and the new Assembly was convened in Port Moresby. In 1953, it seemed that Paliau, the gifted Manus leader, had reached the zenith of his power. In 1964, as a member of the House of Assembly, representing the whole of the Admiralties, he was treated with respect and was looked upon with great interest by those to whom his name had been anathema only a decade earlier. The Territory was caught up in the excitement of this new style of movement toward integration and political education for a population that until recently had constituted one of the few isolated groups of tribes still remaining in the world.

In Peri Village, where, thirty-six years before, I had watched small children playing all day long, there was now a big school. Teaching in English had begun ten years ago, and thirty-three children had already gone elsewhere for further education. While I was in Peri, the school children followed the Olympic games in Japan over the village radio. They were hoping that the people of Papua-New Guinea—this new entity created in their own lifetime—would compete in the next Olympics. Earlier, they had listened to John Glenn's flight and had heard how the people of Perth, Australia, had turned on their lights in greeting.

Not long ago a sea voyage of perhaps twenty miles was so dangerous that people wept when a canoe set out across the reef and wept again when a returning canoe was safely beached with all its crew alive. Now the school children were writing essays—in English—about space satellites. One essay began: "The sputnik is round like the moon." The writer continued: "It stays in the stratosphere like a star and sends signals back to earth which are picked up by tracking stations in places like Manus Island,

Perth in Western Australia, and many other places around the world."

In another essay, a girl wrote: "A sputnik goes round the world faster than an aeroplane. If I could fly in a sputnik, I would be able to see all the countries of the world at once. They would not look the same because I would be very high up, but I would know them when I saw them because Mr. Buckham (i.e. the teacher) has told us about them."

In Peri, education was not a matter for children alone. The whole village was involved. A curfew was set up to ensure that the children did their homework, parents followed their children's grades, and people speculated about where the children's education would lead them.

The transformation I witnessed in 1953 taught me a great deal about social change—change within one generation—and about the way a people who were well led could take their future in their own hands. It helped correct the widely held belief that slow change, however uneven, was preferable to rapid change. The Manus children I studied earlier, in 1928, had taught me about the consequences of the kind of education advocated by contemporary educators. For Manus children, given great freedom, grew up to accept—even though grudgingly—the standards of the adult world. I learned that it is not enough to depend on the next generation; adults themselves must take part in change.

So in 1928 and 1953, my principal learning came from the Manus people themselves. But in 1964, I learned a great deal by watching the young Australians who had volunteered, after a year's special training, to teach in New Guinea villages. This gave me new insight into the process by which more advanced, complex societies can teach the less advantaged—a problem that must somehow be met by the new nations of Africa and Asia, old nations like India and China, and modern nations in their attempts to deal with pockets of poverty and ignorance in isolated rural areas and in urban slums. Watching these young teachers, each one enthusiastically engaged in teaching a whole village, I realized what we lose when we permit education and enthusiasm to be diluted by progressive ignorance, as we set the partly educated to teach those who know even less.

Once I would have agreed with the advice of those who advocated the use of educated Australians only to train teachers who, in turn, would work in the villages where people were illiterate and no one had learned any one of the great modern languages. But what I saw in Manus, in 1964, changed my thinking. There, village children are taught by educated Australians who can give them a sense of the whole complex civilization in which they themselves were reared and who expect their pupils to go on to secondary school and possibly to a university. Contrasting this with the effect of a closed system in which those who teach speak a modern language with difficulty and, knowing only a little, teach their pupils even less, I realized the implications of our failure to bring the best we have to those who need it most.

In Manus, people have not been beaten down to the state of hopeless inferiority that characterizes many peasant and proletarian groups in other parts of the world. Peri children were able to catch the first rapture of widening knowledge as they listened to their Australian teacher who, before he became their teacher, had been an actor, a producer, and a protocol secretary to a diplomat. He talked to them about everything on earth and they listened, fascinated, never daring to look away lest they miss something new. Formerly, the future of Manus children was contained within the limited bounds of their small archipelago; more recently, they could look forward to a constricted life as plantation workers and boat crews. Now they are part of an open-ended system and, potentially, they can go as far as they have the intelligence and drive to go.

The people of Peri have built themselves a new, better village. At Christmas, 1965, they plan to be hosts to all the members of the original Paliau Movement, some 5,000 people, proud that they have succeeded in producing a 20th century statesman. The children who have left the village to go to school in Rabaul, Port

Moresby, and Australia will be coming home for the Christmas holidays. And I, too, will be in Manus to celebrate the occasion with the people of Peri Village and the other Manus villages of the South Coast of the Admiralties who have been collaborators all these years in my exploration of their way of life.

Last year the absent school children were names in my census book and photographs in my collection that evoked memories of the gay, water-loving youngsters I had watched, in 1953, playing by the sea. This year, like Paliau, who is now Mr. Paliau Moluat, M.H.A., they will be coming back with new names. And like Paliau, they will bring to the people of Peri and Bunai, Baluan and Mbuke, a sense of what is happening all over Papua-New Guinea. When they arrive, I shall see their first meetings with the parents who have so enthusiastically backed up and fostered their ambitions. I shall see what they look like, how they walk and talk, these grandchildren of the men and women I knew before they had emerged from their Stone Age isolation.

New York
August 16, 1965

preface and acknowledgements

I have written this account of my return after twenty-five years to the village of Peri, to the Manus people of the Admiralties, in a way which I hope will make it possible for the reader to share some of my own sense of discovery. There are several themes blended together: the simple excitement of returning after twenty-five years and finding children I had known grown to maturity after having had no news of them in between; the intellectual pleasure of realizing how much we had learned in the last twenty-five years about the relationships between the institutions under which people live and the cultural character which embodies those institutions (this emphasized by the discovery of points to which I had been blind then, but which seemed so obvious now); the sheer detective work of finding out what had "really happened" to transform this small cluster of stone-age headhunters into a community asking for a place in the modern world; and finally, the recognition of what one people, in one place, who had made such a leap could mean to our hopes for the world.

In writing, I have woven my way back and forth, between past and present, as the actual course of the return visit went. On the walls of my house in New Peri, I pasted up a whole set of scenes of the water world of Old Peri, where yesterday's children, and myself twenty-five years younger, still looked through an open doorway into the past. The Manus people and I have the same kind of memory, so that each day was doubly lived, each detail of the present illumined by comparison with the past. For those who prefer to know all there is to be known about one set of circumstances before they have to deal with another, the account of twenty-five years ago, *Growing Up in New Guinea*, is still in print.* But in the present book, the past will come and go, as it actually did during the months, from June to December, 1953, that I spent in Peri. Even though I spoke both Manus and the new lingua franca, Neo-Melanesian, and half the people were old friends, still much that had happened remained obscure for many weeks: What was the role of Paliau, the leader who had made their modernization possible but of whom no one ever spoke; what was The Noise, in which everything belonging to the past had been thrown away; what was the New Way, "This new form of thinking"; what was the meaning of the phrases, "inside us" and "outside us"; why had the most worthy heads of the village been in prison; what was the actual position of the men who styled themselves "council" and "committee"?

Each field trip among a new people is always enormously exciting. One wonders what the language will be like, what the people will be like, what intricacies their culture will turn out to have. But this was a different adventure, following the footsteps of those I had known before as they crossed thousands of years to join us, members of the modern world. So, if any point seems strange at first, unexplained, ambiguous, contradictory, it is because that is the way it was presented to me, by the people of Peri, as we became reacquainted across the years.

---

*Mead, Margaret, *Growing Up in New Guinea*, William Morrow, New York (1930); reprinted in *From the South Seas*, William Morrow, New York (1939); Blue Ribbon Books, New York (1933); Mentor edition, New American Library, New York (1953); English editions, George Routledge, London (1931), and Penguin Books, London (1942 and 1954).

This book is a restudy based on a piece of co-operative research in 1928–29. In 1928, Dr. Reo Fortune, as a Fellow of the Australian Research Council, and I, as a Fellow of the Social Science Research Council of the United States, collecting for the American Museum of Natural History of whose scientific staff I was a member, made a field study among the South Coast Manus tribes of the Admiralty Islands, which were then part of the Territory of New Guinea mandated to the Commonwealth of Australia.

Our work was facilitated by the Department of Anthropology of the University of Sydney, under Professor A. R. Radcliffe-Brown, and by the Department of Home and Territories of the Commonwealth of Australia and the Administration of the Mandated Territory of New Guinea, most particularly by His Honour Judge F. B. Phillips now Chief Justice, Territory of Papua and New Guinea, and Mr. E. P. W. Chinnery, government anthropologist. In working up our material subsequently, we were assisted by Columbia University and the American Museum of Natural History.

Three major publications resulted from this early work: Dr. Fortune's *Manus Religion*, published by the American Philosophical Society in 1935; my *Growing Up in New Guinea* (William Morrow, 1930); and *Kinship in the Admiralty Islands* (an anthropological paper of the American Museum of Natural History, 1934). Materials from the Manus research were also included in *Co-operation and Competition Among Primitive Peoples* (McGraw-Hill, 1937), and *Male and Female* (William Morrow, 1949), and in a variety of articles in journals and periodicals.

The 1928–29 field work was a co-operative venture with subsequent specialization in writing up the reports, and I am indebted to Dr. Fortune for the co-operation, in and out of the field, that made the work possible then, and for his help and unpublished notes which he made available to us in 1953.

I am especially indebted to Dr. Ian Hogbin of Sydney University, author of *Experiments in Civilization* (Routledge, 1939) and *Transformation Scene* (Routledge, 1951), who acted as an anthropological specialist in New Guinea during the war, for his insistence that it was important for me to return to Manus and do basic field work on the relationship between the over-popularized "cargo cults" and significant socio-political movements, and for his careful reading of this manuscript.

For the training which underlay the 1928–29 field trip, I am indebted to Professor Franz Boas and Dr. Ruth Benedict, and to Dr. Fortune's previous field-work experience in Dobu; for encouragement in the undertaking, to the chairman of my Department of Anthropology, Dr. Clark Wissler; and for orientation and theoretical criticism, to Professor A. R. Radcliffe-Brown.

The 1953–54 expedition was also a co-operative one, and I am indebted to Mr. and Mrs. Theodore Schwartz of the Graduate School of Anthropology of the University of Pennsylvania for collaboration in the field and in analyzing the research materials, and to Professor I. A. Hallowell and Mr. Charles Schwartz for facilitating their participation. My discussions draw heavily on our joint observations between June, 1953, and December, 1953, and on their field notes during their further six months in the field, January to June, 1954, on Theodore Schwartz' analysis of social change from 1946 to 1954—of which this restudy of Peri is only one part—and on Lenora Schwartz' studies of infants, children's arts, projective tests, play and motor behaviour. As in Dr. Fortune's and my previous collaboration, they have read each sentence of my work, added their suggestions, criticisms, and amendments, giving me the benefit of everything in their material that corroborated, contradicted, and called it in question.

This 1953–54 study was majorly financed by a generous grant from the Rockefeller Foundation, with supplementary grants from the Voss Fund of the Department of Anthropology of the American Museum of Natural History. I am indebted to Dr. Harry Shapiro, chairman of the department, and to Miss Bella Weitzner for the opportunities to undertake and pursue this research.

For facilitation of our work in the field, we are especially indebted to the Honourable Paul Hasluck, Minister of Territories, to the Honourable D. M. Cleland, the Administrator, and Mrs. Cleland, to His Honour Judge F. B. Phillips, to the Department of District Services and Native Affairs, and the Departments of Health and Education, Port Moresby, to the Royal Australian Navy, especially Captain J. A. Walsh, to Mr. Malcolm English, District Commissioner of Manus, and Mrs. English, to Mr. James Landman, Native Authorities Officer, Baluan, and Mrs. Landman, and to Dr. and Mrs. Leo Petrauskis. I have to thank the firm of Edgell and Whiteley, particularly my old friend, Mr. Edgell, Mr. Byrnes of Lorengau, and Mr. Dodderidge of Ndropwa.

Half a lifetime of research, field work, theoretical give and take, lie between my first trip and my second, and all of those from whom I have learned have contributed in different ways to the intellectual equipment which I took on my second trip into Manus. This list is so long that no individual would stand out on it as each actually would deserve. But I feel especially indebted for insights which have contributed directly to this particular piece of work to all my collaborators in the research on *Co-operation and Competition* and in the Columbia University Research in Contemporary Cultures, and especially also to: Theodora Abel, Louise Ames, Gregory Bateson, Alex Bavelas, Jane Belo, Ruth Benedict, Ray Birdwhistel, Gotthard Booth, A. R. Radcliffe-Brown, Eliott Chapple, Edith Cobb, Carl Deutsch, Milton Erickson, Erik Homburger Erikson, Lawrence and Mary Frank, Arnold Gesell, Geoffrey Gorer, Ronald Hargreaves, Ian Hogbin, Evelyn Hutchinson, Frances Ilg, Marie Jahoda, Robert Lamb, Harold Lasswell, Nathan Leites, Kurt Lewin, Konrad Lorenz, Margaret Lowenfeld, Robert and Helen Lynd, Rhoda Métraux, Philip Mosely, Gardner and Lois Murphy, Kingsley Noble, William Fielding Ogburn, Edward Sapir, Erwin Schuller, Frank Tannenbaum, James Tanner, Martha Wolfenstein, Harold Wolff.

We are indebted to Professor W. M. Krogman of the University of Pennsylvania and Dr. James Tanner of St. Thomas Hospital for the loan of instruments, and to Dr. James Tanner and his research staff for the design of our somatotyping and the analysis of the photographs.

For many of the photographs of the earlier expedition, I am indebted to Reo Fortune, and for many of the still photographs and the background materials provided by the tape recorders and moving pictures of the 1953–54 expedition, I have to thank Ted and Lenore Schwartz. For photographic assistance and advice, I am indebted to Josef Bohmer of Vassar College, and Lee Bolton, Elwood Logan, and Alexander Rota of the American Museum of Natural History.

Of the projective materials, I am indebted to Ted Schwartz for the Rorschachs, Lenore Schwartz for the Mosaics, Gesells, Bender-Gestalts, and Stewart Ring Puzzle Tests, and for preliminary interpretation of the Mosaic Tests to Dr. Margaret Lowenfeld and her staff. For comparative materials from the British West Indies, I am indebted to Dr. Rhoda Métraux and Dr. Theodora Abel.

For assistance in preparing and obtaining my medical supplies and advice in their use, I wish to thank Dr. Walter Modell; the Department of Medical Research of Winthrop Stearns, Inc., who contributed all the antimalarial drugs; the Department of Public Health of Papua and New Guinea; and Dr. Leo Petrauskis, medical officer in Lorengau.

For continuing assistance in our tape recording problems, I wish to thank Dr. and Mrs. Francis Rawdon Smith and the Magnecorder Company, who took special pains with the selection of our tape recorder. To the Travel Department of the Bank of New South Wales, Sydney, Australia, we owe the facilitation of transport, and to Grace Brothers, Sydney, the assemblage of a most complicated Christmas cargo.

For criticism of the manuscript, I am indebted to: Theodore and Lenora Schwartz, Geoffrey Gorer, Rhoda Métraux, Ian Hogbin, James Landman, F. B. Phillips, Marie Eichelberger, Isabel Lord, and Constance Sutton.

For help in the preparation of the manuscript I wish to thank Mrs. Barbara Bauman, Mrs. Leila Lee, Mrs. Constance Sutton, and Mr. Fred Scherer.

For hospitality and kindness on the way to the field, in the field, and on the way out of the field, anthropological field workers are particularly dependent and unendingly grateful. In 1928–29, for all the years in between, and again in 1953–54, my period of preparation and readjustment to a world of cold and traffic in Sydney has been mediated by Timothy and Caroline Tenant Kelly. From the first day of landing in Rabaul in November, 1928, to my last day in Port Moresby in December, 1953, I have been indebted to the hospitality and prevailing care of His Honour Judge F. B. Phillips, and for this visit to Mrs. Phillips, who arrived later in the Territory. On the 1928–29 expedition I was indebted to Mr. J. Kramer and Mr. and Mrs. Burrows of Lorengau, Mr. F. W. Mantle and Mr. and Mrs. MacDonnel, District Officers at Lorengau, to Mr. and Mrs. James Twycross of Rabaul, and Mrs. C. P. Parkinson of Sumsum. This expedition would have been much more difficult and cheerless without the friendship and hospitality of Malcolm and Mary English and James and Marjorie Landman.

In the field one is unremittingly dependent upon the other members of the team. Upon their strengths, their skills, their devotion, their tears, and their laughter, the whole tone of the field work depends. To Reo Fortune and to Ted and Lenore Schwartz, I owe two of the most stimulating periods of my life, as we worked, eighteen hours a day, complementing each other's needs and skills.

Finally, it is the people themselves, whom one goes so far to find, who are the heart of the matter. To all the people of Peri, those who were alive in 1929 and are no longer alive, those among whom I lived in 1953, and especially to Kilipak, Michael Nauna, Raphael Manuwai, Petrus Pomat, Karol Manoi, Stefan Kaloi, Peranis Cholai, Penendek Pokanau, Beneditka Ngalowen, and to Johanis Lokus—once one of my household but who became major domo for Ted and Lenore Schwartz—I extend my special and hearty thanks. To Paliau, whom I did not know in 1928, for the spontaneity with which he responded to our research, I owe special thanks also.

Margaret Mead

American Museum of Natural History
New York
August 12, 1955

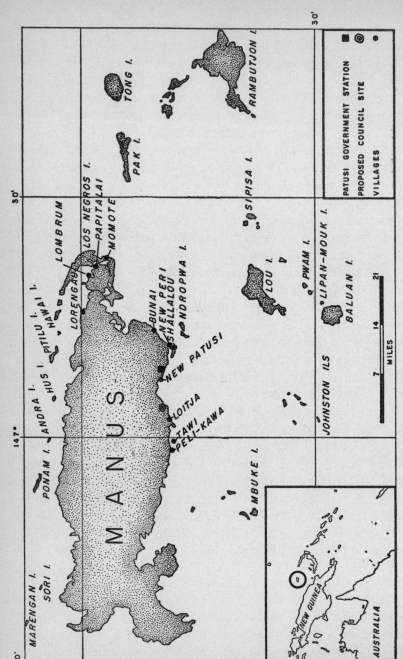

MAP OF MANUS

geographical and linguistic note

The events described in this book took place on the South Coast of the Great Admiralty Island—popularly known as Manus, the name of the people who lived in the lagoons along the shore and near the islands of Baluan, Mbuke, Johnston Island, and Rambutjon before they moved ashore after World War II.

The Manus divided the 15,000 people of the Admiralties into three groups: themselves, about two thousand; the Usiai, bush people of the Great Admiralty; and Matankor, the people of the islands. The new movement, led by a returned police boy, Paliau of Baluan, had included all of the Manus, some of the Matankor (selected parts of different villages), and some of the Usiai who had moved down to the beach and formed composite villages with the Manus. There were thirty-two villages and just under four thousand people who were members of the New Way which had grown out of the Paliau movement.

Paliau had returned to the Admiralties in 1946 and had begun to agitate for modernization, including in his program a new version of Christianity which former versions had hidden from the natives, plans for modern villages, for political unification among the historically hostile peoples of the Admiralties, for economic betterment, and for community organization and community enterprises. Before Paliau's movement had developed very far, but probably partly out of the excitement which rumours of his plans generated, a familiar form of an apocalyptic cult—a "cargo cult"—sprang up among the Manus on the island of Rambutjon, with the familiar paraphernalia of a prophet who preaches that the spirits of the dead will miraculously bring a cargo of the much-desired goods of the white man. Characteristically, this outbreak was accompanied by the destruction of property and by seizures in which individuals saw visions, heard marvellous sounds, shook violently, and fell to the ground. When the cargo did not materialize, the prophet was killed, but not before the religious excitement called The Noise had spread throughout the South Coast. Under the additional impetus given by this religious movement, Paliau's leadership was established and he was able to consolidate his plans.

The movement was exceedingly unpopular locally among Europeans, especially because it involved a break with the Roman Catholic Mission. The Administration was following a policy of working with local leaders; Paliau was taken to the capital, Port Moresby, and councils and co-operatives were explained to him. A local council was subsequently established in 1950, with Baluan as the headquarters, and a council promised to the rest of the South Coast.

When we arrived in 1953, I established my headquarters in New Peri, now a land village situated on the little island of Shallalou, and Theodore and Lenora Schwartz, my young colleagues, were established in the new land village of Bunai, on the shore of Manus Island. This was now a composite village containing the old Manus villages of Bunai and Pamatchau and four Usiai villages which had moved down to the shore. There was a senior Native Affairs Officer on Baluan, Mr. James Landman, and an Australian manager, Mr. Dodderidge, on the island of Ndropwa. Communication with Lorengau, the government station, was by way of Messrs. Edgell and Whiteley's little schooner, which came out periodically to pick up copra, by way of government work boys on the way to Baluan, or by a Manus canoe going into Lorengau or to the aviation field at Momote or the Australian Naval Station at Lombrum.

The word *Manus* is both a singular and a plural and is pronounced with a broad *a*. Words in Melanesian Pidgin, Neo-Melanesian, have been anglicized where the word was derived from English, e.g. *white man*, not *waitman*. In appropriate contexts the term *European,* conventionally applied to all Caucasians, is used to cover—as in New Guinea practice—Australians and Americans also.

The term *the Territory* applies to the Trust Territory of New Guinea, at present administered by Australia from Port Moresby as part of the Territory of Papua and New Guinea. Spelling of personal names conforms to my 1928 usage for ease in relating the personalities in *Growing Up in New Guinea* but place names have been somewhat modified to meet present day usage. In the case of the village conventionally written Mbunai, I have dropped the *m* and written it as Bunai for the convenience of American readers. The word *native* is used in those contexts where it seems appropriate, when discussing the point of view of the Administration, Europeans, etc., and the term *people* is used when the people of the Admiralties are being referred to as part of the modern world.

M. M.

new lives
for old

1

introduction

In each age there is a series of pressing questions which must be asked and answered. On the correctness of the questions depends the survival of those who ask; on the quality of the answers depends the quality of the life those survivors will lead. But first of all it is necessary to identify carefully what questions must be asked. What are some of the crucial questions today? Are they: How soon will the population of the earth outrun the food supply? How soon will all human beings break down under the strain of modern life? How complete will the destruction of civilization be in World War III? But such questions presuppose doom. In the very asking they plunge us into an abyss of hopelessness, or apathy, or the quick snatching violence of despair.

Yet the blandly certain questions provide as little solace: How soon will it take for men to come to their senses, realize that the old ways were best, return to nature and to God, learn to borrow rakes and hoes over the backyard fence, learn to spell as our forebears did, and learn to live a proper life by always deferring consuming until tomorrow anything that we can possibly save today? One has only to listen to this diluted utopia, which is all that it is possible to build out of the ghost of former ways of life, to know that it is a ghost as fleshless and inadequate as the way of life was once full-bodied. Those who rear their children on the nostalgic memories of long-dead lilacs in the dooryard give their children's imagination thinner fare than tiny plastic jet-plane toys which crunch on the new scratch-proof floors with a sound out of which no one has yet written any music.

We are squarely up against the dilemma of whether out of fear and desperation we will seek to prop up a crumbling old pattern or too hastily run up a new one, intent only that the new shall be a bulwark against the destruction of the old—new scaffolding against an old, too-weathered wall—or whether we can believe that we can build a new world suited to men's needs, twentieth-century housing for twentieth-century people. But if we choose the new buildings and yet realize that no blueprint of an unknown is ever satisfactory—that there are always a thousand small adjustments to make until doors and windows and passages from room to room become harmonious and livable—what estimate can we make about how long it will take and what the price will be?

What will have to happen before we have constructed a world which takes into account that instead of near-starvation we can hope for food for all, that instead of the picture, grown unbearable as it has become real to us, of nine-tenths of the world living in poverty and near despair, there is now a possibility for all people to have food and health for their children?*[1] What will have to happen before those who teach learn a new tone of voice so that those who are taught can hear what they say? Or before men learn that machines can be as homely, as fit for human uses, as tables and chairs, loaves of bread, and bottles of wine—all as artificial, as man-made, as a calculator or a carburettor—on the basis of which we now hymn the simple life or even see visions of the essential *itness* of a chair, as if a chair were of the same order as a wild rose rather than a roadster?[2] How can we circumvent the depressing and damaging effects of both those who so lament the old that the new cannot be welcomed and those who so hail the new—claiming that the revolver has forever replaced the rose[3]—that men deprived of traditional

---

*References notes follow page 441.

imagery have no free imagination left to work with loving hands upon the necessary present?

For what we need today is imagination, imagination free from sickly nostalgia, free from a terror of machines bred of mediaeval fantasies or from the blind and weather-bound dependence of the peasant or the fisherman. And yet that imagination must not be empty, for an empty imagination and a free imagination are not the same thing. From a room out of which all the devils have been swept come only meditations about other devils or counter-devils. Then the mind is free only to take horns on or off the frightening face of the future. To be really free one must have good fare to eat, adequate for flesh and bone, one must have tools that one can trust, a horse or a ship or a car or a plane with which to travel swift and far as need be; one must have companions for the task in hand, elders whom one can trust and youngsters for whom the effort is worth the making.

There are a host of voices raised today to say that one or another or all of these conditions cannot be met, that there are no good fare, no tools that can be trusted, no steed to be safely mounted, no companions for the task, that we are hopelessly alienated from the old and only fearful of the fate of the young, and so without faith.

This book is set firmly against such pessimism. It is based on the belief that American civilization is not simply the last flower to bloom on the outmoded tree of European history, doomed to perish in a common totalitarian holocaust, but something new and different. American civilization is new because it has come to rest on a philosophy of production and plenty instead of saving and scarcity, and new because the men who built it have themselves incorporated the ability to change and change swiftly as need arises. This book is based on the belief that Americans have something to contribute to a changing world which is precious, which can be used with responsibility, with dedication. But, like all precious things, this essence of American culture is rare and therefore vulnerable; not rare like some precious stone appropriate to an age of kings and rajahs, but rather rare like some one of the rare elements necessary for the expansion of atomic energy, rare in proportion to the needs of the whole world, our twentieth-century measure of value.

This precious quality which Americans have developed, through three and a half centuries of beginning life, over and over, in a virgin land, is a belief that men can learn and change— quickly, happily, without violence, without madness, without coercion, and of their own free will. For three centuries, men of vastly different ways of life have come to America, left behind their old language, their old attachments to land and river, their betters and their subordinates, their kin and their icons, and have learned to speak and walk, to eat and trust, in a new fashion. As we have learned to change ourselves, so we believe that others can change also, and we believe that they will want to change, that men have only to see a better way of life to reach out for it spontaneously. Our faith includes no forebodings about the effect of destroying old customs and calls for no concentration camps or liquidation centres such as have been used in totalitarian states by those with the desire and the power to change others. We do not conceive of people being forcibly changed by other human beings. We conceive of them as seeing a light and following it freely.

There have always been many who have doubted this faith of ours. They have pointed to the loss in America of all the accumulated civilization of Europe, ignoring the fact that most of the Europeans who came to America had not shared that civilization but instead had eaten the black bread of poverty outside the gates of the palaces and opera houses where that civilization was housed. They have pointed to conformity in the United States, to sensitivity of individuals to the opinion of their present peers, and somehow made the willingness to change one's mentors as one changed one's job into something hideous, while contrasting it with the dignity of living all one's life in a distinctive setting, even though in mortal terror of the gibes and jeers of one set of neighbours, gibes and jeers which kept one firmly fixed and so secure in the position in which one was born. People accustomed to lands where men have nourished a sense of difference—and traditionally a sense of the absolute incompatibility among those with different political ideals—where differences in the architecture of the main streets of small cities and little villages were a matter of pride, find the sameness of American towns soul-destroying, and never see that this is also a form of liberation—to be able to move so far and yet find the familiar and the trusted just at hand.

Today this doubt is very deep indeed. It has been fostered by the presence in America of refugees who did not come freely, but who were driven out from countries which they still prefer. It has been fostered by the moves and countermoves inspired by Communism, which has incorporated the standard Russian myths about European civilization. It has been manipulated by the leaders of non-European countries who confuse the retention of various outmoded forms of feudal power with a defence of ancient civilizations against the "vulgarities" of the American way of life, a vulgarization which makes it possible for a simple labourer to buy articles of good design in Woolworth's. So today there is a great doubt in the land, a doubt of our own distinctive heritage, a doubt as to whether we have anything to give to the rest of the world, even a fear that we may be—as our ready critics, especially the ready critics within our doors, are so quick to tell us—offering nothing to the world except the cheap and the destructive, or soft drinks seen not against a poverty which could afford neither bottled drinks nor shoes for their children, but only as beverages lacking in genuine intoxication, fit only for children, and containing sugar which will destroy the teeth.

In accepting this negative image of America, we often feel we are getting closer to, reaching a better understanding with, our sophisticated and cultivated European and Asian friends. Actually we are depriving them of finding something here to value, something that they, who are searching rather more busily than we for ways of change, could use. And we deprive them either way, whether we slavishly agree that America is a dreadful country in which drugstores and conformity contrast in sorry fashion with the ubiquitous culture of the Old World, or whether, still reacting to their negative image, we insist that everything in the United States is better, brighter, and nearer perfect than any-where else. American complacency and bumptiousness was born of just such doubts two centuries ago. It is the voice of the immi-grant assuring the relatives he left behind, and himself, that America is better than Europe. So, in every foreign capital today, the emissaries of American diplomacy, the Point Four men, the journalists, jostle one another in their laments and counter-laments, seeing America through this smoke screen of the feared judgement of other, older countries, in turn denying and trucu-lently defending our institutions.

Meanwhile, our genuine heritage, our personal knowledge of change, is denied and forgotten, as false prophets seek to change our priceless inheritance of political innovation and flexibility into some untouchable fetish of unchangeableness.

This book—the record of a people who have moved faster than any people of whom we have records, a people who have moved in fifty years from darkest savagery to the twentieth cen-tury, men who have skipped over thousands of years of history in just the last twenty-five years—is offered as food for the imagina-tion of Americans, whom the people of Manus so deeply admire. It is no accident that a people who represent a civilization built on change should catch the imagination of a primitive people intent on changing. Every mile of both my voyages to Manus is relevant to the whole problem of what American civilization—a civilization dedicated to the proposition that all men are created equal, created with a right of equal access to all that men have learned and made and won, a civilization made of men who changed after they were grown—has to give, to Americans and to the peoples of the world with whom we work.

The Manus people know this is so. They know that I would never have come back again if they had not changed. They know that when they beat the death drums for my departure a quarter of a century ago, I had recorded all that they could give, all that the modern world could learn from them. They know that I came back because I had heard that they had changed more remarkably and more drastically than any other of the peoples of the Pacific, that after twenty-five years of no contact between them and myself it was their act and our need that took me back. They understand that we need to know *how fast a people can change* and they know they have contributed to the answer. "Remember," said Samol to his village council, "that for these months everything you do will be recorded, filmed, put on tape . . . and *all America* will know whether we are succeeding in our new way of life." During the war, as a million of our troops poured through the Admiralty Islands, a mere thirteen thousand people were the audience, weighing the behaviour of one American to another, building what they learned into the background for a new way of life. Now, we in turn—or so they understand from what I told them—become an audience, many to their few, weighing their behaviour, one to another, as they attempt the fastest change in modern history, and learning from them something about the nature of change itself.

Twenty-five years ago, we had learned, just learned, that we could gain much from the disciplined study of primitive people, that here was a priceless laboratory in which we could investigate the possibilities inherent in human nature. Exploration of the ways of life of savages, as materials for art, for philosophy, for history, was not new. But the calculated use of a primitive culture to throw light on contemporary problems was new. Institutions which made it possible for an American anthropologist to travel the nine thousand miles to Manus, to remain there seven months, to learn the language and record the life of the people, were also brand-new. Anthropology was just beginning to be taught in universities. The Social Science Research Council which gave me a fellowship to go to study the Manus had recently been set up to channel foundation funds into research relevant to an understanding of society. In faraway Australia, three weeks by boat in those days, a Department of Anthropology had been set up in an Australian university, chaired by an English professor, who in his theoretical approach represented the best of

English and French social science thinking, just as my own Professor Boas represented the best of American and German thinking. It was a grant from the Rockefeller Foundation which had helped establish the Australian National Research Council, through which anthropological work in that part of the world could be administered.

The choice of the Admiralty Islands was made because no modern ethnographic work had been done there. From a district officer attending the course for New Guinea administrators at Sydney University, we learned a little about conditions in the Admiralties. He recommended the South Coast Manus because it was possible to go everywhere by canoe. The problem I went to study was whether children in other societies, in contrast to adults, were as animistic and magical in their thinking as they were said to be in Europe. This problem came from the combined theories of Freud, a Viennese, Lévy-Bruhl, a Frenchman, and Piaget, a Swiss. The use of primitive cultures to test psychological theory was a very new approach, prefigured once before in the psychological explorations of Rivers in the Cambridge Torres Straits Expedition of 1898 to the same part of the world.[4]

For the first time since the development of science, the scientific world of the West was ready to use constructively and imaginatively the priceless and so-soon-to-vanish living behaviour of those people who had not yet come to share in any one of the great streams of civilization. Instead of armchair ruminations on the records of travellers and missionaries and colonial officials, problems could now be set up which could be answered—not by turning human beings into experimental animals, but by scientifically controlled observation of the living stuff of history. We realized that if we could go and study carefully the diverse ways of different groups of human beings, like us in body and brain, strangely unlike us in all of their learned behaviour, we could add enormously to our knowledge of human potentialities.

The research that I did in Manus twenty-five years ago depended upon this new climate of opinion, the compounded theoretical work of Europe and America, and the beginning of organized financing for work in the social sciences. Without the record of twenty-five years ago, this present study would be a shallow record of a people who appeared to be interested in modernization or, less charitably interpreted, of a people who had been upset by contact with World War II armies or, romantically seen, of a people whose beautiful primitive civilization with its own style and dignity had been ruined by contact with the modern industrialized world. Without a knowledge of what life had been like before, we would be unable to estimate whether, when the people say that life is better today, they mean that there is less quarrelling and more co-operation, less fear and more friendship, whether they are describing a reality or speaking from an ideological platform.

In 1951, I decided it was time for those of us who had given up our major task—studying primitive peoples as a way of throwing light on the processes of human society—for wartime work on problems of morale, communication among allies, and psychological warfare against totalitarian forces,[5] to go back to our laboratories in the jungles, on the small islands, around the arctic fringes of the world. Our practice had outrun its theoretical base; we were over-drawn, just as Rabi[6] described the physicists who returned after the war from their atomic ventures to their laboratories.

But where to go first? What was the most pressing problem? Here the discipline of the war years, when there were so few of us that allocating anthropologists to problems had to be conducted like a major deployment in which an army of a hundred confronts a thousand-mile front, came to my aid. It was no longer possible to live comfortably inside a developing science, choosing one's problems by what seemed to be a pleasantly logical next step. When I selected the problem of animistic thinking in 1928, this was just a next step. We needed to know more about the relationship between the child's unfolding mind and the culture within which it unfolded, and the study of adolescence had taught me that adolescence was too late an age at which to start.[7] So I proposed to study, in the quaint ethnocentric language of the twenties, "the thought of the pre-school child in a primitive society." And a responsible research council gave me the funds to do so.

Five years later in Germany, Hitler had taken power, and those who presented evidence of man's capacity to learn and change, regardless of race, were exiled or silenced. In the Soviet Union, those who presented the complementary evidence that heredity was important in the individual descendants of family lines—as in the Soviet twin study—were liquidated or silenced. The interdependence between political climates of opinion and the ideas which came from research on human behaviour was vividly pointed up. We became more articulately responsible for the problems we chose to concentrate upon. With Nazism the immediate threat, it became important to explore very thoroughly just how individuals, born into a particular society, became members of its culture regardless of their initial racial inheritance or the culture in which their ancestors had lived. Unless we could spell out, step by step, how a human baby, capable of learning any culture, learned completely to be a member of one culture, the racist myth, with its dangerously appealing and glib generalizations, its easy reliance upon the comforts of physical similarity, its irresponsible disposal of three-quarters of the world, might prevail.

So the study of personality in culture, of character formation, of the relationship between political institutions and the character structure of those who developed them, or lived within them, became a focus of research. The emphasis was on human plasticity, on how, what, when, under what conditions, a human being learned to be a savage, a peasant, an urban worker, European or Asian or African, to speak and act and feel like his fellow members of one society instead of another. In this emphasis on learning, the problem of the importance of individual innate capacities, or the possible innate capacities of certain physical types which are found in all racial groups, was not forgotten, but it was kept in the background unemphasized. A responsibility to the democratic ethic, created by the threat of Nazism, dictated a muting of one set of problems and the energetic pursuit of another set.[8]

Twelve years of asking what anthropology had to contribute to the concrete problems of World War II, and of the cold war that followed it, disciplined us still further. It was now no longer a question of *not confining* ourselves to the next steps in research which were dictated from inside our own or related disciplines, nor was it a question of emphasizing in the choice of problems for study those problems which seemed *least* likely to bolster totalitarianism and most likely to underwrite the intellectual faith and practice of its enemies. Faced with the world of the nineteen-fifties, it was necessary to try to match what was needed, not this time simply in a war but in the world, with what one could do. Keeping even the most humble talent wrapped in a napkin becomes the more reprehensible the greater the emergency. And the mid-twentieth century is an emergency for humankind. In anthropology, previous field experience is a precious asset.[9] Our capacity to see, to recognize, to isolate significant variables, is a function of which other peoples, how many other cultural groups, we have seen and studied, for how long and with what conceptual and practical tools. Deciding what one ought to do next is tied tightly to what one has done before—in the social sciences, long experience is the analogue of the rigorous formulations essential in the natural sciences.

The most pressing problem, in the range of problems which anthropology was equipped to attack, seemed to me to be how change occurred within a single generation. All the world was on the move. Peasant, feudal, and primitive economies were crumbling before the onset of new ideas and new technologies. Traditional faith and traditional practice were disappearing. What was happening to those who were asked to skip centuries in the way they ordered their lives? How did these rapid changes inevitably involve those individuals who lived through them in disturbances of personality which would leave their mark on society for many generations to come? So often in history, those who killed their rulers have turned to killing each other, or those who have rejected their fathers also, in turn, have rejected and have been rejected by their children. If we knew what happened in the lives of given individuals involved in such changes, we would be better able to plan for a world that was taking on new shape before our eyes.

So when I made a trip to Australia to reconnoitre the general condition of my "laboratories," the primitive peoples of the southwest Pacific, my sense of the urgency of this problem was accentuated by the insistence of my anthropological friends in Australia that I go back to the Manus of the Admiralties, to the people I had studied as children twenty-five years ago, to a people who were changing so fast, so unaccountably, that no one knew quite what was happening. An investigation of this rapid change, they argued, might provide a key to our understanding of similar, less rapid changes all over the world.

Because this task would make full use of my particular experience, my previous study of small children who had now grown to leadership, and was relevant to a problem of world-wide urgency, I chose it to do next. The account of what I found is presented here to a generation on whose imagination the future of civilization is most intricately dependent.

But if this record of what happened to one small group of people on one small island is to feed our imaginations, it must be seen in the whole context of history, as men have come into contact with each other, face to face, through trade, or caught a glimpse of some strange other way of life from a drifted bowsprit or a floating coconut which could be planted and grown into a tree the shape of which no one could guess. Since the dawn of history, human culture has grown through such contacts, through the stimulation of strange ideas, the interchange and interlocking of human inventions, and the interaction between neighbouring tribes, between trader, missionary, explorer, and the people among whom they went, between conquered and conqueror, colonizer and native, displaced rulers and the new class who displaced them.

This whole process anthropologists have called "culture contact,"[10] emphasizing that men may experience the organized shared experience, the *culture* of other peoples, in many ways, by seeing or touching, dissecting or analyzing a strange object, planting a new seed, redomesticating an animal new to their shores— as the American Indians did with the horse—savouring the spices and silks bought through the China trade, listening to the tales of crusaders to the Holy Land, or whalers sailing the Seven Seas. The more dramatic forms of culture contact take place directly between peoples whose ways of life are strongly contrasting, between English colonist and American Indian or Australian aborigine, between Belgians, Frenchmen, Germans, Englishmen and the jungle inhabitants of Africa, where there is a clash between a complex old way of life and a new, with such differing reactions as Japan admitting the West, or old China shutting it out. But these are all part of the same process by which men, temporarily isolated enough to develop distinctive ways of their own, share in and modify the ways of other men, and the strength of human culture is enriched thereby. The contact of this handful of Admiralty Islanders with German colonists, Australian expeditionary forces and administrators, German, Australian, and American missionaries, Japanese conquerors, American and Australian forces, and finally post–World War II Australian administrators—all bringing a knowledge of the technology, law, and religion of the civilizations of the West—to be appreciated must be seen within this world-wide, history-long context.

So also the sequence of particular events which occurred is important—the submission to German force before 1914, which left the Manus still savages, but afraid any longer to be warlike; the adjustment to British Law under an Australian League of Nations mandate and to a system of orderly indentured labour and limited trade; the enthusiastic acceptance of the Mission in the thirties; and finally the response to the catastrophic changes of the war with political plans and mystic "cargo cult."

All through history men have greeted periods of change, the challenge coming from foreign ideas or other peoples, in many ways—by imitation, incorporation, rejection, transformation, trance, manifesto, retreat. In Manus, two of the most significant types of social response took place. One, the movement led by the native leader Paliau, which attempted to understand and incorporate the values and institutions of the Western world, to build a real modern culture of its own, complete with democratic government, schools, clinic, universal suffrage, money, individual and community responsibility, was the stuff out of which abiding, steady social change comes. Counterpointed to this, facilitating and retarding, was a nativistic cult, a "cargo cult" called in Manus The Noise, in which men shook like leaves in the grip of a religious revelation that promised them all the blessings of civilization, at once, without an effort on their part except the destruction of everything they still possessed.

From one point of view the whole of history, and very particularly our own American history, can be seen as a struggle between those who seek a utopia here on earth and those who feel that the life of man is made better by ever-changing institutions carefully shaped and daily renewed by human effort. These are history-old, these alternations and conflicts between the proponents of apocalyptic cults, who deny the need for continuous, partial human effort and wait for a supernatural event ushered in by prophet or party, and those who feel it is the duty of priest and prophet, statesman, artist and scientist, to "cherish and protect the lives of men and the life of the world." And no civilization is immune from the temptation to desert the considered pursuit of a better life for the easy, drastic solutions of apocalyptic cults or revolutionary and counter-revolutionary panaceas.

The Admiralty Islands are a small, isolated, remote part of the world. Once, for a brief period, the lives of a million Americans touched directly these faraway people. But their experience is part of our experience as we learn to draw for our inspiration not with outmoded snobbery only on "all the best," but on *all* that has happened in the world.

part one

2

arrival in Peri, 1953

So this is an account of how a people only recently correctly called "savages" have traversed in the short space of twenty-five years a line of development which it took mankind many centuries to cover. It is the story of the particular tribe of the Admiralty Islands—the Manus—whom I saw in 1928, a mere two thousand nearly naked savages, living in pile dwellings in the sea, their earlobes weighed down with shells, their hands still ready to use spears, their anger implemented with magical curses, their morality dependent upon the ghosts of the recently dead. It is the story of a people without history, without any theory of how they came to be, without any belief in a permanent future life, without any knowledge of geography, without writing, without political forms sufficient to unite more than two or three hundred people. It is the story of a people who had become, when I returned to visit them in 1953, potential members of the modern world, with ideas of boundaries in time and space, responsibility to God, enthusiasm for law, and committed to trying to build a democratic community, educate their children, police and landscape their village, care for the old and the sick, and erase age-old hostilities between neighbouring tribes.

In 1928, with training as an anthropologist, I was equipped to understand their tight limited little world, their language spoken by only two thousand people, their complex system of kinship relationships and exchanges of shell money and dog's teeth by which they maintained their precarious hold on the fringes of their island world. I learned to talk with them, to observe their taboos, to ask the right questions when an adult was ill: "Which ghost is striking him down? Why? Has he confessed yet? Has he paid expiation yet?" I learned to interpret sudden sullen periods of silence or angry outbursts of outraged virtue. As long as I spoke in their language and used their ways of thought, they understood me. But they had no way of understanding my ideas and values, even of making the simplest assumptions about why I would tie up their wounds and still less of why I should have left some faraway country and come to live in their village, except for simple crude gain. When I left, every house in the village sounded the death beat on the house drum, quite appropriately, for I was dying to them. No one could write me a letter, and no letter that I could write, if it could have been read, would have had any meaning beyond the mere statement that I was alive somewhere in a world stranger than the abode of their dead.

In 1953, I stepped ashore on the new site of Peri village, now built on land with "American-style" houses, all of the same design, in careful rows. While a clamorous crowd gathered around me in the darkness, I was greeted by a man in carefully ironed white clothes, wearing a tie and shoes, who explained that he was the "council," one of the elected officials of the community.* A few minutes after I arrived a letter was handed to me. Translated, it read: Dear Missus Markrit: This letter is to ask you whether you will help me teach the children.

It was signed by the name of the present locally chosen schoolteacher, who had been a baby in arms when I went away. A few days later, the owner of the house which had been put at my disposal, responsibly, because he was an elected official, brought

---

*Actually, he was only councillor-elect, for the people did not yet have a government-approved village council. In local usage the words "council" and "committee" are used for officials of the New Way rather than "councillor" and "committeemen."

in an exercise book in which he had written down a long set of rules for modern child care, feeding, sleeping, discipline, etc. He said they hadn't promulgated these rules yet but that they were the best they could write from their memories of what they had been told, or what they had seen the wives of Australian officials do in the ports. Would I check them? When I explained that my comments would be in terms of the latest thinking on the matter as developed in an International Seminar on Mental Health and Infant Development, held at Chichester, England, in 1952,[1] under the auspices of the World Federation for Mental Health and the World Health Organization of the United Nations, *he understood what I was saying.* Each item in this complicated modern sentence could now be made meaningful to him.

I had thought I knew what the word *literacy* meant. I had helped prepare the material for the Unesco report on fundamental education;[2] I had listened to impassioned accounts of teaching ordinary Chinese troops to read and write in a simple phonetic script; and I had lived with and spoken the languages of peoples without writing, and had lived with one people, the Balinese, among whom only a few priests and scribes could write. But this was different. The experience of opening this letter, of reading the "notes" written by people whom one had never visualized as becoming part of the modern world, with whom, in truth, there had been no hope of real two-way communication, had a quality that I had never imagined. I felt almost as if someone—and I was not quite sure who it was, they or I—had been raised from the dead. Someone who, not knowing it, had been dead, and lived again.

I realized for the first time the relative emptiness of my earlier years among the people of Peri, among all the non-literate peoples of New Guinea, who had been so deeply comprehensible to me while I remained incomprehensible to them. I had known, of course, that this lack of understanding existed, that, while I could count on affection, on loyalty, and even on some intellectual enjoyment when we discussed some intricate point of kinship terminology or dog's-teeth exchange, I could never hope to be "real" to them in the way they were "real" to one another. Only as they gave me a name—"Piyap" (pronounced *peeyap*), "Woman of the West," or "Markelita" (which I had told the children to call me, as a familiar thread from my previous work in Samoa)—and saw me living in a house over which the ghost of Pwanau presided, could they talk with me as a person. They were clearly very intelligent, more sophisticated than we in knowing that other men had other ways—and were more tolerant of such other ways. They also knew that basically none of this temporary sharing in their culture was "real." The anthropologists in the new house which had been so oddly designed did not belong with them; they were part of the intrusions from the world of the European, intrusions some of which were useful, some troublesome, and all strange.

Quite suddenly that June evening in 1953, I knew that today all of us, the people of Peri and of the other Manus villages, the boys who had run my house for me twenty-five years ago, now tall, mature men with households of their own, and the weight of office on their shoulders, lived in the same world. Twenty-five years ago Kilipak, then a lively fourteen-year-old, had tried hard to follow the feel of what was said. Someone had sent me a picture of a meeting at which I had spoken, and I explained to the boys that these were a group of important men to whom I had been asked to speak. Kilipak had suddenly given me a resounding whack on the shoulders and shouted, "Hurrah Piyap!" applauding some uncomprehended feat, perhaps because it was by a woman. But today I knew I could explain to them just exactly how I had come, that in the councils of the United Nations and in the governments of great nations there were those who were concerned with the problem of how men changed, what were the quickest and safest paths, what happened when people planned the change themselves—as they had done. I explained that I had come back because I had heard how much they had changed, how they, of their own expressed desires, had taken their old culture apart piece by piece and put it together in a new way, in a way which they, who knew it best, thought would make it work to achieve their new goals—a modern Manus way of life, bounded and responsible, in which all were citizens, to which all were proud to contribute, within which those who had been reared to a different way of life would do their best to educate their children to full membership.

I found that they—these people whom I had known as young men and adolescents and children, impervious and uncaring about any issue beyond each immediate, inwardly warring household—were facing decisions as complex as any that face us, and facing them with full consciousness. What was the relationship between the way children were reared, even the way in which they were carried, and the kind of people they would become? How should the authority of school and home be divided? How did a society get rid of old ideas about exclusive private property, and yet make the new users of resources, now opened to all, responsible instead of wanton wasters of what was owned by all? How was collective responsibility to the community to be reconciled with the new rights which had just been accorded to individuals, the right to choose a spouse, to work as one wished, to keep what one earned instead of being under the control of older kinsmen who held one in economic bondage? How were stable marriages and ideas of individual sexual freedom of choice to be combined?

It seemed incredible, for after all these were the same boys who twenty-five years before were an integral part of their native system in which elders exploited the young for gain, in which ghosts punished with illness and death the slightest infraction of the rigid sexual code set up to protect the property invested in marriages—the dog's teeth, gallons of oil, tons of sago—and where the air of the village was blue with imprecations over the breaking of a small cooking pot. This was the same village in which Korotan, the blind old war leader, had invoked magical death upon the children of his economic rivals; the same village in which women were mercilessly driven into marriages which they continued to fear and hate; the same village in which there had been continuous exacting observation of taboos between all in-laws, and in which there had very recently been living in the men's house war-captured prostitutes whom the men had had to take fishing with them to protect them from the short obsidian daggers of the jealous, joyless women.

Was this Manuwai who now fetched me to come and taste the residue from making coconut oil—for all the world like an invitation to scrape the cooky tins—and, as we scraped the pot, said, so happily, while his young wife knelt by the fire and her mother stood beside her, "*You* know what this means. *You* know that before I could never have spoken to her mother, that I could never have called her, my wife, Elisabeth, by name, never have sat down with her or eaten with her. Now it is different!"? In the flickering light, for kerosene is still costly and more precious than ever now that it is used for fishing, he beamed affectionately at the two women toward whom he would have felt shame and aversion before.

Was it really Manuwai the day there was a village vote on an important question, and the vote was taken by having people on each side of the question stand at opposite ends of the village square, was it Manuwai who went and stood with the weaker side, saying, "They will lose, but they will not mind it so much if I stand with them"? And on that night when I was worried by a message from the village where my two student assistants were located, which informed me that one of them was very ill, was it Manuwai who said that if I was bent on going at once, which meant going through a crocodile-infested, mangrove-impeded river passage, for the tide was out and the wind was up so the sea route was not practicable, he would be one of the boat crew, for this was no work for inexperienced boys? For if Manuwai had grown up to be a man like his father, Pwisio—and in temperament he was very like his father—he would have haggled and haggled before he would even have rented me a canoe for someone else to punt for such a journey. Was this gentle massive man, so uxoriously devoted to his young wife, so protectively possessive

toward my comfort, really Manuwai, whose ear-piercing I had
described twenty-five years before when I wrote:

> Or take the feast for ear piercing held in Pwisio's house.
> The house is full of visitors, all the relatives of Pwisio's wife are
> there, with laden canoes to celebrate the ear piercing of
> Pwisio's sixteen-year-old son, Manuwai. In the front of the
> house all is formal. Manuwai, in a choker of dog's teeth,
> painted and greased, sits up very straight. His father's two sis-
> ters are waiting to lead him down the ladder. But his mother is
> not there. From the curtained back of the house come sounds
> of weeping and the low-voiced expostulation of many women.
> In the front sits Pwisio, facing his guests but pausing to hurl
> insult after insult at his wife whom he had caught sleeping
> naked. (There were strangers in the house, and during the
> night an unwedded youth, a friend of her son's, had stirred the
> house fire into a blaze.) So Pwisio overwhelms his wife with
> obloquy, fearful to beat her while so many of her kin are in the
> house, and she packs her belongings, tearfully protesting her
> innocence and angrily enumerating the valuables she's taking
> with her. "This is mine. I made it, and my sister gave me these
> shell beads. These are mine. I traded the materials myself. This
> belt is mine; I got it in return for sago at the birth feast last
> week." Her little adopted daughter Ngalowen, aged four,
> stands aside in shame from her mother whom her father
> brands thus publicly as a criminal. When her mother gathers
> up her boxes and marches out the back door, Ngalowen makes
> no move to follow. Instead, she slips into the front room and
> cuddles down beside her self-righteous and muttering father.
> After the long confusion, the ceremony is resumed; the ab-
> sence of the mother who would have had no official part in it
> receives no further comment.

Today Manuwai's name is Raphael, after one of the better-
loved angels, and I found that I was grateful for the new name,

that, although it was so easy, so simple to trace the features and voice of the boy I had remembered, and although both of us recalled vividly many tiny details of those distant years, still the conflict in my mind between the man he would have become under the old system and the man he was now was eased by the two names, Manuwai for the boy of old Manus, Raphael for the responsible citizen of the new, "RM" in my notes as a final resolution of the two images. Yet Raphael was, in another sense, just what I might have expected—very bright, a leader, his anger under control, patient and persistent, and for vices, a little greedy and more than a little vain. His essential personality had not changed, only the values within which he exercised it had been radically altered, so that when his voice was raised in anger it was for other reasons, and when he spoke of being a member of the ranking clan of the village he remembered to add, "but we don't speak of such things any more. Now all men are to be treated alike, we make no distinctions between men of rank and 'rubbish men.' "

Pomat had been one of my group of young adolescents, and even then I called him my "butler" because he took his duties with such pomp and seriousness. His great shock of Melanesian hair, the kind of hair which stays up when it is combed so that it can be combed into any shape, seemed more unruly now that he wore European clothes than it had seemed as a small boy. He was the son of the most brilliant woman medium in the village, Isali. Isali, ruthless, forceful, a member of the ranking clan, had manipulated her solid, reliable commoner husband into becoming a rich man, presided over and helped by the ghosts of her kin who did her bidding. Petrus Pomat, and here again I was glad of an added new name to relate the present to the past, told me, in tones which relived the tragedy, of how his mother had been shot down by an American machine gun as she had tried to hide under a house, holding his child in her arms. "Even though she was dead, she held the child fast until I could come and take it from her dead arms. Then I took the child, and she floated away." And then he added bitterly, all his anger for the Japanese, the deeply admired Americans completely exonerated, "The Japanese lied to us, they said they were their planes."

Petrus was, as he had promised to be, a pillar of society. When I asked him to describe his principal weakness, he gave me a long list of sins of envy, hatred, and malice. But when I pressed him, he added, "No, of course I do not do these things, these are just the things all human beings do." But still I had to rub my eyes to be sure I wasn't dreaming the night we sat on my verandah, after a protracted town meeting had finally dispersed and the square was empty in the moonlight, and discussed, as two group dynamics leaders might have, "what had gone wrong with the meeting." I pointed out that he had counted on the young clerk making a longer speech, and that when he had not, Petrus had not got to his feet quickly enough, that he was getting too accustomed to making the summary speech always. The village was very quiet, people had stumbled sleepily home to bed, the fishing boats were still out, with their tirelessly reworked "American" pressure lamps providing the light for fishing. For a brief moment not one of the village babies had wakened crying lustily for absent fathers or immediate food, and Petrus sat, his great uncouth brow wrinkled, worrying and questioning me about the unfamiliar processes of democracy.

My memory went back to his father's raging speeches against his daughter, who, although betrothed, had been seduced by an unstable boy, long since dead. I remembered how his father's wrath had been turned in a storm of threat, fury, and inconsistency against his daughter, against his young unstable cousin, the seducer, against the angry men of the clan of his daughter's fiancé. I then swung back again to 1953, to the careful but too tardy speech that Petrus had made that evening, to the speeches that he, as councillor-elect and "judge," habitually made to calm the disputants in a local case:

> "All of us are human,
> All of us are weak,
> You do wrong and come before the court,

> I also do wrong and if I do I must come before the
>   court,
> None of us is without blame,
> This is the fashion of humankind."

Petrus, too, was explicable, trusted by all the village, as his father, a man of solidity and integrity, had been.[3] Only the ethical system and the political processes were changed almost beyond recognition.

Kilipak was Petrus Pomat's cousin. Petrus' dead mother, Isali,[4] was his paternal aunt, and one of the first stories he told me, as I struggled to fit together his haunting, hawklike features with the lively expressiveness of the most gifted of my small boys, was how Isali had nursed him for months the time that an epidemic of dysentery had hit the village in the late nineteen-thirties. "They put me in an empty house, and I left my wife and children in my house. Petrus nursed me at night and Isali in the daytime; one of them was always with me. Whatever I fancied, Petrus would scour the hills to find. When we came ashore, he and I built our houses close together, side by side in the front line [i.e., facing the public square]."

I caught a glimpse of what had happened back in 1946 when the New Way had swept through Manus villages, and under the heading of the new brotherhood of man, Kilipak had realized that he and his cousin Petrus Pomat would remain friends in their mature years and would never have to become harsh, demanding, goading rivals, essential to each other only because on the repayment which one made to the other depended one's financial status, health, and the very life of one's children. Under the old system, Petrus Pomat and John Kilipak—he had been christened Johanis in 1932 and when the Americans came he had changed it to John—once they were important men would have been tied together like the presidents of rival banking houses in a small city, but they could no longer have been friends. They would have visited each other's houses only formally, for a preview of important payments of dog's teeth and shell money. Now

John could say, "We built our houses side by side." The devoted friendships of the small boys, who had no property about which to quarrel, had been part of the repertory of young Manus men, a base out of which the new, more friendly social system had been built.

John Kilipak went on to speak of Kutan, the one boy of their age group who had not worked at my house. Kutan would also have been an economic rival under the old system, but had instead remained John's close friend. Later on, John and Kutan built the canoe to fit our outboard motor, working in the old style but with "American tools," grown worn through the years since the Americans went away, working together, without ever looking up, in rhythm, one at one end of the canoe, one at the other. In the past a man might work so with a younger brother or a young dependent relative, or some economic hanger-on, but not with a man of equal strength and worth. Now John could say, "I am bored tonight. The village seems empty when Kutan is away."

For all of them, the past was still vivid, but bathed in the lights and shadows of moonlight. After I had been in the village only a week, a volcano came up out of the sea, and, in fear of a tidal wave which would have swept the sea-level village quite away, we were ordered to evacuate the village. After we had watched the volcano all day with a sense of unbelief, for no one had ever seen an active volcano in Manus, a police boy appeared near midnight with the District Commissioner's letter. I sent for the village authorities, and they beat on the large pendent iron gong, made of a United States acetylene tank, which had replaced the old wooden slit gongs that had been thrown away at the time of the religious outbreak in 1947, The Noise, when all the old things had been thrown into the sea. The people gathered quickly, quietly, and a plan was made to go to the heights behind Patusi where there had been, so the people said, a police patrol station during the war. It was high ground where I had never been. There were some old houses there, they said. As my contribution to the well-planned exodus, I had all the lamps in the village brought and filled, and we all set about packing essentials. Not one child cried. It was bright moonlight as canoe after canoe, still of the traditional single outrigger type, laden with all the food on hand, left the village. I did not know then how little food people kept on hand. In the old days, people used to pile up hundreds of pounds of food for use in the great economic exchanges, and such food was safe against the importunities of hungry kinsmen or would-be borrowers or buyers. But today, with all of the big exchanges abolished, there are no such protections. People live a more hand-to-mouth existence, keeping just as little as possible on hand, building up longer-term credit with the landspeople with whom they still trade in biweekly markets.

In the emergency, people's roles stood out sharply. Stefan Kapeli, the fourth of my boys, the odd, stubborn child of a ramshackle family, the only one who years before had never run away, and who had now come back to work for me, was a mixture of leechlike possessiveness—officiously supervising the packing of all the native foods I had in stock—and worried dependency: Should he take his four-day-old baby with him? If so, could he have a special lantern? Old Polkanau, who had been our best informant in 1928, and who was now the grand old man of the village, alive beyond his life expectancy and spoken of as the "lawyer man within the village" or just "our grand old man," announced he would stay till the very end and be the last to leave the new village about which he was as enthusiastic as were the younger men.

The big canoe which had been built for the use of the schoolboys was commandeered for my heavier stores—typewriter, tape recorder, camera box, medicine, rice, tinned meat, and kerosene—and loaded with women and children, including Stefan's newborn baby. John Kilipak, still lithe and quick as a boy, stepped forward to "captain" it. He had sent his wife and children ahead with his younger brother. Petrus Pomat, Simeon Kutan, Raphael Manuwai, with little flotillas of small canoes or single large ones, had gone on ahead. On the shore, looking worried and anxious, stood another councillor, Karol Manoi, the New Manus man,

professional official of the New Way, who had rented his house
to me to live in. What if something should be stolen from my
house, which was impossible to lock, built as it was with walls of
leaf and split sago palm? He would stay and watch, especially
since a few of the unattached young men had announced their
intention of also staying, and he didn't trust them. If I would
leave him a flashlight, he would stay and guard the house. So with
Karol Manoi, the immigrant from the next village of Patusi, exhibitionistically devoting himself to Law and Order, new style, and
Pokanau standing, a rocklike representative of all the villages he
had seen his people build—seven in the sea after inter-village
quarrels, and now this new one on the land—John Kilipak, heir
of the aristocratic clan, punted the great canoe out into the shallow lagoon.

In the moonlight we passed first the mouth of the mangrove-
arched river along the banks of which the people of Peri village
had hidden during the war while the Americans were bombing
out the Japanese. John told stories of the way in which the people
of Peri had fled together, silently mourning the slain Isali, and
slept that first night without lights or shelter on land haunted by
the demons of the land people, with whom they were still at odds.

Then we came to the site of the old village, where the grey,
weathered tops of the old house-posts made a diagram on the surface of the water of the position of the houses as they had stood
right after the war, just before the village was abandoned forever.
The tiny, steep islands were heavily covered with foliage now; the
three flat coral feasting platforms, beside one of which my old
house had stood, were overgrown with trees. Halfway between
Old Peri and Patusi, as we followed the route that the canoe had
taken and that had carried me away from Manus twenty-five
years before, we passed a single broken-down house, where one of
the two men who had remained loyal to the Mission had lived
until very recently, when he, too, had gone ashore. (It was not
until much later that I learned he was the man who twenty-five
years before, as a returned work boy, scornful of the local ghosts,
had experimented with hanging charms on the backs of young

female cousins and then seducing them hoping he would be immune from supernatural punishment.)

Leaving the old village behind, John Kilipak spoke gently of the big men of the past, of his grief in thinking of them, and I heard for the first time what I was to hear whenever the old village was mentioned, the felicitous mixture of unrepudiating regret for old scenes and individuals, whose role memory had softened, and a forthright and complete repudiation of the old way of life, where property was the greatest value, where men were separated from each other by economic rivalry and social prejudice (lagoon people against land people, village against village, aristocrat against commoner, husband against wife), where no one was free to choose his own course. For the household ghosts who had been dethroned long ago, in the early thirties when the people as a group embraced Catholicism, there was almost no emotion left, except amused laughter over particular episodes. People spoke of them like the outgrown bugaboos of nursery days, and indeed that was what they were, for these men, as children, had given little heed to the ghosts who presided over their elders' lives. Later, when I showed them photographs of their old way of life, these grown men of forty would laugh in an amused, almost tender fashion at pictures of their dead elders, wreathed in leaves and dog's teeth, brandishing obsidian-pointed spears and shell-bedecked phalluses at each other. It was only the very young men, under twenty-five, who disliked the pictures of a past which they had never known. For them the old ornaments and flamboyant hairdressing, now all abandoned, were symbols of pure evil, for their first vivid social experience had been seeing the elders inaugurate the new order by pitching strands of dog's teeth, spears and daggers, ornamented baskets, and carved slit gongs into the sea. But where there was memory, the tenderness remained, sometimes with a little puzzlement, as when Kutan said to me, "We had lost these [pictures of the old village and the big men of the past] and you bring them back to us. You have had them all the time, but for us, where have they been?"

So, in the bright moonlight, without a ruffle of wind, our

canoe, the last of the exodus, drew into a narrow cleft in the high bank, and we clambered up the steep cliff. The infertile red clay which promises Manus so slight an economic future was already reduced to slippery mud beneath the many feet which had climbed that way this night. At the top of the cliff, we found an empty patrol post, falling to pieces, but still habitable. There was a large European-style house and four large police houses, rickety, leaky, but all far better than no roofs at all. The single guard, a member of a nearby village of the land people, had left that morning to walk overland to Lorengau to accuse some children of having stolen a mislaid knife; so the whole post was empty.

The people were uneasy at first, for trespass on the police post was forbidden by the government. But with their flexibility in translating from one situation to another and their enhanced sense of their own dignity, they easily accepted my statement that this was like war, an emergency in which ordinary rules were laid aside and one thought mainly of the safety of people, especially the young, the weak, and the old. I settled some seventy people in the big house with me—the others were already billeted into the police houses—and distributed a very little of the stick tobacco, which had to serve both to buoy up flagging spirits and as currency to buy food from the land people.

While waiting for news from the district office, we spent five days on that hilltop, with a clear view of the great spout of white smoke from the volcano, looking directly down into the green pattern of the old village. We watched an airplane circle out over the volcano and go away. Each day some of the older men went to the markets or into the hills to trade for food with the land people, who were becoming more and more frightened, until finally on the fourth day they abolished markets altogether until the emergency was over. The people were very quiet. They organized the distribution of food and the conservation of water, and waited. "We know," said Petrus, "about war, about war planes, about bombs and machine guns. This fire is something which we do not yet understand. Until we understand it, we will wait, and not go fishing, for we do not yet know what the danger is."

It was not until the third day that old Pokanau, the historian of Peri, the man most interested and best versed in the past, remembered that his grandfather had told him that once, when Pokanau's father was a little boy, there had been a great wave which had destroyed the villages located out near the smaller islands, and had—how laughable it was to think of it even now!—left a woman of one of the coastal villages stranded in the top of a mangrove tree. On the fifth day, when there was still no word from the district office, a big meeting was held. The whole community was seated on the ground in little clusters, young men here, there a group of older men and children, or a group mostly of women, while the leaders moved about, disappearing without fanfare from one spot, appearing noticeably in another, occasionally rising to make a speech to which everyone listened without looking toward the speaker. I saw how the old speech-making habit of each man speaking from his own house platform or canoe—shouting out into the lagoon while people listened from their houses—still held, although the words were so different.

They had asked me to speak first, and I spoke in Neo-Melanesian, the lingua franca which was like Manus in grammar but with a large vocabulary derived from English, for my Manus, unused for twenty-five years, was not yet trustworthy enough for such speech-making. I went over the conditions of the evacuation, what the District Commissioner had said, what we had done, what the dangers were. I enlarged on the danger of a tidal wave in Peri to anyone who could not climb a coconut tree and said that I thought it would be safe for grown men and strong women, who weren't either pregnant, old, or encumbered by small children, to go back during the day when they could keep a sharp lookout, but that I proposed to stay until I had word from the District Commissioner that it was safe to return, and that I could keep the older women and the women with young children with me, holding fast to some reserve food for them. At my request old Pokanau repeated what I had said in Manus.

Then, in the speeches that followed, I had a chance to watch

the new democracy at work. Karol, the New Manus man, who
had come from Peri after a couple of days of guarding my goods
had convinced him they were safe, spoke first, emphasizing that it
was "our own business, what we do. No one is forcing us to do
anything. It is we ourselves who must decide." He sat down. No
one answered. Several people changed their position. Poli, who,
next to Pokanau, was the most respected older man, stood up and
suggested sending a canoe to Lorengau to ask the District Commissioner's will. Then Kaloi spoke, a heavy white-haired man of
perhaps forty-five, already moving into the ranks of the old. He
emphasized that the "talk" had come from the District Commissioner and from me, and that he would follow my lead; his children were too young to escape from the volcanic fire. As the
speeches wore on, the elected leaders stressed that it was their role
to follow the will of the people; if half the people returned and
half stayed, then they must divide their time between them.
Pokanau repeated the whole story of our recent evacuation over
again, point by point.

And from two men, one, Lukas Kapamalae whom I had
known as an eleven-year-old, the other an illiterate, very powerful
immigrant who acted as a lay preacher, I heard two notes that
contrasted with the orderly, dignified sense of decision and
responsibility in the voices of the others. Kapamalae, a huge man,
resembling his dead father—whom I had once nicknamed "the
Australian aborigine" for his un-Manuslike appearance—spoke
in Manus in a loud, raucous voice, "I want to go back to the village. God sent this [volcano]. This is God's affair. This fire came
because our thoughts were evil. If I go up into the mountains, the
fire can burn me. If I go back to the village, the fire can burn me.
I want to go back to the village, and I am going." The other
speaker, Lukas Banyalo, spoke with a note of high hysteria,
describing himself as a man of no importance, just a rat, a little
rat. He couldn't give decisions, he would follow my advice; he'd
stay. If others wanted to go, they could, and people shouldn't be
angry at him. He'd help those who remained to worship. The

whole speech rang out of key, but I knew too little of the New Way, of the struggle between slow measured effort and apocalyptic vision of immediate world-shaking events to understand its special tone. Kapamalae's violent self-assertion was explicable, even though I did not know that in the early days of the mystical movement he had been a principal dreamer of prophetic dreams. But the other man, Lukas Banyalo, a man of early middle age, with great physical strength, who called himself a rat, made no sense at all, for I did not yet know what being consciously illiterate could do to such a man's sense of self-esteem, especially to a man who had twice had religious seizures of great violence, whose source he distrusted.

Petrus Pomat spoke, and, although the listening group did not look up, I got a feeling from their bowed heads, "Finally, there will be sense." Petrus spoke quietly:

> "Yesterday, you discussed, I did not,
> I kept my own counsel, and now I would like to say
>   what I think.
> Should I bid you all go back—that I cannot do.
> The Europeans understand this thing,
> Presently word from the Commissioner will come to
>   take us all back,
> But this thing is still here.
> I [as an elected official] cannot tell you to go back,
> Nor can I tell you to stay.
> This [volcanic fire] whether it comes from God or not
> Is something I do not know,
> But I do know that we have a District Commissioner,
> If he says to go back to the village, I will go.
> If he says for me to go down, I'll go down,
> I cannot flout the Commissioner's word.
> This is my thinking in the matter.
> For myself, I would stay here.
> If I see all of you return
> I will follow you."

Then Alois, whom I had known twenty-five years before as a particularly miserable bridegroom, made a brief, confused speech:

> "I'd like to go back to my village,
> I'll ask my wife and children,
> If they wish to return, all right, it's up to them.
> Now the government has made the *luluai* the boss,
> Now the council[lors], together with the *luluai*, together
> with the committee[men] are all here,
> The Commissioner has spoken, I will stay.
> You [councillors and committeemen], you tell us,
> If you want to stay, I'll stay.
> I say, if my council[lor] goes, I'll go,
> If my council[lor] stays, I will stay."

Here a diversion was created by a high-tempered old woman who made a comic speech about her husband, already returned to the village—a matter over which she had loudly and publicly lamented earlier. People laughed; women were now supposed to be fully emancipated. Only Stefan, perennially disgruntled, muttered rudely, "We heard you. Sit down!"

There was a general rustle in the crowd; the talk was really over. Karol, the other councillor-elect besides Petrus, stood up to say that if people decided on two courses of action he would try to divide himself between them. Then young Peranis, the schoolteacher and clerk, the grandson of the old war leader, who represented both the pride of Old Peri, the traditional lore of Pokanau, who had tutored him, and the weight of record keeping and teaching in the New Way, stood up:

> "Now all the officials have spoken
> And they have all said the same thing.
> There is no man with power enough to make me go
> back to the village.
> There will be no possibility for trouble later
> [over their having given mistaken orders],

If the volcanic fire sends a tidal wave to kill us
Men, women, and children,
All of it is written down [in Missus Markrit's book].
It is recorded.
As for me, I want to return to the village,
But my wife, if she wishes, may remain here.
As for me, I will go."

Then John Kilipak stood up:

"There are now two ways of life here among us,
Among us, the people of the ground.
One is the way of the present,
The other is the way of the past, an evil way,
A way of anger and quarrelling and flouting of law.
I wish to stay here. . . .
Those we have elected can follow the style of the captain
    [of a ship],
They cannot go ahead of you,
They must come behind you.
My thought is that none should go
Lest trouble come up.
I myself am afraid,
I do not wish to perish in this fire.
I have a wife and children,
My children are too little to flee from the fire.
I wish to follow the advice of the government,
When they say I can return, I will return,
Meanwhile, my wife and I stay here.
I have work in the village.
During the day I will go to the village and work,
At night I will return here and sleep."

John sat down beside Petrus Pomat. Here the matter might
have rested, with the decision left to each individual, and the officials pledged to trying to do the best they could with the decision.

But Karol, the immigrant, Karol the New Manus man, anxious
and didactic, could not let well enough alone. He stood up to
deliver a long scolding harangue, the type of harangue which I
was to hear so often in the months to come:

> "Where is the road of [true] thought?
> Where is the road of knowledge?
> Where is the road of good men?
> Where is the straight road?
> You men and women and children,
> You talk of your bodies,
> You can attain to nothing thinking thus.
> You cannot attain strength.
> You men and women and children,
> You have no more thought of the Way.
> You have forgotten the Way again,
> In the future, you will find knowledge where?
> You and I think of our bodies, that is all,
> All the good ways of thinking are lost,
> No good way of thinking appears,
> All of us are lost.
> Today no single good way of thought came up,
> We walk about in confusion,
> We no longer understand anything.
> However, let this discussion finish.
> I just mention this,
> Whoever wants to return to the village
> Let them return,
> It is their own business,
> Not ours."

And old Pokanau, "the lawyer man" within the village, said in a
loud voice, *Kinepwen* [It is finished].

The short tropical twilight was closing down on us. The people stood up and went back to their temporary houses. In the big
house we carefully divided up the food that had been bought dur-

ing the day. As they had done before, the officials all returned their shares to the common pool. I counted over those who had gone back to the village and those who had remained. This proved to be, although I did not know it then, an accurate index of social responsibility. Those who had gone back, sure that God could destroy them anywhere so that mundane precautions or attention to governmental warnings were unnecessary, were those who had originally been active in the mystical "cargo cult," and those who were to be active in it again.[5] The responsible core of the village leadership remained on the hilltop.

That afternoon I had had a preview of the principal themes in the life of New Peri: old Pokanau, still acting with sureness from the past; John Kilipak, speaking with the assurance of his rank, his native intelligence, and his arrogance; Lukas, the lay preacher, anxious from his mixture of illiteracy and instability; Peranis, the clerk, mixing fanaticism, attention to record keeping, careful obedience to the new rules about the freedom of women—"If my wife wishes to remain here, she may"—and nevertheless returning to the village; the older men, Poli and Kaloi, showing their old habits of dependence upon authority; and Alois, who was to muddle so many issues, finally muddling this one by saying, "I'd like to go; I'd like to ask my wife and children; I'll obey the officials whom the Commissioner has set up" (so lumping together in confusion the elected officials of the village and the old appointed constable, *luluai*). Petrus Pomat had given the clearest statement of the carefully considered position; a woman had asserted her right to speak and had been laughed at; Raphael had not talked at all because it wasn't necessary; and the New Manus man, Karol, the immigrant, had laboriously and pedantically tried to distil some new ethico-political point (he wasn't quite sure what) out of it all.

That night we talked over again the new beliefs about sin and illness. Again they asked me: Did I agree that medicines could not work if a man's thoughts were wrong? As I answered them, from a framework of psychosomatic assumptions, I realized that this would be a new kind of field trip, in which people would

demand the "truth," would demand to know the Western thinking about the way of life which Westerners knew how to live. It
would no longer be possible to answer questions with "Some people believe one thing, some people another," for the questions
would come, not from the mere curiosity of people who were
fundamentally sure of their own set of answers, but from people
who passionately wanted to know how to keep their babies alive,
how to make their new system work. "The trouble with the Mission," they said, "was that although they told us the truth, they
did not show us the way" [how to be like Europeans]. Here were
a people who were going to be as eager for the analysis of a new
law as for medicine to cure an illness. Twenty-five years ago it had
been possible to spend six months in the village and leave its owners unchanged, their memories of me and mine of them blended
inextricably together, but their way of life, their judgement of
themselves, as it had always been. This time that would not be
possible. Just as no anthropologist could refuse Western medicine
when it was asked for, so also it would not be possible to refuse
this new kind of help.

The next day a letter came from the District Commissioner
saying the volcanologist who had been flown over the volcano
had pronounced it "benign," and had said we would certainly
have two days' warning of a possible tidal wave, so we could all go
home. That evening three sets of young people completed their
plans for a triple wedding to take place the following day. Before
dawn on Sunday morning, the next day, a whole flotilla of canoes
proceeded under full sail—for this time we had a wind that was
with us, a tide that was in, and a deep lagoon—back to Peri, passing the old village, which looked as lonely in the morning sun as
it had looked ghostly in the moonlight. As we neared the new village with its neat rows of houses, I realized how scattered and
broken the community, which I had come nine thousand miles
across twenty-five years to seek, would have been if a tidal wave
had destroyed that new pattern, those rows of houses built by
shared labour, that village square where matters were settled by
discussion, that gateway which showed them to be a self-

governing community, proud members of the New Way. One of the carved filigree scrolls in the gateway was broken. All of it, their fantastic attempt to bridge millennia, my finding them again, had trembled on the edge of extinction, and survived.

But I had begun to get glimpses of how very difficult it was to be for them to stay in this new world into which they had come with such vigour and imagination. It was not only volcanoes, which might be interpreted by religious fanatics as acts of a vengeful God threatening their new efforts, but the very precariousness of an effort to enter a wider world, an effort which depended, almost entirely, on how that wider world interpreted it. I knew now that the councillors were only councillors-elect without any authority from the Administration, and indeed subject to discipline if they exercised authority. I could see how fragile the new integration was, unbelievable in its intensity and extent, but terribly vulnerable. And this recognition was blended with my knowledge that all new-won freedoms are vulnerable and in need of cherishing.

3

Old Peri: an economic treadmill

Manus culture twenty-five years ago had distinction and style. The way of life of this handful of people, only about two thousand in all, differed from the way of life of other Admiralty tribes, of other Pacific peoples, and of every other group of people on earth, differed clearly, identifiably, systematically. It was not simply that their houses had a certain shape, as they stood with their dome-thatched roofs, supported on substantial house-posts in the shallow salt lagoons along the South Coast and among the southern islands of the Admiralties. Nor was it due to the shape of their spears and war charms, the high proud filigree handles of the feast bowls they used, or the wooden beds on which their brides stood. These ornamented objects they shared with other Admiralty Island tribes—with the Usiai, the agricultural people of the main island, and the various Matankor peoples of the smaller islands. All of these objects were distinctively Admiralty Islands in conception and execution and could have come from no other spot. Experts have, in fact, less difficulty in identifying as Admiralties a small obsidian-bladed dagger with handle made from paraminium gum, painted red and hafted to a pointed piece of wood, than they have in deciding on the authorship of some complicated unsigned picture from mediaeval Europe.

But it was not the distinctiveness of every object of use and every ornament which stamped the Manus as having a culture of their own so much as the way in which act and object, use and habit, words and feelings, were fitted together into a whole, in which each Manus shared and shared beyond question. Such distinctiveness, such individuality of institutions and beliefs, has survived in out-of-the-way places in the primitive world long after the spread of great civilizations has eliminated it for most of the world. Such a culture has the aesthetic, almost the moral, appeal of the genuinely original.

No artist could have failed to respond to an old Manus village—water ways between the houses (themselves perfectly designed to withstand sudden gusts of "round winds") filled with small canoes, each canoe, except the play canoes of the smallest children or the old canoes used merely to go from door to door, designed with a single outrigger attached to keep the balance, a small platform on which people and food could be carried, sometimes a carved prow showing a crocodile swallowing a human being. On the canoes and in the canoes, the Manus adults sat or stood with unmistakable posture—a taut, tense adjustment displayed by every muscle. When they punted, they moved lightly with the impetus of the canoe.

Some canoes sped swiftly along with shouts and laughter coming from them. These were the canoes manned by at least one pair of joking cousins. For in this highly patterned primitive culture such matters as whom to laugh with and whom to treat with painful respect were not left to the accidents of personality. A boy or a man might jest with his father's sister, and with her husband and her son and daughter and daughter's husband, and these persons in turn were expected to jest with him. Between such joking relatives the tensions of a people who were both actively lustful and obsessively puritanical could be temporarily resolved. The ghosts—and each house had its own guardian Sir Ghost, the spirit of the most recently dead male—punished all obscenity, even the lightest, except for such joking as that occurring between the proper relatives. Cousins would attack and

counter-attack one another with rapid sallies on the most intimate details of their future or present sex lives, while they kept
the punts or paddles flashing in the sun, or whiled away the long
hours of sailing voyages when the boat crew worked as a unit
holding the halyards, responsive to the least turn of the wind.

The costumes, too, made each canoe crew stand out sharply
against the pale flat lights of the shallow lagoon. The women
stood so tautly, their crinkly grass aprons falling, for all their careful crimping, into straight severe lines; arms and legs, breasts and
waists, confined by black bands of netting covered with black-
ened rubber nut, so tight that it took a good hour to work the
breast bands on. The women's heads were shaved, balanced on
each side by heavy earrings. The men wore their hair in great
knots, or combed up like halos around their heads, ornamented
with combs the handles of which were modelled from rubber nut
painted red. Both men and women wore wide black anklets and
armlets ornamented with beads and, on feast occasions, bristling
dog's-teeth collars, strands of shell money. For brides and just
nubile girls, there were whole costumes woven of shell money—
grey tubular shell beads—which dragged heavily as they walked.

The costumes had style and outline, but were also strangely
unpleasant. The sense of tightness, of weight, and of lack of
buoyancy with which they were worn was confirmed when one
learned that all the more conspicuous ornaments—those elaborate bead aprons, those big-beaded squares which hung under the
arms, the long pendant of dog's teeth falling from the nose, the
black upper-arm bracelets edged with bright beads—were all
either money or mourning packets or supports for the bones or
the hair of the dead. So the bride was dressed from head to foot in
money, over each item of which the bridegroom's relatives would
quarrel shrilly and vituperatively. They would go so far even as
scratching her skin, which for this occasion at least had been oiled
and polished, as they tore their separate shares off her arms and
legs. Every feast day the bodies of the women bristled with bones
of the dead whose funeral payments were not yet complete.
Sometimes, rather rarely, the young men—to whom all contact

with the girls of the village was prohibited—would break a few green branches or twine leaves in their hair and parade their canoes noisily around the village.

But objects of dress and ornament which had been made with care and devotion—by women for their brothers and nephews, not by wives for husbands—were designed for more serious purposes than any simple expression of gaiety. The laughter of joking relatives was heard only rarely, while the background sounds of village life were the sullen, carefully guarded words of those who must remember which names, which words, must be avoided. This comparative quiet was punctuated by angry, upbraiding voices complaining about a broken pot, a borrowed canoe, an unfulfilled debt, and rising into crescendos of abuse and defiance lasting far into the long moonlit evenings. Meanwhile the young girls were hidden indoors, safe behind the leaf curtains in the long houses, and the older men stamped on their house platforms bickering, accusing, affirming, or denying some matter of financial obligation.

Those who delight in things made by hand would have taken pleasure in any one of these houses: the characteristic landing platform, the ladder leading up into the house, the shelves suspended above the four oblong fireplaces, near which beautifully turned black and red pots stood, set in three meeting stones. The women knelt before the stones to cook sago with the sureness of a traditional gesture. The tipped shallow pottery vessels, their hollow surfaces so perfectly adjusted to the sago-meal pancake, were placed against the hottest surfaces. Here the handles of the soup ladles, many of them beautifully carved, there the oil containers, made of two coconut shells of incised and painted rubber nut, caught the firelight. On the hanging shelves fish were laid out to smoke; on the floor, leaf mats holding little mounds of beads were spread, while dogs lay quiet, schooled against touching the fish; and babies, just learning to creep, had already learned not to disturb the beads. Every gesture the women made was made with the assurance that comes to those who have been handled from babyhood by those who move in the same way, a way that has

been steadily disciplined through many generations. Their movements were sharp and angular, but surely patterned. This was the
sort of scene against which those who lament our modern age like
to place our way of life, based as it is on mass manufacture and
mail-order catalogues, where objects clash in colour and form
(unless they come from that rare modern store which has a colour
expert to match the handle of the potato masher with the edging
of the dustpan). Here could be found that lovely thing, handicraft production for human use with each object fitted to its task
and each human being actively alert and aware of the meaning-
fulness of utensil or tool. Or so it looked if one only glanced into
one of these houses, lit by fitful unrevealing fires at night, but
where during the day the baby's only sunlight came from the narrow strips of light reflected up from the water through the slatted
floor.

But it took only very little investigation to shatter this picture. These objects, so perfectly fitted to use and need, were not
made by these people who were using them. No Manus hand was
skilled in making the basket bases for the rubber nut vessels, in
carving the filigree handles of the soup ladles. Nor did the Manus
know how these things were made, and, in some instances, they
did not even know where they were made. They were objects of
trade, bought in the biweekly markets where fish also was
exchanged for vegetables and fruit, lime for betel nut and pepper
leaves, salt water for fresh. Or, more often, these objects were
obtained in more elaborate long-term trade relationships between
individual lagoon and land-dwelling families. Carved and decorated objects were valued, not because they delighted a Manus
eye, but because they could be retraded at a higher value than an
object on which less work had been expended. The Manus
regarded them all in terms of durability, scarcity, and, especially,
resale value.

For this Manus way of life, these adequate, well-stocked
houses, these large sea-going canoes, carved and painted with so
much style, were all the products of a highly commercial society,
of a people who pitted unremitting labour, supported by a driv-

ing and relentless religious system, against the poverty and uncertainty of an existence in which they owned nothing and had no skills, except those of the rough shipwright and carpenter, the fisherman and the navigating trader. House sites were staked out in the salt lagoons, safe from attacks from the land people, but the logs for house-posts and canoes, the leaves for thatching shingles, the rubber nut for caulking, the materials for grass skirts, the bark for fish-line and bark cloth G-strings for the men, the pots and baskets and coconut shell containers, and the starchy foods—sago and taro, which formed their staples—all these came from the land. By unflagging diligence, by long, chilly, dangerous voyages, by all-night fishing to catch a handful of fish to take to the market at dawn, and by a complex economic system in which every man was caught in a chain of obligations such that each debt paid plunged him into a new indebtedness, they prospered. Their house shelves were stocked with other people's manufactures; their brides stood on carved beds made by Matankor carvers, their big men rattled beautifully carved lime sticks from the northern islands against the sides of lime gourds delicately incised by the nearby hill people and decorated with little tassels of shell money, worked laboriously with hand drills from the thin spouts of conch like shells, by still other peoples. They were the richest, proudest people on the South Coast. With the fewest natural resources they had the most material things and the best diet, for they supplemented their supplies of fish and shellfish with large quantities of purchased coconut oil, sago, and taro.

Property so laboriously acquired was highly valued. Small children were taught respect for property before they could walk. They were taught never to touch, always to handle carefully anything entrusted to them. The smallest accidental breakage was a matter for endless recrimination from the owner, and on the part of the parent of the offending child, fury toward the child and angry apology to the owner. The material appurtenances of life combined two characteristics: they stood up in rain and sea water, and they were fragile, easy to break. The pots were thin and brittle, the rubber nut chipped off easily, the rattan fastenings dried and loosened.

Beneath the almost unflagging respect for property—a respect which taught children that it was "theft" to pick up a half-rotten piece of food floating in the water between two houses—there was a kind of reckless instrumental impatience which was more congenial to the Manus personality than the meticulousness with which they forced one another to care for the products of people who had some standards of craftsmanship. Canoes made by Manus men were roughly functional. Houses, although of good designs, were often left unfinished or out of repair. A frequent reason for ghostly chastisement was a house floor so dangerous that the babies fell into the lagoon, or posts so rickety that it only needed a sudden death and the running onslaught of some two dozen mourners throwing themselves on the corpse to send the whole house crashing into the sea. A more congenial attitude toward objects, one that accorded with the Manus impatience, activity, and speed, could be seen when people impatiently chopped a punting pole in half to make drum sticks.

The way of life peculiar to the Manus could be seen also in the stylization of the dances which they shared with the rest of the Admiralty Island people. It took many years of practice for little boys, first practising without phallic shells, then later with them, to learn to fling non-erect phalluses capped with white ovalis shells, expertly, athletically about, symbolically scorning their ceremonial opponents. The women's dance, a stylized hop from side to side, was equally formal. Beneath these learned and only partly meaningful movements, there was the dart-hurling style of warfare in which, protected by no shield, everything depended on ability to dodge. People sat and lay and slept in any position; children slept flung over the side of a slit gong. Beneath the imposed style and the heavy sanctions that supported the respect for property and the correct ceremonial act and posture, there was a reckless, confident trust in one's own body, in the moment. There was an improvidence based on a supreme self-confidence in one's own ability to deal with anyone and anything, and a complete fearlessness before wind and sea.

This reckless confidence was most conspicuous in the smallest children, who had been taught with unrelenting patience and care just how to manage their precarious water world, how to hold tight around the parent's neck so that both parental arms were free to punt or paddle, fasten the canoe, or fling a spear after a suddenly appearing fish. The children were taught how to swim almost before they could walk, how to climb, how to carry a lighted cinder carefully among the highly inflammable mats and grass skirts, how to balance on the rim of a large canoe and how to climb in and out, how to bail and right the little toy canoes not much bigger than platters, which were made for them. Children were not entrusted to the uncertain hold of older children, but were cared for firmly by adequate, strong, sure-footed adults. The child who fell through the house floor was rescued almost as soon as it touched the water, and everyone who participated in such an episode saw that a broken house floor didn't really matter if everyone was quick and alert, if one's own body could be depended upon to react appropriately, immediately, and unselfconsciously. And children's toys, the children's canoes, made only for immediate use, were outside the system of exchange and validation.

So the Manus lived, as it were, two lives. Underneath there was the active, zestful physical immediacy of people who trusted their own muscles and their own eyes, trusted themselves so completely that the moments before action—waiting for spawning fish to come over the reef, or for the moment when the sail must be lowered or the canoe would capsize—were the most relaxed moments. They lived with the complete physical self-assurance and the certainty that it was always possible to construct what was needed, partly with their own bodies, partly out of anything that came to hand, breaking a house rafter into a drum stick or a punt. But overlaid on this vigorous optimism was a second system, respect for property, careful observance of word and gesture toward whole categories of relatives, anxious, worried economic

effort, first in response to the demands of one's elders who had
financed one's marriage, later in response to the demands of
ghostly elders who sent misfortune, sickness and death, all of
which were attributed to sexual or economic laxity.

Their economic system was one of the most elaborate of
primitive systems for which we have any record. They had real
money. Dog's teeth and shell beads, as handled by them, met all
the requirements of a modern definition of money. It was small,
durable, the units were interchangeable. It was scarce and valuable and had a separate use—ornament. It was a form of liquid
wealth usable for any type of need. It was always possible to manage straightforward money transactions, to pay for anything in
dog's teeth or shell money or—after tobacco was introduced—in
sticks of tobacco. Thus the value of pots of taro, lime gourds, or
containers of oil could be accurately stated. But there was no
dependence upon a monetary system to ensure the manufacture
of the objects that people needed. Simple money incentives were
as ineffectual as they become in wartime in modern societies
when peasants refuse to sell food because there are no manufactured goods to buy with the money, and when newly booming
defence towns lack necessary services like laundries or dairies for
which people would gladly pay, but for which the ordinary
supply-and-demand mechanisms have broken down. In the old
Manus economy, money was not a sufficient incentive because it
could not compel others to make or transport the particular
objects which one group or one person needed at a given time.
There was reliance instead on a kind of compulsory barter, fish
for tubers and sago, lime for the betel nut and pepper leaves. Each
trader compelled others to bring him, either to the market or in
terms of individual trade friendship, the particular objects which
he needed. This compulsion was carried even further by the fishing people who dealt in goods such as fish, turtle, dugong, the
supply of which was partly unpredictable. When they did make a
big catch, their usual market partners were forced to accept it on
credit, and the land people then had to go away and work sago to
repay it. Between more ceremonial trade friends, there was both

the gift which could not be refused and the request which must be acceded to if the relationship was to continue. Here also barter was the rule—one kind of desired object in return for another kind of desired object, without dependence upon money to stimulate the necessary production.

Within a Manus village, there was a day-to-day, hand-to-mouth subsistence economy. It was necessary that someone from every household should fish, daily, nightly, to trade for fresh supplies of taro, pepper leaf and betel nut, and to supply their own pots. Additionally, the big men, the entrepreneurs, organized the longer-term overseas trade as well as the hundreds of exchanges that were pivoted on each event in each marriage—the betrothal as children, the girl's menarche ceremony, the series of marriage ceremonies, the announcement of pregnancy, the birth, the return of the new baby to its father's house, and finally the deaths that ended this particular cycle of exchanges between the kin of the bride and the kin of the groom.

Such exchanges are a commonplace of marriage arrangements in many parts of the world, and the special economic device through which one side to the exchange gives perishable goods—mainly food—and the other side gives imperishables, which can be passed on in other exchanges, is also widespread. But the Manus had developed these customs to a particularly high point of efficiency by adding several distinctive touches. In the first place, they freed themselves from any dependence on the vagaries of age and sex. Rich entrepreneurs were no more likely to have children who were appropriate mates for the children of those of the same financial stature than was anyone else. Nor were their traditional vis-à-vis—their cross-cousins*—whose children were supposed to marry their children, guaranteed to be men on high finance. All the intractabilities of age and sex which of enterprise and substance. But high finance battens only bedevil

---

*Cross-cousin is a technical term for first cousins who are the children of a brother and sister.

the lives of less enterprising primitive peoples were well under control. Men of means and ambition simply invested in marriages of young people of the right age and sex, using well-worked-out legal fictions to "make the road." In this way they could balance the inflow of thousands of dog's teeth and hundreds of fathoms of shell money against the outflow of tons of sago, great flagons of oil, pigs, pots, grass skirts, wooden beds, canoes, and large fish nets. The young men so financed worked for their financial backers and, together with a fringe of older dependents who had never taken the trouble to assume responsibility for their own lives, formed an adequate supply of labour— boat crews for long overseas trading trips, fishermen and market messengers for daily needs, assistant carpenters, shipwrights. So, in each generation, a few men grew important, rich in terms of the goods which passed through their hands, in the variety of services which they could command, in the number of enterprises of which they were a part.

About once in each short generation one or two men from each village would organize a further elaboration of the system— a big exchange in which whole villages participated. Huge quantities of oil were manufactured and redistributed, and the whole area was set up in oil supplies for several years. These were focal events toward which effort was directed for years before, and about which boastful tales were told years afterward.

This was the system through which the daily supply of food was brought in by the daily toil of fishermen—matching the daily toil of land people as gardeners and sago workers—and the system in which unremitting industry, planning, scheming, trading, sailing, combined with the tireless handiwork of the women— stringing and combining beads and rubber nut into ornaments, bark into cloth and string, leaves into thatch, clay into pots— made possible the provision of the materials for the more permanent needs of life. But what were the incentives and the human costs of the system? It is possible to emphasize how well such a system functioned economically, permitting the sea-dwelling

Manus people to wrest from their limited resources of sea and seagoing skills an adequate living, without explaining at all why the Manus were willing to work so hard, to fish and sail and worry to add one dog's tooth to another until they died, in early middle age, their power passing on into other hands before their eldest son's first children were born.

In some societies men work for prestige, spurred on by the acclaim which is given the man who has given many feasts, distributed many pigs, floated many companies, or underwritten many patents. In some, power over others or freedom from daily toil may be the incentive. But among the Manus both leaders and the led were driven to their ceaseless economic endeavours by a persistent fear of illness and death. Each man was expected by his Sir Ghost to keep up an appropriate amount of economic effort, in return for which his Sir Ghost would prosper his enterprises and protect his household from misfortune, illness, and death. Active economic effort and safety from supernatural penalties were tied tightly together.

In many primitive societies the system is adjusted to the weaknesses and differing skills and abilities of different men. So, in Bali,[1] a man who rose by sheer age and survival in the local hierarchy of the village of Bajoeng Gedé had to feast the entire community. If he was a poor man, the village lent him the money with which to give the feast. In Manus, differences in wealth and ability were compensated for by two devices. In the first place, there was no system of communal responsibility through which the lazy and inefficient could escape doing their share. In the great exchanges where hundreds of thousands of dog's teeth changed hands, individuals on each side were not only individually responsible for their co-operation with their own financial leaders, but also were individually responsible to men or women on the opposite side of the exchange. So each individual's contribution and responsibility were calibrated and people who could not meet big debts acquired only small credits. The second device was the religious sanctions—enforced by the Sir Ghosts—behind the initiation of new enterprises as well as behind the payment of debts, so that the more successfully a man conducted his affairs, the more he was impelled to undertake.

In practice it worked out like this. When illness struck a household, the cause of the illness most likely to be first divined—by a male diviner well versed in the affairs of all that household's members—was some type of sex offence, often only a word or touch but regarded very seriously because it endangered the stability of the entire investment system in which each repayment was calculated on the dates of other payments involved in other marriages, and one broken engagement or marriage had an effect like a sudden bankruptcy on the whole group. But if no sex offence could be found, then some economic laxity was invoked; a debt had not been paid, an obligation not discharged, or if no such defaulting could be located, then some new enterprise which *should* have been launched had not been launched. In long protracted illnesses, or in the too frequent illnesses and deaths of young infants, other, more fanciful ghostly explanations had to be invoked—the malice of the ghosts of other houses rather than the moral disapproval and chastisement of one's own Sir Ghost, black magic used in open anger by rivals, contact with the property protecting magical charms of the land people. But primarily the religious system, with the ghost of the last dead male member acting as the Sir Ghost of each household, kept the economic system rolling (sometimes adding insult to injury by taking the soul stuff of a man's fish net as a form of chastisement for not working hard enough) on the spiritual principle that "from him that hath not shall be taken away even that which he hath."[2]

This system ensured that the rich and enterprising were punished if they paused for a moment in pursuing their far-flung enterprises, and that the man who had elected to remain a dependent of some entrepreneur was chastised for not fulfilling his simple dependent role. The man who had elected to stand aside from the complexities of high finance and simply fish was chastised for not returning some very small debt or for letting his house floor get too dilapidated. No one was permitted by the oracles—male diviners and female mediums who interpreted the will

of the ghosts—to keep his one talent wrapped in a napkin once
he had even peeked at its possibilities for exploitation, and no one
having started an enterprise was permitted to turn back.

There was one additional factor which welded the young
men to the economic treadmill. This was the circumstance that,
after a long and carefree childhood, their services were demanded
by their financial backer in the name of the wife for whom not
they but their backer had paid. They had been high-spirited,
high-handed children, flinging angry taunt for angry taunt back
at their elders. As adolescents they had been restive, unruly, and
unindustrious, prevented from seducing and raping the girls of
the village only by incessant vigilance and the accusations of
bringing illness and death on all their kin when they did trans-
gress. But the dependent position of being paid for, above all else,
having one's future sexual life, which was regarded with enormous shame, paid for by someone else, was intolerable. It struck
at the centre of their sense of their own autonomy, which had
been cultivated throughout all their upbringing. So young men
worked for older men, their uncles and elder cousins, hardly ever
their fathers, in a sort of sullen, driven anger, working to get out
from under, to take over the control of their own households, to
have houses of their own.

There were three ways in which they could do this. They
could elect to become permanent economic dependents by working long enough to establish themselves in that permanent status;
they could take over the financing of their own marriages at the
birth of first or second child and then withdraw from the competitive economic arena; or they could combine taking control of
their own marriage payments with modest investments in other
marriages, financed largely still by others. In time the latter course
would lead to their becoming entrepreneurs in their own right.
Thus the period of angry, economic servitude was relatively
short. It provided a spur to choice. Each individual was free to
take his own road, toward co-operative dependence, or modest
independence, or large-scale entrepreneurship. Once this free
choice was made, the normal sequence of misfortune, illness, and

death, interpreted as punishment for laxness or deviation, would drive him on.

This highly adaptive system in which natural resource and human capability were exploited to the limits was, however, a specialized development on a social system which also embodied a whole series of status principles—the relationships between kin, the obligation any wife owed her husband, any brother and sister owed to each other, the obligations which men owed others of their patrilineal clans, and the obligations between men of rank (*lapan*) and their "people" (*lau*). Within a Manus village clans were ranked; in Old Peri there were two clans of entirely *lapan* rank, one an offshoot of the other. Each clan was divided in a series of lineages called "houses," and within the lesser clans there were "houses" of *lapan* rank. These lesser clans were formally subordinate to the *lapan* clans, and spoken of as the outriggers of their canoes. Commoners could become *lapan* with several generations of conspicuous effort. Moreover, if a man of very low economic activity was known to come of a *lapan stock*, or to have a recent ancestor who had ever claimed to be *lapan*, he might have his unvalidated status, of which he was showing himself unworthy, thrown in his teeth.

In times of peaceful economic functioning, actual wealth and economic leadership overshadowed these status relationships based on kinship, and the man who would become a leader in the next generation was most likely to be the child of his father's or adopted father's successful early middle age. In times of crisis clan and rank would come to the fore. After a death, a man might sever close co-operative ties with a man of another clan, or even abrogate an adoption in early childhood, and return to the clan site—a patch of lagoon perhaps entirely empty of houses—of his true father's clan. Furthermore, clan and rank membership, by birth or adoption, provided a thread on which beads of inalienable pride could be strung. So the children of ranking families heard about behaviour appropriate to an aristocrat, heard pretentious men without rank, who defaulted on an obligation, taunted as upstarts, learned the names of several generations of their most illustrious ancestors. And when the past was recalled in descriptions of "big men who handled big exchanges," rank and clan status were blended.

For individuals in *extremis* actual blood ties became paramount. If a woman of little importance had been ill for months, it was her own kin who tended her, not the wife of her husband's economic backer. And a man of importance also turned in the end to his own sister or his sister's son. Beneath the bustle of entrepreneurship, of chosen economic role and autonomy, of alliances formed for business reasons, of adoptions carefully planned to meet deficiencies in the economic strength of a household, of effort proportionately rewarded and failure roundly punished, there lay this area of intrinsic belongingness, of clan ties that could never be denied, and of blood traditions which could never be altered from what they were, proud or shameful.

Out of these varying heritages, the children of aristocrats took pride and developed a thin but recognizable sense of responsibility for their clans and their village. After all, the ranking family provided the war leader for the whole cluster of clans who lived together; it was *lapans* who could speak in public—first, because of their rank, and secondly, because they behaved like men of rank; it was *lapans* who had ancestors to talk about; and it was the children of *lapans* who learned to look down on the "rubbish," people without rank or tradition who ranted without style and conducted their affairs in a slovenly fashion. But in counterpoint to this, people of low degree, or the people who had failed to live up to the possibilities of their rank, could fall back on the active sense of autonomy which all Manus shared. They could stamp and shout that what they did was their own affair, they could sneer and gibe at and undercut the man of rank who showed any sign of resting on his laurels. So, while the absoluteness of rank and clan membership did provide a little security, a little tempering of manners, and a slight basis for responsibility to kin and village, it became important mainly in personal or communal emergencies. Between the emergencies, people formed their alliances along economic and practical lines. The only

approved reason for divorce was a business one, when both husband and wife proved to be "too stupid for trade."

Along with the emphasis on individual effort was the fact that
old Manus life was anchored in no seasonal calendar,[3] fixed
against no point of origin, moving toward no climax of possible
holocaust or day of judgement. The other Manus villages were
said to be offshoots of Peri village, and tales were told of the quarrels which had led to this clan and that clan moving away. But
before that time, conceived of as quite recent, there were no
genealogies to refer to, no moment of migration or of creation.
Their mythology was a series of inconsequential tales, centring
around folklorish birds with human or supernatural powers and
clearly more related to the life of land people than to that of the
Manus. The southeast and northwest monsoons gave two seasons
to the year, and the waxing and waning of the moon subdivided
these seasons, while approximate daily counts from the new
moon and the full moon gave the time of the spawning of the
fish, when large catches could be expected inside the reef. It was
fully realized that the season for the ripening of some fruit on one
of the islands was a period which was recurrent, but no festival
was named for this occasion and no calendar came to be based
upon it.

Although market days were fixed, every third day here, every
fourth day there, they were not named nor arranged in any fixed
sequence. When men made plans they counted ahead, beating on
a slit gong with the opening announcement pattern, "I am going
to announce how many days from hence I mean to make an economic exchange," then a slow count of days. The people who had
paused when the first beats rolled out listened and counted,
thirty, or thirty-six, as far off as forty-two. Then would come the
signature, the drum beat of that house, and the listeners would
supply the plot—ah, that is when Nane will make his *metcha*.
Smaller exchanges were keyed to larger ones so that the same sago
might change hands very rapidly, validating one event after
another, before the dozens of ten-pound packets were shared out
to be eaten.

For any large event, there were scores of small events: overseas voyages to be made to visit trade friends, soliciting and collecting trips to relatives on distant islands, appointments to be kept at the market, with special large supplies of fish to be paid for by future special supplies of land food. All of this planning was endlessly subject to disruption, by the weather, by ill luck in fishing, by good luck in fishing—for the sudden catch of a large turtle might make it possible to speed things up—by events which could only be imperfectly foreseen, birth and menarche, and, most of all, by illness and death. When a baby was born, there was a sudden demand for feast foods, especially for coconuts. Coconuts in large numbers also had to be husbanded for every other sort of feast; great piles of them were needed to throw publicly in the sea when a girl reached menarche. After everything looked perfectly planned, there might be a major disaster at sea— a whole set of bridal finery being taken from a Tawi bride to a Peri bride to wear, or from a Peri bride for a Mouk bride to wear, might go down, or a canoe-load of people might be shipwrecked so disastrously that two drowned and the remainder, when they reached their village, had to be slept with and mourned over by all their kin as if they had been dead. All this would mean a need for more food and the withdrawal of many people from hard work to days of hovering over the rescued relatives.

It was a world in which planning was crucial and also almost impossible, in which one man's plans depended not only on another's word, on another's industry, but also on all the vagaries of fish, of wind, and of weather. Each entrepreneur drove ahead with his own plans and drove all those who were necessary to his plans, and every man's course was subject to direction and deflection from the initiative or failure of some other. One had to be continuously on the alert, to send a message by a passing canoe, to work out the implications of a Mbuke canoe's stopping at Patusi in terms of gallons of oil or extra packets of sago which would follow from that Mbuke canoe's stopping at Patusi. Would it be worth while sending someone to Patusi to find out just how much oil had been left there, or to whom the sago had been sold? Each person was forever trying to reorganize events so that his special series of plans would succeed—plans which he must complete under pain of ghostly anger on the one hand, and human anger and recrimination on the other—because other people's plans depended in turn upon his. The faster the system moved, the faster it could move; the more it jammed, the more it was slowed down by an unexpected cluster of events; the stronger the pressures were to speed it up, the more were illnesses laid to the door of unfulfilled obligations and angry speeches made from creditors to debtors. As long as a man could keep his planning ahead of these pressures, he could have a certain sense of initiative. But planning meant contracting debts and feeling convinced that one could pay them, in time, and so one of the points on which men of affairs were able to mobilize the greatest anger and indignation was in refuting the accusations: "This is more than you will be able to pay back."

In 1928, we soon found that the surest way to obtain anything we needed was to give loans. The individual in our debt was completely vulnerable to our insistence on repayment. A debt of ten sticks of tobacco could send a man out fishing in the rain, while the offer of ten shillings in money, or its equivalent of thirty sticks of tobacco, might have no effect whatsoever. Men were adjusted to keeping ahead of pressure; they did not respond to offers of wealth as such, unless it in turn was needed to help them meet some other pressure. And when a man died he bequeathed to his heir a place in the system, the "good will" of a set of hereditary trading partners scattered on other islands, a whole structure of collectible unpaid debts, and a chance to be as driven, as harried as he had been.

This system not only lacked a calendar for scheduling future events, but it lacked any clear way of scheduling different series of events in relation to each other. As a result, one man's purposes were always partly in accord with—through one set of expected payments or repayments—and partly in opposition to the purposes of others. Each man crashed ahead, deflected from his own intentions by events and not by any desire to defer to the will of

another. Any purposely initiated change in another's affairs had
to be done with truculence. One had to present oneself as an
immovable obstacle, a thing, not as an individual asking for compliancy. This attitude stood in sharp contrast to the initial solici-
tation of help from others, when one might plead one's dire
extremity to persuade a sister to start making bead belts or grass
skirts.

As the past was not anchored down on any point of origin or
migration, and the present was not situated in any determined
time scale in relation to any agreed-upon event, all life was a
cross-section viewed from the position of the speaker in time and
space. A man spoke of "my grandfather's time," "the year I married," "the time, two moons ago, when my net took three hundred fish," "ten days from now when I am going to Mouk." Only
in arguments as to just when something had happened were
events other than those of one's own life brought in, such as the
year the Kalo canoe was shipwrecked, but in a minute the conversation would revert to "And I know it was then, because my
sister had just had a baby and she was living in our house; her
husband was late with his sago, so it was more than thirty days
after the baby's birth," etc. So old Pokanau at the time of the
evacuation finally remembered that his grandfather had told him
that in his son's time—Pokanau's dead father's childhood—there
had been a tidal wave. The event remained placed in relation to
Pokanau's idiosyncratic past, for his father had died soon after he
was born and he had been adopted by his grandfather, who then
told him stories about his father's childhood.

Beliefs about the soul and life after death only accentuated
this cross-sectionalism in time. Individuals came into being as a
result of copulation. During life they had soul stuff, a divisible
and alienable substance. Bits and pieces of this soul stuff might be
carried away, by angry hostile ghosts of various sorts or for discipline's sake by one's own Sir Ghost. When this happened, the
individual sickened but when all the soul stuff was returned, the
patient recovered. After death this material became a ghost, a sin-

gle entity, attached to the skull, or, if the skull had been lost in
war or at sea, a coconut properly charmed and bedecked could be
substituted. So something which either was the skull, or did service for it, had to be kept ceremonially in the rafters of the house
where the recently dead presided now with more power than he
had had in life.

During his brief rule, which would come to an end as soon as
some other male member of the household died, thus proving
him impotent to protect his own, the Sir Ghost demonstrated his
superior strength in various ways. In the world of ghosts the prohibitions of mortal life could be flouted; adultery and polygamy
flourished, and ghostly sanctions against ghosts were reflected
back on the mortal plane, as ghost angry at ghost made mortal
wards ill. But everyone knew that this period would be brief; and
after that, unhoused, his skull, or its coconut surrogate, thrown
out into the sea, his bones no longer adorning the bodies of his
female relatives whose mourning finery was now worn for those
who had died later, the recently powerful ghost degenerated into
a sort of half-life, hanging around the edges of islands, until
finally he became a sea slug and disappeared altogether. His name
might survive in the genealogical chants used over adolescents,
but this was a survival in name only, and the ghostly children
whom he was sometimes reported to have begotten on the spirit
plane survived only as long as they were necessary dramatic
devices to explain some mortal tangle.

Real events in the past and reported events on the other plane
were treated by the Manus in the same way. The memory of each
dimmed as the validating events surrounding them receded: the
debt to be repaid, the bone ornament which had been donned for
a mourning ceremonial, the property charm which had been fastened to the central post of a house, the memory of the laments
which had been composed at their deaths, the names of their children who had died without issue, the long voyages they had
made, the battles they had fought, and the feasts they had given.
Nowhere was there any mechanism of immortality or time record;

no court where records were kept so that people could measure time backward by reigns or the names of chiefs or the dates of famous battles. The intensely crowded life of any Manus generation existed between two voids; it was tacitly assumed that the past had always been like this and the future always would be, and at the same time both past and future were seen as continuously unpredictable because each depended upon combinations of events, and no combination could be accurately predicted. Where a people with a calendar could assume that corn had been harvested at the same time for a thousand years, the Manus could never assume that a shipwreck, a quarrel at the market, and the first menstruation of an important man's daughter had occurred or would ever occur together again.

This attitude toward time was repeated in their attitude toward space. The known world was the world in which they lived—the South Coast of the Admiralty Islands, each small creek mouth and bay accurately known. When people spoke, they spoke of going either *up*—toward the open sea—or going *down*—toward the nearby shore—or going *along*—parallel to the shore. The open sea surrounded them in every direction, stretched up to an indefinite horizon, unbounded, unnamed. Other places, such as New Guinea, where a few of the men had been, Australia or Germany, where no one had been, simply existed somewhere unmapped, uncharted, unguarded even by mythological sea serpents or gods of the sea, on that vast watery rise, just sea water and more sea water—an element they trusted and knew.

The contours of the lagoon villages made this picture of themselves as living at the bottom of a giant shallow saucer seem real and palpable. The mountains of the surrounding main island and the far-off peaks of Baluan and Lou stood high against the horizon, and from the sheltered lagoon in which Old Peri stood, one literally looked "up" to the pounding reef on the edge of the open sea. Within the reef, the light was paler, the tides moved less stormily, the winds blew less dangerously; and beyond and up, beyond the islands which ringed the South Coast waters, was a sea for which larger and larger canoes were needed if one were to venture forth. There are European records of Manus canoes having traded off the coast of New Guinea, but the memory of these long voyages had already faded, even as the Administration forbade them.

When one asked about any custom, it was described not as something that had always been or something that had been established by some mythical culture hero, a Prometheus or a marplot, who set the pattern for man's way of life. The answer was simply, "This is our custom," "This is the way we do it now," coupled sometimes with statements of a golden age which had existed some time before the present and in which the ghosts had not concerned themselves so vigilantly with the sexual peccadillos of mortals and men were free to be as immoral as the Usiai and the Matankor. True to their puritanical style, the Manus regarded all the differing customs of their neighbours as saturated in the darkest sin, and looked upon their women as legitimate game for Manus men who upheld, albeit against their wills, so much higher standards for their own womenfolk. The land peoples, who, in 1928, were beginning to know something about missionaries, retaliated by accusing the Manus of being "just like indigenous missionaries."

So the present, with its exacting moral standards, its escalator economics that drove men mercilessly on and on, was suspended somewhere in an undated time, set against a past when licence had been possible and an immediate ghostly future in which licence, combined with capriciousness, would again be briefly possible. The eleven Manus villages, among which there were many ties of kinship and economic obligation, the nearby villages of other tribes with whom contacts occurred almost daily, the more distant villages where a single household harboured distant kin or trade friends, and the other parts of the Territory far away over the horizon from which New Guinea natives came, constituted the world, a world in which the European was an intrusion hard to comprehend. This small universe was not only based on islands, but also might be said to float, unanchored in time or place, without known origin or glimpsed destination. Taken at its face value, as it was, and giving no indication of its age, this small universe of the Manus might have been five generations or fifty old—there was no way of knowing by any examination of the form of their culture.

4

the wider context in 1928

In 1928, the people of Manus, as a part of the Mandated Territory of New Guinea administered by Australia, were just coming under the influence of Western civilization.[1] During the German period of occupation which ended with the Australian conquest of New Guinea in World War I, a few Manus natives had been persuaded to work for the Germans. Missions had been set up at the other end of the island, a day's journey away, and a few Manus had become police boys. After the end of World War I, indentured labour became more common; many of the young men were away at work, while in the village there were several returned work boys who had worked formerly for the Germans. All of the Manus islands were technically under control, although there were still Usiai hill villages with doubtful reputations on the main island. The people paid a head tax. Each village had its appointed headman, called a *luluai* or *kukerai*, chosen from among the older men, and its appointed interpreter, a younger man who could speak the lingua franca of contact—the language that in those days we called Pidgin English, but now is more accurately referred to as Neo-Melanesian.

In each village there was a "doctor boy," who, like the other two officials, was excused from taxation and provided with a hat. Usually he had been taught a little simple medicine. A few articles of trade had found their way into the villages: steel plane blades to replace stone in adzes and axes, steel fishhooks, canvas for canoe sails, cloth under which women could hide their heads from their male relatives-in-law, fine coloured beads to supplement the yards and yards of shell beads made by laborious native manufacture. Tobacco, in the trade form of "Louisiana twist" at three sticks for a shilling, provided small change and continuous moral conflict between smoking and saving for trade. Work boys brought home camphor-wood boxes and women wore big, practically interchangeable iron keys to these boxes swinging loosely between their bare breasts. Some dependence on European trade was already part of the system.

Warfare had been forbidden and there had been no real warfare for fifteen years. The old war leaders danced up and down with rage, clattering upon their house platforms, reaching for spears which they no longer hurled. Younger boys no longer practised dodging spears in earnest, and these boys would have been mown down for lack of dexterity in dodging in any real warfare in which shields were not used. But the sounds and gestures and costume of war were still familiar to everyone: the elaborate face paint, the obsidian-tipped spears, the picturesque feathered war charms which protruded horizontally from a man's shoulder blades and "nudged" him into bravery. All of these were still used in the mock hostilities which accompanied any large exchange of property.

People told stories of warfare, of splits within a village, of how all the different Manus villages, then eleven in number, had originally budded off from Peri, of how there were still small islands among the thirty-two small bits of rock and rubble which constituted Peri's only land that were named for these clans which had broken off from Peri. Fighting used to be initiated by quarrels within a village, between two alliances of lagoon (Manus) villages and their trading-partner villages on the main island (Usiai) or in feuds with one of the small outlying islands (Matankor) against other lagoon villages and other trade partners. Sometimes warfare was undertaken as a form of vengeance reminiscent of the practice of head-hunting after a death. The Manus were proud, however, of the fact that they themselves had never been cannibals, although they were unashamed of selling war captives to their cannibal neighbours. This invidious distinction between those who merely sold human flesh and those who both killed and ate, and bought and ate, remains until today to haunt the relationships between those Usiai—agricultural people of the main island—and those particular Matankor groups—agricultural canoe-using people of the small islands who were cannibals—and the Manus. Occasionally war yielded real spoils—a prostitute for the men's house, who sometimes died and sometimes was returned laden with gifts, or even a title to sago land along the shore. Most importantly, warfare provided an outlet for the energies of the young men, who could not marry until the elaborate economic exchanges necessary to ratify their marriages were completed, and to whom all Manus girls except the war-captured prostitutes were forbidden. The effort to maintain rigid sexual standards, supervised by Sir Ghosts who watched from where their skulls hung in the rafters and objected to a suggestive glance or word, among a people as active and lustful as the Manus, kept the community in continual turmoil. The insubordinate young men responded to the challenge of the forbidden. They were aided and abetted by the girls betrothed to men whom they had never seen and were not allowed to think of and by widows who were alert to the chance to foil the self-seeking plans of their male relatives and relatives-in-law. But adultery was almost unknown among married women.

Warfare had been abolished under threat of imprisonment. Its abolition was additionally enforced by the hostility between villages, only too anxious to report the forbidden activities of rival villages, whether these were capturing prostitutes, fracases in which no longer was anyone killed, or the other forbidden old activity of prolonged washing of the bones of the dead and no burial instead of a burial in a designated graveyard on land.* In 1928, with warfare outlawed by the Administration, men fought with words and open threats of black magic which would kill off one another's infants. The major sanctions were the ghost-sent illnesses, misfortunes, and deaths. The increased strain on the little puritanical communities from roistering unmarried young men who chafed under economic servitude and sexual deprivation alike was in large part relieved by the indentured-labour system. Boys ran away to work at fourteen or fifteen, usually still with the disapproval of their elders, and if, when they returned after three years of work, a very long period of economic servitude to their elders was still to be expected before their marriage arrangements were to be ready, they often ran away again. Their elders appropriated the rewards of their years of work—money, tools, cloth, boxes—in lieu of the work they would have done had they remained in the village. Where, in the past, warfare had provided only a temporary diversion, because after the battle was over a young man was still dependent on his elders for financing his marriage, indentured labour was beginning to provide a permanent alternative. There was always the possibility that a man might stay away forever as a work boy, enlisting for many years in the police corps or becoming the trusted captain of a white man's schooner. Manus boys were much in demand as police boys, house boys, and as crew members on boats. With a relaxation in the governmental policy which attempted to prevent lifelong indenture, expatriation into a faceless, illiterate labour force, without any heritage except that of the New Guinea work boy

---

*It is ironical, if not otherwise significant, that graveyards, which were such a focus of governmental anxiety wherever Australians governed so that in both New Guinea and Papua conflict over burial customs was central to native-Administration relationships, should in 1954 have become the focus of a brief rebirth of the mystical nativistic cult in which anti-Administration ghosts were mustered in the graveyards, now enjoined by the Paliau movement as they had once been enjoined by the Administration.

with his impoverished view of himself and his employers, might have become the rule.

In 1928, one factor in the return of the young men to their villages was the friendship among boys of one group, who used to go away to work at the same time, take out work papers of the same length, and plan to return to the village together. The pull back to the village was very strong, even though they knew they would be hard-driven there, first by shamed dependence upon their financial backers, later by the restless goading of their ghosts toward their putting out more and more economic effort. It was important also that there was no real New Guinea–wide society for them to join. If they wished to marry and have children their choice lay between returning to their own tight little villages, where the older men still held supreme power, under the Sir Ghosts, or marrying girls from other tribal communities and becoming absorbed in some alien system of savage controls, and one, in all probability, even less rationalized, more sorcery-ridden, than their own. Alternatively, a man could remain a dependent employee, fed and clothed by a master, either the government or an individual European, and subject to penal sanctions for disobedience or insubordination.

The administrative style was inaugurated by the Germans whose colonial policy had been to abolish all resistance by heavy punitive sanctions, and continued by the Australians under the humane rules established by the old League of Nations mandate and the general traditions of responsibility of British colonial government. It continued to be shaped in large part by the character of the Europeans who lived and worked in the Territory, and the lack of political cohesion or understanding among the natives. To the natives, the governmental system was heavy but essentially external. Manus men wanted some of the things they could buy or trade or work for; Manus boys wanted the adventure of sailing the seas in European ships, of working with engines and driving motor cars in Rabaul. The Manus obeyed certain government rules because they were afraid not to. They approved of the present times, as compared with the time when there had been warfare, because more widespread trading was possible without danger of being ambushed and because the new lingua franca facilitated communication. They were in a European-controlled world, but distinctly not of it. Perhaps the closest parallel to their position is that of the gipsies in the early twentieth century in Western Europe, who would use our society but would recognize none of its values; or, from a different angle, the Greek peasants who came to America to earn dowries for their sisters, and then returned to Greece, often without having understood anything of American values.

The Manus were being brought into a world which would slowly destroy all of their old values and ties, their distinctiveness as a people with a name and a way of life of their own. The steps by which they were to enter and live in this new world were already determined. Even the anthropologist—approved by high Administration policy, distrusted by the practical man on the spot, who alternated between accusing him of paying no attention to people who really knew something, and, if the anthropologist did ask questions, of picking the local man's brains and living a parasitical existence—was becoming a recognized part of the group of occupational specialized invaders of native life.

The style of trade and planting, also well defined, was a system through which owners of the old coconut plantations, planted by the Germans and sold to Australian veterans after World War I, maintained trade stores where indentured labourers and nearby village natives could buy the standard low-grade trade goods, paying a higher price than was charged to the white man. The main missions of the area were already well established. The Roman Catholics built great economically self-supporting mission centres, which made it possible for them to adopt a policy of slow education. The Protestant missionaries came as individual family groups, supported by stipends, with less continuity of support on the spot and more urgency toward immediate conversion. And varieties of smaller sects like Seventh Day Adventists, extreme evangelical groups, established small missions here and there. In some parts of the Territory there had been attempts at gentlemen's agreements, so that certain areas were left to Catholic missionizing, and others to Protestant. In many areas, as in the formerly German-owned Solomons, there were numerous incidents as catechists and native teachers, unschooled in the proprieties of Christian peacefulness, egged their converts on to burn down one another's churches. Gold had been discovered in the mountains of New Guinea, and the first mining machine was a huge dredge flown in in ton sections to the gold fields. All of these patterns of government, labour, trade, missions, mining, were sifted through the talk and experience of the work boys who returned to their small, relatively untouched villages to gossip about the ways of the modern world and the white men who peopled it.

It was against this contact situation that the natives of Peri perceived us and were both able and unable to communicate with us. But before I try to discuss finer points, it is important to specify just how they saw this outer world. At the top was the government, arranged in a hierarchy of "Number Ones" of different levels. The Neo-Melanesian words "Number One" were used for a concept which did not exist at the native level. Villages or groups of villages might have one big man, the only big man, even a single war leader, but there was no paramount leader until Europeans brought in the idea. The district officer, the "Number One" of each area, had to be obeyed under penalty of the "calaboose," which the Manus, unlike many other Oceanic peoples, regarded as a terrible disgrace. "Calaboose" carried few heavy punitive measures, the food was adequate and the work reasonable, but the Manus readily caught the moral taint in the world of European values which accompanied imprisonment, and became preoccupied with the moral question of whether an imprisoned man had or had not deserved to be imprisoned.

So, in 1929, an unstable youth from an unstable family background in Peri was apprehended on the government station as a Peeping Tom—a frequent enough occurrence in a world where full-grown males, coming from tribal backgrounds with a great variety of sex mores, become house boys of European women with very limited experience in handling castelike situations, who vacillate between going about in dressing gowns or bathing suits and sleeping in rooms in which every window has been barricaded with several layers of chicken wire. The people of Peri were very much disturbed by the arrest; they felt that because the boy was unstable mentally, he was not accountable for his behaviour. Here they attributed to the white man's court more fairness in assigning responsibility than they attributed to their own ghosts, who could be very much offended by the behaviour of the unstable. They asked us to record—in genealogical form, which they had already grasped from our note-taking methods—just how bad this boy's inheritance really was, and to send it into court as evidence. The idea of law, of a set of rules over and above the caprice of any individual, was taking hold, despite their experience with the caprices of actual individual junior and senior officers on patrol, annoyed and unreasonable about the floor of the government rest house being out of repair. Government officials in turn were already branding the Manus as "bush lawyers," sensing that the Manus would be ready enough to turn their articulate high standards against them.

Experience of, and as, police boys provided another side of this view of European life. A native police boy was drilled and disciplined; he accompanied government officials on patrols, enforced their requests, and, when his superiors' backs were turned, battened on the relatively helpless natives, bullied, blackmailed, raped, and robbed. He also, in uncontrolled areas, risked his life for the safety of the patrol, was the effective executor and the essential link between the single government official and native peoples, just brought into an uneasy state of obedience. Manus were early choices as police boys. In the village of Peri in 1928 there were two police boys trained by the Germans who had served in New Guinea. Up to 1929 the Manus constituted a mainstay of the Territory police force. The government had found that their puritanical reformist behaviour, scorn of danger, activity, industry, and high intelligence were all assets. But in 1929, in Rabaul, the capital of the Territory, with its handful of

Europeans and its thousands of indentured native labourers, an event occurred which ended the Manus dominance in the native police force. This was the Rabaul "strike," an event which, although deadly serious at the time for the local officials who had to deal with it, had elements of a Gilbert and Sullivan opera about it.

Rabaul was a pretty little tropical town, with wide streets which had been laid out by the Germans. The European residents, government officials, traders, missionaries, and functionaries of the small bureaucracy, lived in bungalows set high on stilts, each house with its complement of house boys, usually from some distant land, indentured for several years. The big firms, which combined trading with the ownership of steamship lines and plantations—Burns, Philp and Co. Ltd., W. R. Carpenter and Co. Ltd., and the Melanesia Company—maintained work-boy groups or labour "lines" of several hundreds. Hundreds more were employed on surrounding plantations. The local natives, the Tolai, cultivated their gardens, celebrated a mixture of their old pagan masked dances—the Dukduks—practised Christian burial within walking distance of the capital, and had their women bring native food to the market. Outside Rabaul, Catholic and Methodist missions had large headquarters for training and furnishing their staff. The native police, recruited from all over the territory, also guarded the township, challenged work boys out after the curfew hour without a pass, and stood on guard before the doors of the courtroom. Very smart they looked, naked except for uniform hats and very short bright uniform *laplaps*, abbreviated kilts of red and blue, with well-polished rifles over their shoulders. They were under the control of native sergeants, and their shiny arms were stacked at night in the guardhouse, the key to which was kept by a native.

At Christmas time, 1928, feeling ran so high among the various groups of indentured labourers in Rabaul—"Solomons," "Sepiks," "Manus"—that it was found necessary to post guards at the approaches to the Recreation Reserve, to seize the various sorts of weapons that natives were carrying. Commentators on native life shook their heads, remarking that these natives were quite incapable of ever organizing beyond the narrowest tribal borders, overlooking the fact that terms like "Solomons," "Sepiks," or "Manus," when applied in Rabaul, already blanketed many tribal differences. A Manus in Rabaul might be a *Manus true*— one of the lagoon people about whom I am writing—or an Usiai or a Matankor—with the lagoon people setting the style, and all Manus natives acting as a unit—outside Manus.

So the Europeans in Rabaul went to bed, shaking their heads over the "fortunate" inability of the natives to organize. This was in spite of some recognition that inability to organize, to feel loyalty to large units of their own people, might well be related to an inability to develop loyalty to their new employers and administration.*

Then, on the morning of January 3rd, Rabaul woke up and shouted for tea and hot shaving water, the customary harbingers of the new day which native servants should have been preparing in the kitchen. In all but one house, that of the judge who was Acting Administrator, no one answered. People began telephoning to each other. "I say, have you any boys?" It soon became evident that there were no boys, anywhere. They had all vanished like mist "leaving Master Paatzsch to put on his own shoes," as the natives themselves characterized it later, Master Paatzsch being a German of such vast proportions that he could no longer reach down to lace his boots.

Slowly the plot unfolded of a strike which had been organ-

---

*A very popular story in those days was the one about the sentimental white man who had a devoted native servant, and who, probing for reassurances, asked Tapo, or Bimbo, or Yami, "If the time came when all the New Guinea people planned to kill all the white men, you wouldn't kill me, would you?" And, in the story, the native would protest violently, "Oh no, Master, I wouldn't think of doing such a thing. Never! I'd get another man to do it." Such stories alternated with tales of extreme loyalty and devotion, of natives who performed unheard-of feats of courage and endurance to save the lives or even the possessions of their masters.

ized by a company boss boy, Sumsuma, with the help of the Manus Sergeant Major Rami. A Manus type of intervillage organization for big oil-distributing feasts was used. The boss boys of the different labour "lines" each organized his own "line" and the police got out the scattered house boys, using police wards as units. On the appointed day, all had left, except the faithful small servants of the judge, who however hadn't worked out how to reveal the plot from which they had disassociated themselves. The idea of a "strike" had come from American Negro seamen who had expressed shock at the small wages reported by the local natives and had suggested a strike as a solution. A "strike," as the natives understood it after listening to the tales of the visiting seamen, meant that you, all together, planned secretly, absented yourselves from work, and refused to come back until higher wages were promised. As for places to assemble, they had decided that all the Catholics and those who were someday to be Catholics should go to the Catholic Mission, and all Protestants and future Protestants, to the Protestant Mission. (There were informal agreements among the missions, with spheres of future influence mapped out in terms of which many of the natives knew that in some future day, when everyone became Christian, the people of their village, or tribe, or language group, would be "Catholics," or "Talatalas.")

Consternation and fury reigned among the "masters" in the little capital. The exodus had been orderly enough, but it had been organized by the police, the police who had complete access to loaded weapons which they had, but might not have, left behind them neatly stacked in the ordnance room. Small groups of dissident vigilantes demanded various harsh measures against the natives, or against the white officials, who, they asserted, ought to have foreseen and forestalled the event and ought to be taking quicker action against the natives. The strikers sold their underpants to the surrounding bush people for food. When they were told to come back to work, they replied they wanted an increase in wages. They were told they wouldn't get it, and to come back to work at once; and they did. No one had told them what strikers did when those against whom they struck didn't accede to their demands.

The major punishment fell on the ring leaders and native police officers, caught, as happens the world over, between two poorly resolved loyalties; for they had taken an oath of allegiance to the government and had betrayed this oath. The period in which the Manus dominated in the police force was over, but the importance of the police was not. A generation later, after World War II, it was police boys who had returned to their villages who became the organizers and leaders of native movements in various parts of New Guinea, overactive yeast in situations of rapid change.

A Manus native who had been a police boy combined in his person a demand for order and organization, for swift military obedience associated with the posture of the drill ground and the snappy salute, and a vivid sense of ways in which order could be contravened. These ways included clever argument in the court-room, where the highest canons of British justice were adhered to and the Crown inexplicably appointed white men to defend, by the most cunning arguments they could muster, murderers and those who resisted the government itself, and the use of high-handed blackmail and chicanery behind the scenes. Learning to be a police boy meant, from 1928 on, learning to manipulate a system and make quick, accurate judgements about power, when it could be abused, when it must be obeyed.

The village native's view of government officials was medi-ated by the police, with some additions from the boys who worked as house boys for government officials, especially those boys who became the professional cooks of a given type of official such as the chief medical officers. Doctors came and doctors went, but the boy who knew how to work the intractable stove, and the even less tractable shower in the doctor's quarters, remained and became a lively commentator on the differing char-acteristics of "Number One doctors." In great part this meant close intimate association, on the government station and on patrol, with Europeans who had not chosen the New Guinea service as a way of life, but had instead, in most cases, been catapulted into it by accidents of war and were always leaving it to become traders or planters, recruiters or prospectors, or because of some piece of bad luck which was summarized by the folklore statement on the danger of kicking a boy who had a swollen spleen, and who would then die on you—most unfairly. Masters were judged in terms of their tempers, their predictability, their generosity, and, where Manus were concerned, their respect for one as a human being. The Manus reacted to undeserved blame or unfair reproaches with sulky depression, smouldering anger, and running away, the devices that they had always used when passions boiled within the household or within the village. Government, seen through the eyes of police, of house boys, and of appointed native officials in the village, was a power structure with a set of rules which could be learned and used *against* those who wielded the power.

An adventure of Kilipak's in the thirties illustrates individually what the Rabaul strike demonstrated on a larger scale—the Manus ability to engineer an activity against authority. Kilipak and Kutan had gone to work on a plantation on a small outlying island where the supervisor was a Malay and said to have subjected his labourers to many cruelties. The boys' response—and they were still youngsters of only seventeen or eighteen—was to plan a combined "runaway and complaint." They stole a native canoe and with four other boys set sail for Rabaul, believing it was hopeless to take their complaint to the local district officer, who was believed to be a beer-drinking crony of the trader for whom their tyrannical Malay overseer worked. Thus a detailed piece of gossip about the actual relationship between those in power was translated into a four-hundred-mile canoe trip, in a small, poorly equipped canoe, by six teen-age boys, seeking justice.

Trading, planting, and recruiting native labour were overlapping occupations in the late twenties. The German properties had been taken over by a governmental board and sold to ex-soldiers, most of whom had very little ready money and so survived only with the help of large extensions of credit from the big firms. Their copra shipments had to cover all purchases and payments on their debts. Ready cash came from small trade stores and bonuses paid by future employers for recruiting native labour. Sometimes small pinnaces or small schooners were added to the stock-in-trade of the planter. The Manus tended to see these various activities as one: a master who "bought" boys for himself and others, maintained a labour "line" and sold trade goods at a high profit, and—this especially from the point of view of the lagoon peoples—also had a boat which was worth sailing. The planter was seen by the native as someone who lived away from home in order to make money, and whose life pivoted on the arrival of cargo which represented all the good things of life, food, mail, beer. "When the cargo comes" was the phrase oftenest on the lips of white men, many of whom made a practice of consuming all the beer as soon as it arrived and, after recovering from the inevitable bout of malaria, living on gloomy savourless fare until the arrival of the next cargo.

Cargo came in boxes which seldom betrayed their contents. When eagerly broken open, they might or might not contain hoped-for articles. The big firms had been known to pack all the delicacies—the spirits, the Worcestershire sauce and catsup, the coffee, the special pickles and jam—in one box, to be taken back on board in case the amount of copra didn't come up to expectations. Additionally there were always shortages, so that no cargo contained everything one hoped for, and trade goods which had been missing from the store shelves for months often were searched for in vain. The ways of "the cargo" assumed in native eyes an aspect as inscrutable as the ways of Providence itself. Only God knew, in words sometimes profane, sometimes religiously prayerful, as some planter's wife, struggling to make sago pudding without any vanilla, prayed that the next cargo would contain at least "something to flavour with"—only God knew what would and what would not come in the "next cargo."

Even the missionaries, with whom the Manus still had only indirect contact as schooner boys or work boys on the other islands, fitted into the same picture of people who had left home for gain. Here there was a distinction between those missions, mostly Protestant, which made no attempt to establish trade stores or plantations and supported their missionaries with funds from home or with contributions from the converts to the collection plate, and the Catholic missions, which followed the traditional Roman Catholic practice, so highly developed at the time of the Spanish settlement of the New World, of setting up religious communities which were as self-supporting as possible, with brothers well versed in such crafts as shipbuilding and carpentry, sisters who established hospitals, with coconut plantations and trade stores which relied on recruiting labour and customers among the converted natives. This meant that in both cases the natives, without any understanding of motives of altruism, saw the missions as interested in money—the Protestants in collections, the Catholics in trade and labour.

Only very slowly did New Guinea natives come dimly to understand the kinds of motivation which led a government official, in spite of poor pay, endless bureaucratic difficulties, and a poor future, to stay in the Territory trying to make the government service be something that Australia could be proud of. It was a much slower task to gain any understanding of what made a slender beautiful young nun travel the long thousands of sea miles from her home—a place called Germany—to spend her life ministering to people she had never seen, or of what made a young Protestant couple risk getting yaws and malaria and leprosy for themselves and their brood of small children in order to carry the Christian message to a group of natives who had not asked to hear it.

Most of the natives of the Territory still lived in small groups, loyal only to close kin, caught in endless raids and counter-raids, alliances and betrayals, with the people around them. The fear of hunger and the fear of death, death from ambush, death from a raid, death from shipwreck, death from sorcery, death from disease, hung over them all. Whether their views of illness and death were more dominated by fear of ghosts or live enemies, sorcery or spears, or whether their food supplies were more likely to fail from black magic or enemy actions, differed from one tribe to another. But these were their fears. All human love and loyalty was compromised by this background of fear. In marriage one feared one's relatives-in-law; courtship, passion, and marriage outside the group tempted death in many forms; friendship meant involvement in affairs not one's own when the burden of dealing with close kin was enough. Pride and self-esteem were usually a matter of pride in one's immediate ancestors, one's clan, and only to a slight degree in one's village or one's tribe. Religion was a severely practical matter, a way of preventing illness, restoring the sick, obtaining immunity, gaining protection for warfare and dangerous journeys, and obtaining success in growing food, catching fish, or carrying on trade or courtships.

Or, put another way, man's potentialities for wide and continuing social responsibility, for delight in knowledge and order for its own sake, for a sense of communion with his God, were undeveloped. These very simple cultures provided a framework within which an artist could paint or carve, a mother could play with her baby, children could weep for their parents, men could feel that they were fully human beings because they felt they knew who they were and why they lived. No one had yet raised any questions which were too difficult to answer.

The new government brought by the white man, the new religious concepts brought by the missionaries, the new machines and ships, which were the product of white civilization, were all lumped together by the native peoples as parts of one system, obviously technologically superior but not necessarily preferable to their own way of life. And a New Guinea culture grew up mediating between hundreds of tribes, each with its local distinctive style, and the missions, the government, the traders, the prospectors. Each group saw the other in generalized and oversimplified terms.[2] The Europeans came to like some natives and dislike others, sometimes to like villages, or even whole tribes or areas, and to develop certain expectations from them, based originally on felicitous or unpleasant contacts with a few individuals, as house boys, police boys, catechists, etc. The natives did the same thing, preferring one mission to another, or one occupational group of Europeans to another, overgeneralizing from the individual. Then sometimes, if they were sophisticated, as the Manus were, they learned something about the occupation itself, knowing what kind of generosity you could expect, for example, from an employee running a trade store as compared with the owner.

It was only very slowly that any understanding of what we call "spiritual values" or "higher things" began to penetrate. By 1928, in some parts of the Territory, especially New Britain, New Ireland, and a small part of the Solomons, there were native peoples who knew what sorts of choices the Mission would make, who knew how to get the Mission to defend them against exploitation as carriers, who realized that an up-and-coming missionary could be involved on their side of an argument against a trader or government official. There were village groups that had supplied labour to a responsible government official or a trader for ten years who developed a trust in that particular individual.

But the principal area of communication between Europeans and natives was food and medicine. Little as they understood why any white man when presented with an open wound would open his supply of medicine, no matter how slender, and bind it up, they knew this so well that when patrols or prospectors penetrated into areas where the people had never seen a white man, it was only a matter of minutes before sores and wounds were brought for bandaging. The Christian ethic of binding up the wounds of the suffering, expressed in the person of the devoted nuns riding in their voluminous habits many miles over bad trails, and the government medical assistants going up and down mountains to give shots of NAB, was the one situation where the natives, as a group, came to identify an ethic different from their own, an ethic stemming from a belief in a God whose children they also were, and an ethic which somehow was independent of the particular virtues of individual white men or white women.

Food was a less reliable medium. Whereas every white man responded to the request for medical aid, many of them could not respond to the requests for food. Out on patrol with only a shot-gun, a few rounds of ammunition, and a few tins of meat for an emergency, the isolated white man couldn't afford to feed a hun-gry village, although with luck he might shoot them a bush pig. The isolated trader waiting for "cargo" now three weeks overdue had to conserve every cup of rice for his indentured labourers. So generosity in terms of food, of blankets, of mosquito nets, were matters which were again referred to individuals. *Only* in medi-cine were the will to cure and save and the ability to cure and save joined together in a combination of a Christian ethic and drugs which worked wonders—NAB for yaws, argyrol for infected eyes, calomel and castor oil and iodoform for intestinal disorders, qui-nine for malaria.

But for the Manus people in 1928 even this single line of understanding was almost completely cut off because of their own religious theories of disease, which were one degree more "spiritual" than the usual New Guinea theories—in which men sickened from vengeful ghosts or the black magic of vengeful, envious, disgruntled men—but still considerably less "spiritual" than the ideas of service to God and to man, the religious values of the white men with whom they came in contact. In 1928, the Manus did not want medicine; they hid their sick from govern-ment inspections; they resisted our occasional attempts to deal with an extreme case. Only in the case of a wound were they will-ing to accept our help. Catgut and iodine were seen as technolog-ical improvements not involving religious principles. So, when the adolescent Ngaleap fell from the rattan swing—the same kind of swing in which the angry wives of an earlier generation used to swing and laugh and shout tauntingly at their husbands congre-gated in the men's house nearby with their captive prostitute—and cut a great gash in her knee, they were willing to let the anthropologist sew it up. Otherwise, when we held aromatic spir-its to the noses of people who had fainted and they came to, coughing and spluttering, the people said, "Take it away, can't you see that they dislike it."

So there was no way at all to communicate to them why an
anthropologist would come all the way to Manus, live in a native
house so inferior to the kind of houses in which Europeans lived,
with a floor made of slender split betelpalm trunks through
which fountain pens and keys and chair legs slipped, sleep on narrow uncomfortable camping cots, eat food prepared by untrained
little boys who burned the fish as they argued about who had
more burnt tattoo marks on his upper arms, work eighteen hours
out of the twenty-four in the steaming heat. There was only one
way they could understand this—it must be for money. After all,
they themselves laboured long and late, made dangerous voyages
on the open sea, took unpleasant trips into the hostile, magic-
infested land in order to trade, for by trading they could under-
pin their way of life so that their children could live and become
traders in their turn. Someone in Rabaul had seen a white man
with pictures of nearly naked women in his pocket (probably pictures of Matty Island girls which formed the stock-in-hope of so
many lonely white men who had never been to Matty Island,
where the girls were said to be attractive to a European eye, com-
plaisant to foreign advances, wore only a green leaf as pubic covering, and had long, beautiful flowing hair). Someone had heard
that white men would pay money for pictures like that. And
weren't we in Peri taking pictures of their women whose breasts
were uncovered—and every native who had seen Christianized
natives knew that uncovered breasts were salacious? And did they
not have a ceremony, their best and most worthy of being photographed ceremony, performed in warfare and during big economic exchanges, in which men took off their breech clouts,
donned white ovalis shells, and engaged in vigorous athletic phallic displays of assertion and aggression? We were taking photographs; photography indeed was one of our most conspicuous
activities; when we developed the film at night we had to develop
in the dark of the moon with scouts posted around the village to
keep fishing canoes with torches far from our house. So, obviously, our motivation for coming to Manus was to take porno-

graphic pictures which we would be able to sell for large sums of money.

In this context, the presence of a white woman was somewhat aberrant. Most of the women, and many of the older men, had never seen a white woman. Those white women the work boys had encountered had been unpredictable in the extreme. In general they were willing to assimilate me into the picture of one of their own female relatives, who gave food and time to her sons and brothers, while they admired Dr. Fortune to the extent that he was dominating, definite, and trustworthy in trade. Men stood and stamped on our house floor as they demanded large loans, daring us to say they would not repay them. People took journeys into the port for us if they wanted what we paid them. Once we lived on canned tomatoes for days, although we had forty pounds in silver, because the "cargo had not come" and we had no tobacco to trade. The men who had been willing to sell us fish every morning—I used to climb wearily out of bed before the sun was up at the urgent salesman's voice saying, "E ni! e ni!"—now had no fish to sell. We were convenient trade partners, to be treated as trade partners only as long as we had something to trade.

The construction of a house to live in had been one long nightmare of economic exploitation; every ten shingles had to be bought separately from somebody's Usiai trade friend, who would then beg for something additional, in the manner of trade friends. Everyone who entered the house tried us out by asking for something; anything would do, the teakettle, a flashlight. Sometimes a sophisticated work boy would begin lower down on the scale and end by asking for the one thing he knew we would not refuse, however slight his claims—ten grains of quinine. Each house-post, each of the floor "planks," had to be bargained for and haggled over, and the transactions were never complete. The Sir Ghosts of the two rival men who were trying to use us in different ways—Pokanau, the intellectual, who was trying to use us as a direct source of payment for his work as an informant, and

Lalinge* the entrepreneur, who as the contractor building our house was accumulating large credits and trying to use us as a focus of trade—tangled over the question of where we were to live. In the end, as a result of the mediums' interpretations of my illness and the illness of three of my boys, we were forced to move into the new house before the verandah and the ladder were finished, although I knew that if we moved the house would never be properly finished. Pwanau, the Sir Ghost of Lalinge, would never have permitted Lalinge to leave his own house in such a state. But after all, we were strangers, to be traded with, nobody's kin, even though Lalinge had once, in a mixture of legalism and sentimentality, said he would be my brother so that, in case the master beat me, I would have a house to run away to, for the position of a woman with no one to depend upon but her husband was too horrible to contemplate. And true enough, the house never was finished. Two months later the temporary ladder broke; I fell and broke a bone in my foot, and spent many weeks on crutches.

This incident crystallized for the Manus as some error, some miscarriage in the way our system of values and theirs had met. In 1953, I found that the house that had been remodelled for me had a heavy double railing on each side of very well made steps, and the bridge which led to the latrine, built out over the sea, was made not of one but of two enormous smoothed logs, with a railing on both sides. During the first week after my return to the village, one person after another took me aside and asked me how, those many years before, I had hurt my foot. No one could remember, search their memories as they might, just how it had happened. They remembered that Ngamel had made my

---

*Lalinge is an alternative name for the man whom I called Paliau in earlier publications, then the *tultul*, government interpreter, in Peri, and a former police boy under the Germans. I will refer to him throughout this book as Lalinge, or by his Christian name of Josef, to prevent confusion with Paliau of Baluan, the leader of the New Way in contemporary Manus.

crutches, whittling down native wooden pillows and fastening them on the ends of punting poles; they remembered the long canoe trip into Lorengau to see if the Australian medical assistant could set my foot; they remembered who went, what we said on the journey, where we stayed in Lorengau, what everybody ate, how the glass of the Tilly lamp got broken when we all slept in the canoe because it was too rough to make Lorengau that night, but no one, no one at all, could remember how I broke my foot. This single piece of forgetfulness, a forgetfulness of which everyone was conscious, about which they worried, a forgetfulness of an event against the recurrence of which they took such explicit precautions, stood out from the detailed memories which we shared of the so-much-that-had-happened in those six months twenty-five years ago. They could remember with pleasure the master's anger and determination, especially when directed against people from a neighbouring village, but my foot which had been broken because they had been self-interested and non-protective worried them. It became a slender link between the old kin-limited standards of the tribal past and the concepts of today, when all human beings must be seen as brothers and sisters, regardless of race, no matter whether they are "black or white, red or green."

The submission to government orders of a people such as the Manus were in 1928 was a submission to greater power and a cool-headed appraisal of the values of the Pax Britannica in promoting trade and of the uses of government courts in pursuing trade disputes. Their approach to the mission was equally as practical and equally as lacking in any sense of real difference between their system of governing ghosts and the God of the Christians. In fact, they knew nothing yet about God, although they knew quite a bit about missions. There were two kinds, Protestant and Catholic, between which it would be possible to choose, and the choice was like the choice offered between working as a house boy or on a plantation or on a ship. If natives were wise enough they could drive a bargain with the recruiter and say what kind of work they wanted to do; if they weren't, they simply let themselves be "signed on" and ended up where the recruiter could get the best premium.

So the Manus were deciding, in 1929, to become Catholic, for three carefully thought-out reasons: because the Catholic missionaries taught their converts to read and write in the lingua franca of Pidgin English instead of in a local language and this would give them access to a wider world; because they did not collect as much money; and because they practised auricular confession. The idea of confessing one's sins to a single person instead of to the entire community was an alluring one to a people whose confessions were now blazoned forth with drum beats to the entire community. Of Catholic doctrine or practice they knew very little. A few Manus-speaking people living as immigrants in the faraway village of Papitalai had already become Christians. Once while we were in the village a Catholic convert who was visiting nearby was brought in to add his curing powers in a case which was proving very stubborn; all the known sins and trespasses on magically protected property had been confessed and atoned for long ago, but the illness still lingered.* So the Catholic convert was allowed to add his religious powers to those of practitioners of the local religions.[3]

But before they became Christians, they wanted to give one last big feast, for they knew that not only would they have to give up their ghosts, but they would also have to give up many of their distinctive customs, and especially their phallic dances, which were conspicuous targets of white disapproval. So first the villages on the South Coast, led by Kisekup of Bunai, would give a great *tchinal* feast, a feast named after the carved dancing pole on which the phallic dancers stood, and then everyone at once would throw out the skulls of the dead, thus deposing the Sir Ghost that gave each household protection against the ghosts of other house-

---

*In other parts of the Admiralties where conversion was a slower matter, notably on Baluan, accounts state that a favourite method of making converts was for the catechist to help to cure someone who, if cured, was then bound to become a convert.

holds. For any household without a Sir Ghost was in a desperately vulnerable state, but if everyone threw out their Sir Ghosts
together, then it would be safe to welcome in the new order. The
Pater would come, people would be christened, receive Christian
names, and there would be a resident catechist and a school, and
they would learn to read and write and keep accounts. This last
item was the most important of the three, for the Manus combined an ability to deal with very large numbers with a lack of any
method of keeping records. If they only knew how to keep
records, then they would be able to settle their financial affairs
without endless bickering, shouts, and insults and vituperation.
So they planned, in 1929. In one or two years they would give the
feast, and then end all feasting, a final individualistic triumph for
the middle-aged leaders who under normal circumstances would
have had to face their power passing into other hands because of
their early deaths, or, if they lingered on, their blindness or illness
which made them no longer able to lead. "After me the deluge"
was translated by the leaders of the South Coast as "After we
make a feast there will never be such a feast again," and "We will
have the satisfaction of throwing out the ghosts who have driven
us all our lives."

For the old system had lain heavily on them all. Dr. Fortune
has described it, speaking of the way in which confession was
extorted by detective work on the part of mediums and diviners
with a "persistent interest in others' affairs, in secret gossip, and
an eye for appearance of guilt." He continues:

> It might be thought that in this way confession would be more
> readily secured. But a great shame is at work to hinder it, and
> further a great fear of reproach. For if his ward's eldest son sins,
> Sir Ghost does not necessarily take his ward's eldest son's soul
> stuff. Sir Ghost takes the soul stuff of the next person in that
> household to fall ill. By the laws of probability, sin is much
> more likely to hurt someone else than the sinner. Then the eld-
> est son must confess to secure the cure of one dear to him. If he
> conceals it, he is assured that a death will likely result from his

concealment. He will then commit patricide, fratricide or the like. But if he confesses, the family reproach will fall on him for having caused illness and danger of death to one of those who love him and who care for him. This type of reproach, which is made with the utmost severity, is feared and dreaded by all. So much is this the case, and so great is the shame felt at public exposure of sex offence or theft, for instance, that confession must usually be wrung from the sinner by the oracles, even though the oracles by virtue of their communications from the ghosts have "more than mortal knowledge" of secret sin. In the end the more than mortal knowledge usually wins, but the oracles may be hard put to it. They have normally to know enough to appear to be able to substantiate their profession of multiscience.[4]

For the people of the South Coast, becoming Christians thus meant, in prospect, a safe and successful revolution against their old religious authorities, the substitution of a system which would be both powerful and much less heavily weighted with public shame, a much pleasanter form of retirement for the current leaders, and an entry into a world to which writing and mathematics held the key. Religion would deal with matters of health and disease; writing would enable them to deal with governmental power and with natives and white men who wished to get the better of them in trade store or court. Arithmetic would eliminate the need for anger. For the Manus twenty-five years ago saw the anger, the stamping and shouting, cursing and raging, insults and epithets, with which their system was shot through, as due especially to an inability to decide on what was the truth. If one man claimed a debt of two thousand dog's teeth or twenty fathoms of shell money and the other man claimed it was eighteen hundred, who could tell which was correct? True, there were attempts to make every transaction public. Payments which were to be made were previewed, each individual knew exactly which strings of shell money or dog's teeth—this one with "five teeth and then a broken one, blue beads between the [dog's] teeth except in the middle, where there are five red ones, blue and red tags on the end"—he or she would receive in the final transactions, and these would be reviewed again later. But still there were quarrels. If only people could handle accounts all this trouble would vanish!

For even then, the Manus had a touch of social engineering in their thinking. They held parts of the "system" responsible for the kinds of human behaviour which they deplored; if the system could be revised, the desired changes in human behaviour would follow. So Lalinge had caught on eagerly to my account of our old rule of etiquette that presents during the engagement period should be completely perishable—candy or flowers—or completely imperishable—jewelry—to obviate the distastefulness of returning such gifts as partly worn-out gloves. He equated this distastefulness with the position of the consumable items in the Manus system, the sago which one had neither eaten nor yet made a return payment for—how much less quarrelling there would be if this situation could be eliminated. And once at a feast, as Lalinge made the customary speeches and hurled little balls of sago at the house front, he added with contemptuous cynicism, "and after all this fuss I will merely pass these things on to someone else tomorrow." The detachment which made it possible to look at the "system" came, as it has come to Europeans, from a knowledge that their own culture was only one among many; *kaiye e joja* [customs of ours, exclusive of the person spoken to] had been compared for generations with *kaiye e ato* [customs of theirs] and *kaiye e aua* [customs of yours]. People were well aware of differences between their own ways and those of their neighbours; trade friendships and periodic intermarriages kept this sense of the reality of difference lively.

So, in 1929, there lived in Bunai a man who was called "Man of Lorengau," who attempted to alter the functioning of the Manus Sir Ghost cult by suggesting that quarrelling would be obviated if each house had a medium of its own for its own dealings with its own Sir Ghost. This would prevent the quarrelling which came when a medium was accused of mixing worldly advantage with supernatural communication. "Man of Lorengau" had been a work boy under the Germans. He is described in our "Who's Who of 1928"[5] as "A magician of Bunai village; a versatile linguist, speaking German, some English, nearly all the languages of the Admiralties, some Bismarck Archipelago dialects, and a little Samoan. He was born of a Usiai mother and a Manus father; an innovator and an experimentalist, very upset at criticism of his not being radical enough. He liked to try internally medicines (European) usually used externally in order to discover the effects, and similarly with everything; a male medium as well as diviner. He had five wives and had a great reputation for love magic. He had a great many friends on the North Coast, where dugong is to be fished." The "Man of Lorengau" was a type who may have existed for centuries within the Admiralties, his potentialities as critic of the "system" enormously increased by his early participation in the culture-contact situation.

The belief in change, the belief that change could be for the better, the belief that men could themselves decide to make a change in their system, were all there in 1928, as they were in somewhat different form in many parts of Melanesia and New Guinea. In fact, the system of entrepreneurship—in which widespread patterns of exchange between in-laws had been tailored to fit the Manus demand for individual responsibility and accumulation of property—was itself a demonstration of how the Manus felt that social forms could be manipulated for conscious individual or group purposes. So, at the end of World War I, an inexperienced government officer, left over from the military occupation and appalled at the litigation involved in marriages, and also at the possibilities of violence which accompanied the presence of grown unmarried girls in the village, lined up all the unmarried men and women in the village and ordered them to marry. This was in total disregard of all the rules of forbidden degrees of marriages and of the approved ways of contracting a marriage in which one cross-cousin could approach another. It was also in disregard of the local Manus entrepreneurship arrangements which had imposed a further pragmatic structure on this kinship pattern, so that real cross-cousins "made the road" but did not finance the marriages. It took the people of Peri years to straighten out these marriages, work out appropriate "roads" which might have been taken, sort out the half-completed economic responsibilities so that in the end these marriages, imposed by governmental fiat, were legal and working within the Manus system. But they had worked with the resourcefulness and zest of lawyers determined to adjust a difficult corporation law to the realities of a local financial situation, worked with a consciousness of the relationship between formal rules, operating situations, and awkward facts. The circumstance that marriages were arranged, and if possible maintained, irrespective of the wishes of the two spouses did of course make this easier, just as it is easier for a corporation lawyer to manipulate proxies in a joint-stock company than to manipulate the feelings of two angry partners.

So, in 1928, the Manus faced what the white man had to offer with a conscious sense of what they wished to take for themselves. They rejected baubles, cheap perishable goods, or the extravagance of matches. Why buy matches; it was necessary to know how to make fire because matches got wet anyway, and matches were expensive. Hard-earned money should go into good steel, or good canvas, or strong cloth. The government collected taxes, the government forbade warfare; very well, government courts should be made to subserve the conflicts once settled by warfare. There were recruiters who wanted to "buy boys," but the Manus themselves would bargain for where they would work and what kind of work they would do. The mission with a religious system which was superior to theirs wished to "come in," but they would choose which mission and when it was to come, turning the acceptance of Christianity to their own uses—to subserve the pride of their leading men and to depose the Sir Ghosts whose rule was irksome. They were like men with well-filled pockets but no stores of their own, who saw the world of the white man as a combination department store and cafeteria within which they would buy what they chose, subject to the rules of purchase laid down by the owners of the store.

They did not question those rules. The Germans had come and been strong and well equipped and able to subdue local warfare. The Australians, "the English," had thrown out the Germans. The Australians were now in control; armed, to be dealt with, traded with, manipulated legally, but not to be rebelled against. Men like "Man of Lorengau," who had travelled widely, knew the greater complexities of the work-boy world. Two men in Peri, Lalinge and Pataliyan, had been to the mainland of New Guinea. Many men had been to Rabaul, had seen motorcars and telephones. The Burns-Philp ships came "regularly" into Manus, every six weeks or so, and during World War I the older people had seen battleships. The world of New Guinea over which the Australians ruled was a world in which men could travel, trade, work for money and goods, choose their routes—within the rules—in a social framework as external to themselves as the ships on which they would travel, or the shillings which they had learned to equate with shell money and dog's teeth.

If there was as yet no form of communication within which the ideals of a higher civilization could reach them, there were plenty of smaller units of communication. The Manus recognized facts—in the Western sense—that a given event could be said to have occurred or not to have occurred, and that in reporting on it, one either told the truth, that is, made a statement which was an accurate report such as a camera would make, or lied. So, when the German traders first came to Manus, they imported fake dog's teeth to use in native trade. The Manus rejected these fakes. Then the indefatigable Germans imported real dog's teeth from Turkey. The Manus recognized that these were real and accepted them, with a perfectly conscious comment on the inflation of their currency—that now it took twenty dog's teeth to buy what two had bought before.

So the idea that they would take over, in whole or in part, available and superior European institutions, was present, as it was in many other parts of the Territory, but the idea that they would become in any sense *like* Europeans was absent. They did not even have as elementary an idea as the fact that they and their neighbouring Usiai were in any sense the same people, even though when away at work they might use the word "Manus" to cover both groups. Prophesying in 1929, one could say that they showed unusual discrimination and a high sense of autonomy and choice in their dealings with Europeans, but there was little reason to predict that their fate would differ from that of other native peoples who had as little sense of themselves as a political unit within which they could confront the massive institutions of the European. African kingdoms numbering millions, with king and court and law courts, with institutions which predated those of Europe, might reel before the shock of European contact, before firearms and machines, and yet survive as intact political units. Even the Polynesian islands of Samoa and Tonga, with populations which could only be numbered in thousands, but with elaborate political organizations, might survive, slowly reworking and adapting the institutions brought to their doors, advised and protected by responsible missionaries and governmental officials. But the Manus, this bare handful of people, without any political unity, without a sense of the past or the future, without a notion of such ideas as "the state," "law," "tradition," with nothing but a respect for facts, a concern for their children's futures, and an autonomous demand to make their own choices—how could they have a chance of survival, except as mere bodies, shorn of their own system and hardly sharing in the wider one, doomed to perhaps many generations of proletarianism, in which, wearing the trappings of culture contact, manufactured cotton underpants, they would eat polished trade rice out of enamel bowls and combine bits of little-understood Christian rituals with an increasing fear of the sorcery of natives from other islands?

This is how I saw them in 1929, against what we knew then about culture contact over the world. I had seen Samoa, where the people had made a fair amalgam of the new and the old and still lived with dignity, protected by the U. S. Navy.[6] I knew what had happened to our American Indians.[7] I knew something about what had happened in Africa. I knew what happened in the

United States, where peasants came to our large cities and as pro-
letarians learned to be Americans.

New Guinea was a kind of economic backwater, precariously
dependent on copra and gold. Whale oil competed with copra on
the world market. As my old friend, Mrs. Parkinson, who had
helped found one of the first coconut plantations in the Territory, said to me, that summer of 1929, "I wish all the whales that
live in the sea would die, and then my poor children could make
some money again." The price of copra was down, the gold that
supplied the money to run the Territory was presumably not
inexhaustible. A few experiments with cocoa and cattle had had
slight success. Germany was not yet renascent; possibilities of war
with Japan were regarded as the wild fear of jingoist warmongers.
Australian interest in New Guinea was very slight. The League of
Nations' demands for progress could be fobbed off with photographs of somebody's cook boy dressed in a shirt sitting before a
typewriter as a report on education. Americans were so little
interested in that part of the world that I had to call my book
Growing Up in New Guinea (using the name for the Territory),
and then include a bit of Australia in the map so that those Americans who thought New Guinea was in either Africa or South
America could orient themselves. My field trip of 1928 was the
same length as that of 1953, but it took six extra months to get
there and get back in 1928, so that my first quarterly report on
my fellowship was due the week that we finally moved into Peri.

There seemed every possibility that social change would be
slow, and the task of the anthropologist was to advise on how to
slow it down, moderate and modulate it, help a responsible government to suppress head-hunting, cannibalism, and warfare,
without wrecking the fabric of these small native cultures. Our
model in those days was the epic achievement of Mr. E. P. W.
Chinnery, first government anthropologist in the Mandated Territory of New Guinea, who had persuaded the warlike Orokaiva to
substitute a pig kill for a human-being kill in the male initiation
ceremonies, and otherwise left everything as it was. True, a little
later the Orokaiva had been seized by a kind of religious mad-

ness[8]—a quaking and shaking and seeing of visions—but we knew something of these "nativistic cults," too. They had occurred among American Indians. The most notable of them was called the Ghost Dance,[9] and they were an expected accompaniment of culture contact between high cultures and primitive ones.

Initially, the natives would be excited by the superior techniques of the high culture and try to acquire them. Later, they would find out how difficult the road was, the long, long road to civilization. Then would come the "nativistic cult," the promise of an immediate supply of all the desired goods, or the return of the buffalo on the American plains, or the ancestors would return to drive out the invader or turn them into servants. These "nativistic cults," not yet named "cargo cults," had been occurring for years in New Guinea and would doubtless recur. The initiators of the cults—which after all were undesirable, for they excited the natives and led them to kill their pigs and neglect their gardens—would be imprisoned, the promises would not materialize, the natives would pass through the "nativistic-cult" stage, and slow change—the slower the better—could go on.

These were our expectations. What was wrong with them, so wrong that what has happened in Manus still seems almost like a miracle? What has happened in the world that we had not anticipated? What was there in the Manus people themselves that we had left out of our accounting, and was it in the Manus alone of New Guinea peoples?

5

yesterday's children seen today

Now that the change has occurred, now that the Manus people have taken such a different route, have come into the mid-twentieth-century world because they want to live in it not by mystical or magical means, but as part of it, we can ask what was it that we did not know in 1929 that, had we known it, would have made us better able to predict their future course.

We think of science as the study of natural laws which once grasped will enable man to predict the course of any comparable sequence provided the conditions can be kept constant. But in real-life situations we always have to come to terms with sequences of events which are outside the limited little set of conditions within which we have learned to predict. So the bridge builder constructing a bridge in New Guinea may understand perfectly the nature of the steel with which he builds and the principles on which such a bridge must be constructed, and still his bridge may fail to materialize because his bridge-building crew died of an epidemic of Japanese River fever. Today, but not in 1929, he can provide the necessary antibiotics which will combat this fever. But his crew may be attacked by hostile natives concerning whose intentions he had no knowledge. Here again, before starting the bridge he may include an anthropologist to

reconnoitre the native situation, establish rapport, or diagnose
the dangers in the local scene. But then an earthquake may come
and shatter his cement posts just after they are built. If he has
included geodetic surveys he knows that an earthquake may happen, but not when; he may construct his bridge differently to
guard against it; he may have brought extra supplies to replace the
damage. Or he may have to radio for new materials, and these in
turn may go down with a ship sunk in a storm five hundred miles
away, or he may get no new building materials because a war
scare half a world away has cut off the supply of materials. He
may suddenly be ordered to stop building because the island has
become a strategic base, due to the discovery of a use for a mineral
it contains, and consequently a new kind of bridge is needed. Or
the engineer himself, on whom the building of the bridge
depends, may die of an embolism, although he had an exhaustive
physical examination before he left civilization.

So, although modern science can tell how the bridge can be
built, of what materials it can be built, what drugs will give protection against some of the diseases which may be encountered,
under what conditions the natives are likely to be unfriendly, and
whether earthquakes are likely to occur in the area, it cannot tell
us whether that particular bridge will be built. Too many different event sequences may coincide on that small island, and men
do not have as yet, and may never have, devices for correlating far
enough ahead the way in which those sequences of events will be
combined. Even if the political scientists feel war between such
and such powers is inevitable, their predictions may not affect the
bridge-building company whose officials want to build for peacetime uses. The anthropologist may be able to estimate that a
group of natives are friendly and peaceful, but not be able to
reckon with the effect of the news that one of their number, away
at work, has been murdered. The success or failure, the shape of
any one of man's ventures within time, is subject always to such
combinations of sequences, each of which is only partially known
and partially predictable.

But our partial knowledge of each sequence increases, and the areas within which science gives us such knowledge are continually widening. If we look at the present fate of the people of Peri against our knowledge of them and our knowledge of human behaviour which has developed during the last twenty-five years, it will be possible to sort out those scientific advances which make it possible to understand what did happen in Manus. In the human sciences where experimentation is very difficult, so that we have to rely on history to provide us with experiments, we progress by locating the natural experiments, studying them, and from the understanding so gained constructing the theory which will give us our partial predictive control in the future.

So it is possible to ask what understanding, lacking in 1929, we now have which makes what happened in Manus intelligible—the first step toward predictability.

My concluding paragraph in the appendix of *Growing Up in New Guinea*, entitled "Culture Contact in Manus"—for it was still the style in those days to concentrate on the culture as it had been and include statements about the living present in footnotes or appendices—read:

> To summarize, Manus contact with the white man has to date been a fairly fortunate one. War, head-hunting, and prostitution have been eliminated. Recruiting has prevented these prohibitions creating new social problems, the recruiting period and its rewards have been fitted into the social economic scheme; trade with the white man has provided the natives with beads which have developed a new decorative art and furnished new incentives to the production of foodstuffs; the peaceful régime has produced more favourable conditions for inter-tribal trading. The Manus at the present time are a peaceful, industrious people, coping admirably with their environment, suffering only slightly from preventable diseases. Their ethical system is so combined with their supernatural beliefs as to receive great force and intensity from them. They are not taking any measures to reduce their numbers, being apparently ignorant of medicinal abortifacients (as they are

ignorant of most herbal properties owing to their water life),
and seldom resorting to mechanical methods. From the standpoint of government they are making a most satisfactory
adjustment to the few demands which white contact makes
upon them. (*This is quite aside from the type of personality which
is developed by their methods of education and their attitudes
towards family life and marriage. These are subtler points which
government will have no time to deal with.*) [Italics new.]

This closing statement, in parentheses, contains one key to
what has happened in twenty-five years to our ability to predict—
within the limits of uncontrollable historical accidents and cross-
sequences—what the members of any culture will do in a given
situation. Twenty-five years ago we had practically no inkling of
the relationship between *type* of personality and politics. Marx
and his followers had insisted, were insisting, that there was a
congruence between economic institutions and character, so that
capitalists had "bourgeois" characters which showed many undesirable traits which would disappear when bourgeois institutions
were eliminated by the victory of the proletariat. But how bourgeois institutions produced bourgeois character no one had yet
spelled out. Max Weber[1] had emphasized the congruence between
Protestantism and the Industrial Revolution, but again a detailed
study of the mechanisms by which the Protestant ethic produced
or imbued the Protestant character was lacking. Freud had
opened up a whole new understanding of the human personality,
but child analysis was still in its infancy, and the tendency was to
explain personality in what were ultimately biological terms, the
inevitable clash of man's nature with the historical content introduced during his upbringing in any society. There was as yet no
room in psychoanalytic thinking to relate the differences in the
style of upbringing, between classes, between nations, between
periods, to the differing social institutions. This was all to come
later.

Lawrence Frank[2] was beginning to ask what the relationships
were between culture and personality, a question which led Mali-

nowski[3] into his single adventure in psychological speculation, the study of possible changes in the Oedipus complex which might exist in a matrilineal society like that of the Trobriands where power was lodged in one's mother's brother rather than one's father. I had made my study of the relationship between culture and adolescence in Samoa.[4] Lasswell[5] was just beginning to apply psychoanalytical thinking to political personalities. Ruth Benedict[6] was developing her theory of culture as personality writ large, a theory I had applied to my Samoan material.[7] In none of these formulations did we yet have a way of dealing with the question of how changes in economic and political institutions, in religious beliefs and world views, actually resulted in changes in the personalities of those who experienced the change. How did the imputed personality of the Middle Ages, otherworldly, incurious, living in an explicable and limited world, turn into Renaissance man, and then into Reformation man? Neither the statement that such events as the sacking of the library at Alexandria, or the discovery of America, set off the change, nor that religion was a neurosis and could be understood by the same rules used by the physician who studied a patient on the analytic couch, nor the statement that class conflict was inevitable and any change in the means of production brought about a social struggle, nor the insistence that culture was personality writ large—none of these formulations told us *how* the changes occurred within the human beings who had to embody the changes.

In 1929, we did not have the conception of character formation—of the specific, detailed, biologically relevant steps by which human individuals subjected to specific types of transactions with other individuals of specified character will develop character structures which are themselves systematically related to the existing institutional structure. That conception was not to develop until the middle thirties, when it developed from a combination of the work in anthropology, psychoanalysis, economic history, child development—in the thinking pioneered by Lawrence Frank.[8]

The Manus field work of 1928–29 contributed to these developments but was done too early to benefit directly from them. So when I studied the children of Peri in 1928–29, I was able to show *how* they were reared, in what ways and at what stages the way they were reared seemed to be inconsistent with the desired adult behaviour, and to develop the idea of the importance of continuity or discontinuity in child-rearing practices. What I lacked was any clear theory of how the presence within the adult character of this experience of discontinuity would function in implementing change.

We were, of course, on the edge of an understanding of this relationship. Psychoanalytic thinking had stimulated the hope that a different kind of early childhood experience would produce a different kind of adult. If children were told the truth about sex, allowed to see the nude bodies of both sexes, allowed to express their hostilities and aggressions freely, then it was hoped that many of the "complexes" which bedevilled their elders would be eliminated. So schools were founded which would put these principles of education into practice. As there was no theory of the relationship between types of education and types of social institution, very little initial thought was given to the congruence between this type of nursery and school training and the experience awaiting the child in the wider society.

This psychoanalytically based hope of bringing up a child freer from complexes was combined with the dreams of those who wished to revolutionize our society socially, to substitute co-operation for competition. If the most conspicuous socioeconomic evils of American life—identified by the social critics as "competition," "soul-destroying materialism," "keeping up with the Joneses"—were all eliminated from the schools, and children were kept in little prefigurative utopias, then when the children grew up they would reform the world. Neither of the two theories—and often they were combined in the same school, where freedom of bodily behaviour and insistence on sharing, work for work's sake, abandonment of marks, etc., went together—dealt with what I called in my appendix to *Growing Up in New Guinea* "the subtler points which government will have no time to deal with." What would be the effect of the contrast and discontinuity between the experience in school and all the rest of the experiences in the society? How did being allowed to go naked in the garden fit in with contacts in a world in which people snickered at the mere mention of nakedness? Or, how did not competing for marks fit with later competition in the labour market? Children, originally reared by parents and nurses who were part of a competitive culture, were to be briefly and partially insulated from this world, and then returned to it, somehow equipped to change it.

To such social reformers, the anthropologists could oppose their knowledge of culture, of how every society of which we knew managed to rear children who fitted the culture within which they were reared, unless there was a radical change pioneered by adults in some section of the whole society. It could be stated as an axiom, it seemed. Change was brought about by adults. So on the one hand stood the socio-economic reformers and revolutionaries saying, "Change the institutions," and begging the question as to why particular adults should be willing to become revolutionaries and instead invoking large-scale impersonal entities like technical change, or the class struggle, or an expanding market; and on the other stood the educators, the psychoanalysts, the religious leaders, saying, "Change the individuals." Change the institutions, said one side, and the personalities will take care of themselves. When goods are properly distributed, human beings will become generous, non-competitive, friendly. Change the personalities, said the other side. Generous, non-competitive, friendly personalities will have no use for institutions which permit individuals to exploit and oppress one another.

The conditions in Manus in 1928 seemed extraordinarily relevant to this controversy, at least to the educational side of it. For in the lives of Manus children there was, by accident not by design, just such a discontinuity as educators were attempting to introduce. While the adults were a driven, angry, rivalrous, acquisitive lot of people who valued property and trade above any form of human happiness except the maintenance of life itself, the children were the gayest, most lively and curious, generous and friendly, that I had ever known. Without property they had no quarrels over property; without responsibilities they did things happily for the sheer joy of activity; without involvement in the religious system they treated the possibilities of ghostly anger, which drove the adults so relentlessly, with lighthearted casualness.

The adult world was a world of continual economic activity and angry driving tension. I described the children's relationship in this way:

> From this world the children are divided completely by a very simple fact: they own no property. They have neither debtors nor creditors, dog's teeth nor pigs. They haven't a stick of tobacco staked in the transaction. True, the exchange may be made in the name of one of them. Kilipak's father may be paying twelve thousand dog's teeth to his cousin, the father of Kilipak's future wife. This brings the question of Kilipak's one-day marriage before the minds of the other children. They chaff him a little, suddenly stop using his personal name and call him instead "grandson of Nate," the name of his bride's grandfather. Kilipak turns hot and sullen under their teasing but he takes no extra interest in the ceremony, although in the name of it his elders will some day bring him to account. Today he simply goes off fishing with the other boys.
>
> Afterwards Kilipak will feel this payment in which he takes no interest: henceforth he must avoid his bride's name and the names of all her relatives, and he must lie hidden if his canoe goes through her village. So to the child's eyes, the elders have a great economic show which takes up all their time and attention, makes mother cross and father absentminded, makes the food supply in the house less subject to the child's insistent demands, takes the whole family away from home, or separates him from the large pig which he used to enjoy riding in the water. Then there is a great deal of beating of drums, speech making, and dancing. Every ceremony is just like every

other. It may be of huge interest to his elders that for a *kinekin* feast for a pregnant woman the packs of sago are stacked in threes, while for a *pinpuaro* feast after birth the sago packs are stood upright. To the elder such important bits of ceremonial procedure are sign and symbol of intimate knowledge, like the inside knowledge on the stock market which the new speculator displays so proudly. But to the child as to the non-investor, this is all so much unintelligible rigmarole.

His version of the whole spectacle is brief and concise. There are two kinds of payments—the payments made on a grand scale and the small gradual repayments made individually. The big ones may be canoes of sago and pigs and oil, or they may be hundreds of dog's teeth hung up on the islet, in which case there may be dancing. Sometimes, for wholly inexplicable reasons, there is no dancing. At other times a pig changes hands and a drum is beaten about that, most annoyingly. The drum beat may turn one from one's play in anticipation of some interesting event. And it turns out to be nothing but the payment of a debt. Afterwards there are always quarrels, insults, and recriminations. If mother is very much involved in the transaction, so involved that it would be inconvenient to go home—in the children's words, if mother "has work"—father will be especially nasty to her, knowing she won't dare leave him. But if it's "father's work," mother is likely to be extra disagreeable, to weary of it in the middle, and go off to her own kin. The fact that a lot of this "work" is ostensibly in his name only serves to set the child more firmly against it, as a most incomprehensible nuisance. To all questions about commerce, the children answer furiously: "How should we know—who's grown up here anyway, we or you? What do you think you are to bother us about such things! It's your business, not ours."

The parents permit their children to remain in this happy state of irresponsible inattention. No attempt is made to give the children property and enlist their interest in the financial game. They are simply expected to respect the tabus and avoidances which flow from the economic arrangements, because failure to do so will anger the spirits and produce undesirable results.

In the child's world, property, far from being garnered and stored, is practically communal in use if not in ownership. Property consists of small canoes, paddles, punts, bows and arrows, spears, spider-web nets, strings of beads, occasional bits of tobacco or betel nuts. These last are always shared freely among the children. One poor little cigarette of newspaper and Louisiana twist trade tobacco will pass through fifteen hands before it is returned to its owner for a final farewell puff. If among a group of children one name is heard shouted very frequently above the rest, the listener can be sure that that child has a cigarette which the others are begging. Similarly a string of beads will pass from child to child as a free gift for which no return is expected. Quarrels over property are the rule in the adult world, but they are not frequent among children. The older children imitate their parents' severity and chastise younger children for even touching adults' property, but this is more for a chance to start a fight and from force of habit than from any keen interest in protecting the property.

Quarrels which spring up from other causes will be justified in terms of property, if an adult inquires into them. The children know that to say "He took my canoe" will elicit more sympathy than "I wanted to make cat's cradles and he didn't"; and the child is an adept at translating his world into terms which are acceptable to the adult.

The constant buying and selling, advance and repayment in the adult world is a serious obstacle to any co-operative effort. Individually owned wealth is a continual spur to self-centred individualistic activity. But among the children, where there are no such individual stakes, much more co-operation is seen. The boys of fourteen and fifteen who stand at the head of the group organize the younger children, plan races, on foot or in canoe, organize football teams, the football being a lemon; or institute journeys to the river for a swim. Surface quarrelling and cuffing is fairly frequent, but there is little permanent ill humour. The leadership is too spontaneous, too informal, and has developed no strong devices for coercing the unwilling. The recalcitrant goes home unchastised, the trouble-maker remains. The older boys scold and indulge in vivid vituperation but they dare not use any appreciable force. A real fight between children, even very tiny ones, means a quarrel between their parents, and in any case the child always finds a sympathizer in his parents. Irritation over missped plans or a spoiled game takes itself out, very much as dominoes fall down one upon another. Yesa tells Bopau to get his canoe. Bopau refuses. Yesa slaps him, Tchokal slaps Yesa for slapping Bopau and Kilipak slaps Tchokal for slapping Yesa. Kilipak being the largest in that group, the scuffle degenerates into a few wailing or sulking individuals. In five minutes all is fair weather again unless some child feels so affronted that he goes home to find sympathy. These teapot tempests are frequent and unimportant, the consequence of a large number of aggressive children playing together without devices for control. At that, they are far sunnier and less quarrelsome than their elders, more amenable to leadership, friendlier, less suspicious and more generous. Deep-rooted feuds and antagonisms are absent. Among the elders almost every person has definite antagonisms, always smouldering, always likely to break out into open quarrels. But the size and the varying ages of the children make a fluid unpatterned grouping in which close personal attachments and special antagonisms do not flourish.

Although the parents take violent part for their children, their children do not reciprocate. Children whose parents are making the village ring with abuse, will placidly continue their games in the moonlight. If the quarrels between the parents grow so serious that the spirits may be expected to take a hand, the children are warned against going to the house of the enemy, a prohibition which they may or may not obey.

The whole convention of the child's world is thus a play convention. All participation is volitional and without an

arrière pensée. But among the adults casual friendliness, neigh-
bourly visiting is regarded as almost reprehensible. Young men
without position or standing go to the houses of older relatives
to ask for assistance or to render services. Men may haunt their
sisters' homes. But visiting between men of the same status, or
between married women who are neither sisters nor sisters-in-
law, is regarded as trifling, undignified behaviour.

I knew, even in 1928, that I had to find a specific answer to
how these children, permitted a life so at variance with the life of
their elders, developed into men like those same elders. I had
tackled this question once before, in Samoa, when it seemed so
puzzling that the badly behaved toddlers, who tyrannized over
their child nurses, could become later such obedient, well-
behaved children. In Samoa I found the answer in the tasks the
older children were given to perform; a tyrannical, undisciplined
little girl was made in her turn into a child nurse, and had to keep
her charge quiet or invite adult anger. And small boys were
reduced to deferential obedience by older boys who only permitted them to accompany them if they behaved. On the freedom
and indulgence of babyhood, a later period of discipline was quite
adequate to develop the Samoan adult character—quiet and casually conforming to external rules.

So I realized that I must discover how these gay Manus children turned into such unlovable and unloving adults. The children's life was not a sort of holiday which stern adults permitted
them in preparation for a hard adult life. Had it been, this would
have made the whole process more explicable—like the behaviour of self-made men who want their children to have all the
advantages they didn't have, which over time may crystallize into
a culturally patterned indulgence toward children, as it has in the
United States. But the people of Peri, in 1928, were not generously according their children a happy carefree period as a preface
to a later life of driven hard work. It was not a conscious and calculated mitigation of the hardships of life, like special treats at
Christmas; it was rather an accident of the way in which Manus

religious and economic life was constructed. Each household had characteristically only one protective Sir Ghost. Very occasionally some just-deposed ghost might be assigned to a male child instead of throwing him and his skull out altogether. But children on the whole had no ghosts to go about with them. As long as they were within the village, they were beneath the watchful protection of the series of Sir Ghosts of related households. When men went abroad they took their Sir Ghosts with them, but outside, and even inside, the village there was danger, and children were safest when close to their homes. The land was dangerous for children; there were crocodiles, snakes, cannibals, and the *tchinals*, the dangerous magical familiars of the land people. Fear of the land barred children from most of the work at which they would have been most useful—going into the bush in search of rattan, or leaves, going to market, running errands to the next village. The manufactures that went on in the village were mainly canoe building and making fishing devices—highly skilled activities which children watched rather than participated in. In the dangerous water world the care of small children was not left to older children; parents themselves took complete charge of babies and children under three. As the girls grew older and were betrothed, they were drawn into the string-making and bead-stringing, grass-skirt-making and mat-making of the older women, but the boys were left free to play in the water all day long. So Kilipak, reminiscing twenty-five years later, describes his childhood:

> When I was little I thought about a little canoe, a very little one. I was little and so I could not think of anything else. However, I also could think about shooting fish. When I woke from sleep in the morning I had no mind for anything else except going to get my little fish spear, my little arrows—you remember the bows and arrows that Pomat and I made and some of those you bought and took away with you—and then I would hurry out to fish with them. But they weren't big fish, they were only little ones. Sometimes they could be eaten,

sometimes they were too small to eat. These were just thrown away. The day would pass and it would be night. I would sleep with my father and mother, in the morning my thought would be on going out fishing again. Then I would go and shoot fish again. I wouldn't think of anything else. If it was bad weather I would stay at home with my father and mother and other children would come and we would talk of something that we wanted to make. We used to talk about it first, talk about toy canoes, about coconut shell [craft]. If the wind from one direction was right, all of us would go there. All of us would gather [in a crowd] and do it. When play finished and it was night, our minds would not give up thinking about them [the canoes]. At dawn, I would get up and think of nothing but playing with these little canoes. Later, I spent all of my time in the sea. My mother would wait and wait but I would not come home. The food cooked for me would harden. It would be cold. My mother would wait and I would not come. Finally she would shout for me. I would go and eat but my thoughts would still be on the games with the canoes. As soon as I had eaten, back I would go. Now I would play again with my little canoes. Each day it would be the same. My thoughts never turned to anything else. Because after all I was a young boy. Now my mind was set on play only.

And in reply to a question about what he remembered about something he greatly desired in childhood:

When I was a small boy my mother took me to a place where a Peri group and a Patusi group were [having a feast] ornamenting a bride with foreign articles. This was the way the big men used to do. Plenty of desirable little things, playing cards, little pipes, little handkerchiefs, these were all placed in the bride's ornaments.* She was a virgin and being married to a new husband. Now I caught sight of one of the cards, now I wanted it

---

*See Plate XI.

terribly, I cried for it, and my mother tried to silence me, but I did not heed her. I cried and cried because I wanted that [playing card]. I cried and cried. I desired it too terribly. Now my mother knew about [my wanting] this, and she watched out [on my behalf]. Now they showed these cards to a certain man. Then my mother went to that man and asked him for the card and she got it from him, and she brought it to me. Then I stopped crying. If she had not brought it I would have continued to cry. I wanted it terribly, enough to cry until my mother went and got it for me. Then only would I stop crying.

Also, the presence of children at ceremonies made the adults uneasy. The name taboos meant that everyone should be on the alert, so as not to pronounce inadvertently the name of someone else's brother-in-law, or more seriously still that of his mother-in-law or his spouse. There were many such circumlocutions, and the children, particularly the boys, were careless and ill-informed about them. At séances everyone was supposed to stay very quiet, waiting for the ghost to speak through the whistling of the medium, or waiting while a ghostly messenger went off to another island to collect a little of the purloined soul stuff of the sick person for whose sake the séance was held. Here again children could be nuisances. Manuwai still remembers with glee putting up his hand to find his mother's mouth to see just who was doing that whistling. Children brought up differently from Manus children learn easily to sit quietly at ceremonies, but Manus children had been reared in a way which made them intolerant of restraints. It was easier for the adults to shoo them away than to keep them in order; attempts to restrain them usually ended in a shouting match in which the elder got the worst of it.

But the accidental nature of the freedom accorded children, the fact that it was based on no philosophy, was important in the way in which it was reflected in adult personality. In some societies little children or older children or young men are accorded a period of license, and after they are grown can gratify their own memories by re-according this same licence to their children. Bearded men can smilingly say, "Boys will be boys," or alternatively, in those societies in which youth is a period of discipline and turmoil, "It's hard to be young." What was striking in Manus was that the childhood freedom was not such a recognized and integrated part of the whole process of education, not a period when children were sent to their mother's brothers to be lovingly indulged, as among the Ba Thonga of South Africa,[9] or to boarding school to make men of them, as among the English upper classes. It was simply a by-product of the way the economic, political, and religious life was organized, combined with the special type of personality which Manus children developed as little children being reared to live safely in a water world. To understand it, it is necessary to look at the way in which very young children were treated.

In describing in more detail the way in which the character of Manus children was formed, I will follow as closely as possible, without reproducing the whole, the descriptions which I wrote twenty-five years ago, adding nothing new here. By doing this it will be possible to demonstrate that we had all the information necessary but lacked a theory of cultural character adequate to deal with it.

My account of Manus infancy begins after birth because I, as a woman who had not yet given birth to a child, was not permitted to be present at a birth. Because our theoretical understanding was so rudimentary, had I witnessed a birth I probably would not have fully recognized the way in which the very first movements of the newborn and nurse or mother are knit together into a pattern, which is gradually enlarged and expanded as the child grows older.* What I did observe was that, after birth, mother and child were isolated together, and the father took care of the

*See Chapter XIII, "Rage, Rhythm, and Autonomy," and *Growth and Culture*, by Margaret Mead and Frances Cooke Macgregor, Putnam, New York, 1951.

displaced older child. He himself was under taboo; he could neither see his wife nor do any useful work until his wife's brother validated the birth with a large payment of sago. So the new father loitered about the village, indulging the toddler, who hardly knew that its nose was out of joint now that it had its father to itself. Here was formed the strong tie between child and father, from which the child never returned completely to its early feeling about its mother. Little boys remained devoted to their fathers until, at six or seven, they struck out on their own to spend all their time in play groups. Little girls were banished from their father's company when they became betrothed, and so subject to taboos.

Meanwhile the new baby had its mother's complete attention, made the more precious to the mother because this was the only time that she was allowed an undisturbed relationship to her child. A month—or a little longer if her brother was behindhand—after birth, she would rejoin her husband, and he would begin urging her on to economic tasks. As the baby grew older he would play with it during the day after his night's fishing. The busy mother left the baby more and more to itself. The old women were as busy as the young and while older children played with babies they were not required to act as child nurses.

This was how early childhood in Manus looked twenty-five years ago when the people lived in Old Peri, in houses over the sea.

The Manus baby is accustomed to water from the first years of his life. Lying on the slatted floor he watches the sunlight gleam on the surface of the lagoon as the changing tide passes and repasses beneath the house. When he is nine or ten months old his mother or father will often sit in the cool of the evening on the little verandah, and his eyes grow used to the sight of the passing canoes and the village set in the sea. When he is about a year old, he has learned to grasp his mother firmly about the throat, so that he can ride in safety, poised on the back of her neck. . . . The decisive, angry gesture with which he was reseated on his mother's neck whenever his grip tended to slacken has taught him to be alert and sure-handed. At last it is safe for his mother to take him out in a canoe, to punt or paddle the canoe herself while the baby clings to her neck. If a sudden wind roughens the lagoon or her punt catches in a rock, the canoe may swerve and precipitate mother and baby into the sea. The water is cold and dark, acrid in taste and blindingly salt; the descent into its depths is sudden, but the training within the house holds good. The baby does not loosen his grip while his mother rights the canoe and climbs out of the water.

Occasionally the child's introduction to the water comes at an even earlier age. The house floor is made of sections of slats put together after the fashion of Venetian blinds. These break and bend and slip out of place until great gaps sometimes appear. The unwary child of a shiftless father may crawl over one of these gaps and slip through into the cold, repellent water beneath. But the mother is never far away; her attention is never wholly diverted from the child. She is out the door, down the ladder, and into the sea in a twinkling; the baby is gathered safely into her arms and warmed and reassured by the fire. Although children frequently slip through the floor, I heard of no cases of drowning and later familiarity with the water seems to obliterate all traces of the shock, for there are no water phobias in evidence. In spite of an early ducking, the sea beckons as insistently to a Manus child as green lawns beckon to our children, tempting them forth to exploration and discovery.

For the first few months after he has begun to accompany his mother about the village the baby rides quietly on her neck or sits in the bow of the canoe while his mother punts in the stern some ten feet away. The child sits quietly, schooled by the hazards to which he has been earlier exposed. There are no straps, no baby harnesses to detain him in his place. At the same time, if he should tumble overboard, there would be no tragedy. The fall into the water is painless. The mother or father is there to pick him up. Babies under two and a half or three are never trusted with older children or even with young people. The parents demand a speedy physical adjustment from the child, but they expose him to no unnecessary risks. He is never allowed to stray beyond the limits of safety and watchful adult care.

So the child confronts duckings, falls, dousings of cold water, or entanglements in slimy seaweed, but he never meets with the type of accident which will make him distrust the fundamental safety of his world. Although he himself may not yet have mastered the physical technique necessary for perfect comfort in the water, his parents have. A lifetime of dwelling on the water has made them perfectly at home there. They are sure-footed, clear eyed, quick handed. A baby is never dropped; his mother never lets him slip from her arms or carelessly bumps his head against door post or shelf. All her life she has balanced upon the inch-wide edges of canoe gunwales, gauged accurately the distance between house posts where she must moor her canoe without ramming the outrigger, lifted huge fragile water pots from shifting canoe platforms up rickety ladders. In the physical care of the child she makes no clumsy blunders. Her every move is a reassurance to the child, counteracting any doubts which he may have accumulated in the course of his own less sure-footed progress. So thoroughly do Manus children trust their parents that a child will leap from any height into an adult's outstretched arms, leap blindly and with complete confidence of being safely caught.

Side by side with the parent's watchfulness and care goes the demand that the child himself should make as much effort, acquire as much physical dexterity as possible. Every gain a child makes is noted, and the child is inexorably held to his past record. There are no cases of children who toddle a few steps, fall, bruise their noses, and refuse to take another step for three months. The rigorous way of life demands that the children be self-sufficient as early as possible. Until a child has learned to handle his own body, he is not safe in the house, in a canoe, or on the small islands. His mother or aunt is a slave, unable to leave him for a minute, never free of watching his wandering steps. So every new proficiency is encouraged and insisted upon. Whole groups of busy men and women cluster about the baby's first step, but there is no such delightful audience to bemoan his first fall. He is set upon his feet gently but firmly and told to try again. The only way in which he can keep the interest of his admiring audience is to try again. So self-pity is stifled and another step is attempted.

As soon as the baby can toddle uncertainly, he is put down into the water at low tide when parts of the lagoon are high and others only a few inches under water. Here the baby sits and plays in the water or takes a few hesitating steps in the yielding spongy mud. The mother does not leave his side, nor does she leave him there long enough to weary him. As he grows older, he is allowed to wade about at low tide. His elders keep a sharp lookout that he does not stray into deep water until he is old enough to swim. But the supervision is unobtrusive. Mother is always there if the child gets into difficulties, but he is not nagged and plagued with continual "don'ts." His whole play world is so arranged that he is permitted to make small mistakes from which he may learn better judgement and greater circumspection, but he is never allowed to make mistakes which are serious enough to permanently frighten him or inhibit his activity. He is a tight-rope walker, learning feats which we would count outrageously difficult for little children, but his tight-rope is stretched above a net of expert parental solicitude. If we are horrified to see a baby sitting all alone in the end of a canoe with nothing to prevent his clambering overboard into the water, the Manus would be equally horrified at the American mother who has to warn a ten-year-old child to keep his fingers from under a rocking chair, or not to lean out of the side of the car. Equally repellent to them would be our notion of getting children used to the water by giving them compulsory duckings. The picture of an adult voluntarily subjecting the child to a painful situation, using his superior strength to bully the child into accepting the water, would fill them with righteous indignation. Expecting children to swim at three, to climb about like young monkeys even before that age, may look to us like forcing them; really it is simply a quiet insistence upon their exerting every particle of energy and strength which they possess.

Swimming is not taught: the small waders imitate their slightly older brothers and sisters, and after floundering about in waist-deep water begin to strike out for themselves. Sure-footedness on land and swimming come almost together, so that the charm which is recited over a newly delivered woman says, "May you not have another child until this one can walk and swim." As soon as the children can swim a little, in a rough and tumble overhand stroke which has no style but great speed, they are given small canoes of their own. These little canoes are five or six feet long, most of them without outriggers, mere hollow troughs, difficult to steer and easy to upset. In the company of children a year or so older, the young initiates play all day in shallow water, paddling, punting, racing, making tandems of their small craft, upsetting their canoes, baling them out again, shrieking with delight and high spirits. The hottest sun does not drive them indoors; the fiercest rain only changes the appearance of their playground into a new and strange delight. Over half their waking hours are spent in the water, joyously learning to be at home in their water world.

Now that they have learned to swim a little, they climb freely about the large canoes, diving off the bow, climbing in again at the stern, or clambering out over the outrigger to swim along with one hand on the flexible outrigger float. The parents are never in such a hurry that they have to forbid this useful play.

The next step in water proficiency is reached when the child begins to punt a large canoe. Early in the morning the village is alive with canoes in which the elders sit sedately on the centre platforms while small children of three punt the

canoes which are three or four times as long as the children are
tall. At first glance this procession looks like either the crudest
sort of display of adult prestige or a particularly conspicuous
form of child labour. The father sits in casual state, a man of
five feet nine or ten, weighing a hundred and fifty pounds. The
canoe is long and heavy, dug out of a solid log; the unwieldy
outrigger makes it difficult to steer. At the end of the long
craft, perched precariously on the thin gunwales, his tiny
brown feet curved tensely to keep his hold, stands a small
brown baby, manfully straining at the six foot punt in his
hands. He is so small that he looks more like an unobtrusive
stern ornament than like the pilot of the lumbering craft.
Slowly, with a great display of energy but not too much actual
progress, the canoe moves through the village, among other
canoes similarly manned by the merest tots. But this is neither
child labour nor idle prestige hunting on the part of the parents. It is part of the whole system by which a child is encouraged to do his physical best. The father is in a hurry. He has
much work to do during the day. He may be setting off for
overseas, or planning an important feast. The work of punting
a canoe within the lagoon is second nature to him, easier than
walking. But that his small child may feel important and adequate to deal with the exacting water life, the father retires to
the central platform and the infant pilot mans the canoe. And
here again, there are no harsh words when the child steers
clumsily, only a complete lack of interest. But the first sure
deft stroke which guides the canoe back to its course is greeted
with approval.

The test of this kind of training is in the results. The
Manus children are perfectly at home in the water. They neither fear it nor regard it as presenting special difficulties and
dangers. The demands upon them have made them keen-eyed,
quickwitted, and physically competent like their parents.
There is not a child of five who can't swim well. A Manus child
who couldn't swim would be as aberrant, as definitely subnor-
mal as an American child of five who couldn't walk. Before I

went to Manus I was puzzled by the problem of how I would
be able to collect the little children in one spot. I had visions of
a kind of collecting canoe which would go about every morning and gather them aboard. I need not have worried. A child
was never at a loss to get from house to house, whether he went
in a large canoe or a small one, or swam the distance with a
knife in his teeth.

In other aspects of adapting the children to the external
world the same technique is followed. Every gain, every ambitious attempt is applauded; too ambitious projects are gently
pushed out of the picture; small errors are simply ignored but
important ones are punished. So a child who, after having
learned to walk, slips and bumps his head, is not gathered up
in kind, compassionate arms while mother kisses his tears
away, thus establishing a fatal connection between physical
disaster and extra cuddling. Instead the little stumbler is
berated for his clumsiness, and if he has been very stupid,
slapped soundly into the bargain. Or if his misstep has
occurred in a canoe or on the verandah, the exasperated and
disgusted adult may simply dump him contemptuously into
the water to meditate upon his ineptness. The next time the
child slips, he will not glance anxiously for an audience for his
agony, as so many of our children do; he will nervously hope
that no one has noticed his faux pas. This attitude, severe and
unsympathetic as it appears on the surface, makes children
develop perfect motor co-ordination. The child with slighter
original proficiency cannot be distinguished among the
fourteen-year-olds except in special pursuits like spear throwing, where a few will excel in skill. But in the everyday activities
of swimming, paddling, punting, climbing, there is a general
high level of excellence. And clumsiness, physical uncertainty
and lack of poise, is unknown among adults. The Manus are
alive to individual differences in skill or knowledge and quick
to brand the stupid, the slow learner, the man or woman with
poor memory. But they have no word for clumsiness. The
child's lesser proficiency is simply described as "not under-

standing yet." That he should not understand the art of handling his body, his canoes well, very presently, is unthinkable.

While in many societies children's walking means trouble for the adults, this was obviated in Old Peri by the extreme respect for property which children were taught. Linked together by rigorous training were the need never to touch anything that wasn't one's own—to be careful in touching other people's bodies or using their names—and the need to be rigorously careful about excretion.

... Where even the dogs are so well trained that fish can be laid on the floor and left there for an hour without danger there are no excuses made for the tiny human beings. A good baby is a baby who never touches anything; a good child is one who never touches anything and never asks for anything not its own. These are the only important items of ethical behaviour demanded of children. And as their physical trustworthiness makes it safe to leave children alone, so their well-schooled attitudes towards property make it safe to leave a crowd of romping children in a houseful of property. No pots will be disturbed, no smoked fish purloined from the hanging shelves, no string of shell money severed in a tug of war and sent into the sea. The slightest breakage is punished without mercy. Once a canoe from another village anchored near one of the small islands. Three little eight-year-old girls climbed on the deserted canoe and knocked a pot into the sea, where it struck a stone and broke. All night the village rang with drum calls and angry speeches, accusing, deprecating, apologizing for the damage done and denouncing the careless children. The fathers made speeches of angry shame and described how roundly they had beaten the young criminals. The children's companions, far from admiring a daring crime, drew away from them in haughty disapproval and mocked them in chorus.

Any breakage, any carelessness, is punished. The parents do not condone the broken pot which was already cracked and then wax suddenly furious when a good pot is broken . . . The tail of a fish, the extra bit of taro, the half rotten betel nut, cannot be appropriated with any more impunity than can the bowl of feast food. In checking thefts, the same inexorableness is found. There was one little girl of twelve named Mentun who was said to be a thief and sometimes taunted with the fact by other children. Why? Because she had been seen to pick up objects floating in the water, a bit of food, a floating banana, which obviously must have fallen out of one of the half a dozen houses near by. To appropriate such booty without first making a round of the possible owners, was to steal. And Mentun would have to exercise the greatest circumspection for months if she were not to be blamed for every disappearance of property in the years to come. . . .

The departments of knowledge which small children are expected to master are spoken of as "understanding the house," "understanding the fire," "understanding the canoe," and "understanding the sea."

"Understanding the house" includes care in walking over the uncertain floors, the ability to climb up the ladder or notched post from the verandah to the house floor, remembering to remove a slat of the floor for spitting or urinating, or discarding rubbish into the sea, respecting any property lying on the floor, not climbing on shelves nor on parts of the house which would give beneath weight, not bringing mud and rubbish into the house.

The fire is kept in one or all of the four fireplaces ranged two along each side wall, towards the centre of the house. The fireplace is made of a thick bed of fine wood ash on a base of heavy mats edged by stout logs of hard wood. It is about three feet square. In the centre are three or four boulders which serve as supports for the cooking pots. Cooking is done with small wood, but the fire is kept up by heavier logs. Neat piles of firewood, suspended on low shelves, flank the fireplaces. Swung low over the fire are the smoking shelves where the fish are preserved. Understanding of the fire means an understanding that the fire will burn the skin, or thatch, or light wood, or straw, that a smouldering cinder will flare if blown upon, that such cinders, if removed from the fireplace, must be carried with the greatest care and without slipping or bringing them in contact with other objects, that water will quench fire. "Understanding the fire" does not include making fire with the fire plough, an art learned much later when boys are twelve or thirteen. . . .

Understanding canoe and sea come just a little later than the understanding of house and fire, which form part of the child's environment from birth. A child's knowledge of a canoe is considered adequate if he can balance himself, feet planted on the two narrow rims, and punt the canoe with accuracy, paddle well enough to steer through a mild gale, run the canoe accurately under a house without jamming the outrigger, extricate a canoe from a flotilla of canoes crowded closely about a house platform or the edge of an islet, and bail out a canoe by a deft backward and forward movement which dips the bow and stern alternately. It does not include any sailing knowledge. Understanding of the sea includes swimming, diving, swimming under water, and a knowledge of how to get water out of the nose and throat by leaning the head forward and striking the back of the neck. Children of between five and six have mastered these four necessary departments.

As in the physical world where children learned by doing, by endless activity, experimentation within narrow limits set by those whom they were imitating, so they learned to talk by long bouts of imitation, adult and child alternately repeating a word fifty or sixty times. Melanesian languages are extremely repetitive. So to express intensity, it was "big, big, big"; duration, "I had to wait, wait, wait, wait"; and distance, "He walked, walked, walked, walked." All are expressed by straight repetition, and this provided an easy medium for teaching by repetition. Manus crowds have a tendency to pick up any phrase and chant it, and small children spent a great deal of time chanting phrases, and in the case of small boys, Pidgin* phrases, for girls refused to learn the work-boy language.

What is true of speech is equally true of gesture. Adults play games of imitative gesture with children until the child develops a habit of imitation which seems at first glance to be practically compulsive. This is specially true of facial expression, yawning, closed eyes, or puckered lips. The children carried over this habit of repeating expression in their response to a pencil of mine which had a human head and bust on the end of it. The bust gave the effect of a thrown-out chest. The thin lips seem compressed, to a native, and almost every child, when first looking at the pencil, threw out the chest and compressed the lips. I also showed the children one of those dancing paper puppets which vibrate with incredible looseness when hung from a cord. Before the children ceased to marvel at the strange toy, their legs and arms were waving about in imitation of the puppets.

This habit of imitation is not, however, compulsive, for it is immediately arrested if made conscious. If one says to a child who has been slavishly imitating one's every move, "Do this the way I do," the child will pause, consider the matter, and more often than not refuse. It seems to be merely a habit, a natural human tendency given extraordinary play in early childhood and preserved in the more stereotyped forms in the speech and song of adult life. It is most marked in children between one and four years of age and its early loss seems to be roughly correlated with precocity in other respects.

This is an excellent example of how the kind of thinking we were doing twenty-five years ago failed to take account of significant differences in child care. With this easy, facile, meaningless phrase, "merely a habit," I dismissed an essential element in

*Today I would write this "Neo-Melanesian."

Manus child care, and so blinded myself to much that lay right in front of my eyes.

Other activities learned through imitation are dancing and drumming. The small girls learn to dance by standing beside their mothers and sisters at the turtle dance given to shake the dust out of the house of mourning. Occasionally a child is incited to dance at home while the mother taps on the house floor. Six- and seven-year-olds have already grasped the very simple step: feet together and a swift side jump and return to position in time to the drum beats. The men's dance is more difficult. The usual loin cloth or G-string is laid aside and a white sea shell substituted as pubic covering. The dance consists in very rapid leg and body movements which result in the greatest possible gymnastic phallic display. It is a dance of ceremonial defiance, accompanied by boasting and ceremonial insult, most frequently performed on occasions when there is a large display of wealth in a payment between two kin groups connected by marriage. Those who make the heavy payment of dog's teeth and shell currency dance and dare the other side to collect enough oil and pigs to repay them. Those who receive the payment dance to show their defiant acceptance of the obligation which they are undertaking. The smaller children are all present at this big ceremony and watch the men's athletic exploits. Boys of four or five begin to practise, and the day that they master the art of catching the penis between the legs and then flinging it violently forward and from side to side, is a day of such pride that for weeks afterwards they perform the dance on every occasion, to the great and salacious amusement of their elders. Slightly older boys of ten and twelve make a mock shell covering out of the seed of a nut and practise in groups.

Whenever there is a dance there is an orchestra of slit drums of all sizes played by the most proficient drummers in the village. The very small boys of four and five settle themselves beside small hollow log ends or pieces of bamboo and drum away indefatigably in time with the orchestra. This period of open and unashamed imitation is followed by a period of embarrassment, so that it is impossible to persuade a boy of ten or twelve to touch a drum in public, but in the boys' house when only a few older boys are present, he will practise, making good use of the flexibility of wrist and sense of rhythm learned earlier. Girls practise less, for only one drum beat, the simple death beat, falls to their hands in later life. . . .

Singing is also learned through imitation of older children by younger children. It consists in a monotone chant of very simple sentences, more or less related to each other. A group of children will huddle together on the floor and croon these monotonous chants over and over for hours without apparent boredom or weariness. They also sing when they are chilled and miserable or when they are frightened at night.

Similarly the art of war is learned by playful imitation. The men use spears with bamboo shafts and cruel arrow shaped heads of obsidian. The children make small wooden spears, about two and a half feet in length and fasten tips of pith on them. Then pairs of small boys will stand on the little islets, each with a handful of spears, and simultaneously hurl spears at each other. Dodging is as important a skill as throwing, for the Manus used no shields and the avalanche of enemy spears could only be dodged. This is an art which requires early training for proficiency, and boys of ten and twelve are already experts with their light weapons. The older men and boys, canoe building on the islet, or paddling by, stop to cheer a good throw. Here again, the children are encouraged, never ridiculed or mocked.

Fishing methods are also learned early. Older men make the small boys bows and arrows and tiny, pronged fish-spears. With these the children wander in groups about the lagoon at low tide, skirting the small rocky islands, threading their way through the rank sea undergrowth, spearing small fish for the sport of it. Their catch, except when they net a school of minnows in their spider-web nets, is not large enough to eat. This toying with fishing is pursued in a desultory fashion by children from the ages of three to fifteen. Then they will go on expeditions of their own and sometimes join the young men on excursions to the north coast after turtle, dugong, and kingfish.

Small children are also sometimes taken fishing by their fathers. Here as little more than babies they watch the procedures which they will not be asked to practise until they are grown. Sometimes in the dawn a child's wail of anger will ring through the village; he has awakened to find his father gone fishing without him. But this applies only to small boys under six or seven. Older boys prefer the society of other children and of grown youths, but shun the company of adults. Boys of fourteen and fifteen never accompany their parents about their ordinary tasks except when a boy has fallen out with his playmates. For the few days of strain which follow he will cling closely to his parents and be officiously helpful, only to desert them again as soon as friendly relations are re-established.

Little girls do very little fishing. As very tiny children they may be taken fishing by their fathers, but this is a type of fishing which they will never be required to do as grown women. Women's fishing consists of reef fishing, fishing with hand nets, with scoop baskets, and with bell shaped baskets with an opening at the top for the hand. Girls do not begin this type of fishing until near puberty.

Of the techniques of handwork small boys learn but little. They know how to whiten the sides of their canoes with seaweed juices; they know how to tie a rattan strip so that it will remain fast; they have a rudimentary knowledge of whittling, but none of carving. They can fasten on a simple outrigger float if it breaks off. They know how to scorch the sides of their canoes with torches of coconut palm leaves, and how to make rude bamboo torches for expeditions after dark. They know nothing about carpentry except what they remember from their early childhood association with their fathers.

But children have learned all the physical skill necessary as a basis for a satisfactory physical adjustment for life. They can judge distances, throw straight, catch what is thrown to them, estimate distances for jumping and diving, climb anything, balance themselves on the most narrow and precarious footholds, handle themselves with poise, skill, and serenity either on land or sea. Their bodies are trained to the adult dance steps, their eye and hand trained to shooting and spearing fish, their voices accustomed to the song rhythms, their wrists flexible for the great speed of the drum sticks, their hands trained to the paddle and the punt. By a system of training which is sure, unhesitant, unremitting in its insistence and vigilance, the baby is given the necessary physical base upon which he builds through years of imitation of older children and adults. The most onerous part of his physical education is over by the time he is three. For the rest it is play for which he is provided with every necessary equipment, a safe and pleasant playground, a jolly group of companions of all ages and both sexes. He grows up to be an adult wholly admirable from a physical standpoint, skilled, alert, fearless, resourceful in the face of emergency, reliable under strain.

But the Manus' conception of social discipline is as loose as their standards of physical training are rigid. They demand nothing beyond physical efficiency and respect for property except a proper observance of the canons of shame. Children must learn privacy in excretion almost by the time they can walk; must get by heart the conventional attitudes of shame and embarrassment. This is communicated to them not by sternness and occasional chastisement, but through the emotions of their parents. The parents' horror, physical shrinking, and repugnance are communicated to the careless child. This adult attitude is so strong that it is as easy to impregnate the child with it as it is to communicate panic. When it is realized that men are fastidious about uncovering in each other's presence and that a grown girl is taught that if she even takes off her grass skirt in the presence of another woman the spirits will punish her, some conception of the depth of this feeling can be obtained. Prudery is never sacrificed to convenience; on sea voyages many hours in duration, if the sexes are mixed the most rigid convention is observed.

Into this atmosphere of prudery and shame the children are early initiated. They are wrapped about with this hot prickling cloak until the adults feel safe from embarrassing betrayal. And here social discipline ceases. The children are taught neither obedience nor deference to their parents' wishes. A two-year-old child is permitted to flout its mother's humble request that it come home with her. At night the children are supposed to be at home at dark, but this does not mean that they go home when called. Unless hunger drives them there the parents have to go about collecting them, often by force. A prohibition against going to the other end of the village to play lasts just as long as the vigilance of the prohibitor, who has only to turn the back for the child to be off, swimming under water until out of reach.

Manus cooking is arduous and exacting. The sago is cooked dry in a shallow pot stirred over a fire. It requires continuous stirring and is good only for about twenty minutes after being cooked. Yet the children are not expected to come home at mealtime. They run away in the morning before breakfast and come back an hour or so after, clamouring for food. Ten-year-olds will stand in the middle of the house floor and shriek monotonously until someone stops work to cook for them. A woman who has gone to the house of a relative to help with some task or to lay plans for a feast will be assaulted by her six-year-old child who will scream, pull at her, claw at her arms, kick and scratch, until she goes home to feed him.

The parents who were so firm in teaching the children their first steps have become wax in the young rebels' hands when it comes to any matter of social discipline. They eat when they like, play when they like, sleep when they see fit. They use no respect language to their parents and indeed are allowed more license in the use of obscenity than are their elders. The veriest urchin can shout defiance and contempt at the oldest man in the village. Children are never required to give up anything to parents: the choicest morsels of food are theirs by divine right. They can rally the devoted adults by a cry, bend and twist their parents to their will. They do no work. Girls, after they are eleven or twelve, perform some household tasks, boys hardly any until they are married. The community demands nothing from them except respect for property and the avoidance due to shame.

Undoubtedly this tremendous social freedom reinforces their physical efficiency. On a basis of motor skill is laid a superstructure of complete self-confidence. The child in Manus is lord of the universe, undisciplined, unchecked by any reverence or respect for his elders, free except for the narrow thread of shame which runs through his daily life. No other habits of self-control or of self-sacrifice have been laid. It is the typical psychology of the spoiled child. Manus children demand, never give. The one little girl in the village who, because her father was blind, had loving service demanded of her was a gentle generous child. But from the others nothing was asked and nothing was given.

For the parents who are their humble servants the children have a large proprietary feeling, an almost infantile dependence, but little solicitude. Their egocentricity is the natural complement of the anxious pandering love of the parents, a pandering which is allowed by the restricted ideals of the culture.

This, then, was early childhood in Manus twenty-five years ago, an early childhood in which boys and girls were treated almost identically, taught to be independent, assertive, subject to a very few absolute requirements of shame and respect for property. This early training had a relentlessness about it that seemed to bite deeply into the developing personality.

So I described early childhood, giving details where I only partially understood their meaning, realizing that in some way this early childhood must provide the basis for the adult character, but very much puzzled by several things.

I was puzzled as to why the children, both the little children and the older children, were so extremely unimaginative. They were so gay, so curious, so active. And yet, when I gave them paper to draw on, they drew endlessly—I collected some thirty-two thousand drawings—faithfully making replicas of the world around, canoes, fishing, fighting, accurate representations of fish, or, in the case of the girls, accurate drawings of bead work. But not a spark of fantasy illuminated this work. Where were the Sir Ghosts, whose activities took up so much of their elders' attention, or the *tchinals*, the frightening supernaturals of the land people? The drawing remained starkly realistic, there were no houses with eyes, no fish who talked, no animistic tricks of fantasy. In a world where the adults lived in such an awareness of the unseen that the personalities of the Sir Ghosts were as real to me, a newcomer, as those of the living, the children had no imaginary playmates, no traffic with the unseen world.

I recorded how the children's attention was focused on the outer world, and they were never encouraged to use alibis to explain a real failure. The drifting of a canoe was relentlessly coupled with an imperfectly tied knot. When I offered children an alibi for unsuccessful drawing, such as "bad pencil," the children themselves refused it. They had been taught to watch carefully every detail of a process and to take full responsibility for mechanical failure. This gave them their readiness to handle machines, machines that were put together in ways which were lawful and could be learned. Far from being intimidated by new machines of any sort, the smallest children set out to understand them. The adult world of mechanical processes was one in which they fully participated, responsibly and efficiently.

But they did not treat the content of the adult world the same way. The feasts, the spectacular dances in which men approached the village in canoes covered with green boughs, shouting in phallic defiance of their rivals, were not imitated by the children. They did not play at marriages or at funerals, at séances or at war. The units of activity involved, the dance steps, the spear throwing, the stamping and shouting which underlay the oratory, were all there, but in their play there was no loving imitation of the adult world, either its work or its ceremonies.

So I was left with a set of questions. Why did the children, so imitative in matters of movements, of posture and gesture, swimming and dancing, imitate so little content? Why was their child life so contentless and empty? And what was the mechanism by which these generous and curious but undisciplined children were finally transformed into the driven, hostile adults that their parents were and that they were destined to be?

6

roots of change in Old Peri

The central questions seemed to be: How were the insubordinate children made into adults who all their driven lives would pursue goals which seemed to have no prefiguration in the children's experience? How were young men who, as boys of fourteen and fifteen, had been completely unmanageable, to be brought to heel, and little girls, who had ranged free as birds, to be turned into women who were afraid to utter even a word which would offend their husbands' Sir Ghosts? A few years later I was able to ask whether such a degree of discontinuity didn't result in some of the churlishness about life so characteristic of the Manus adults. If adult life was to be hard-bitten, ruthlessly self-interested, with friendship subordinated to economic gain, wouldn't the adults be happier if as children they had been permitted no such holiday from the demands of the real world?

In the early thirties, after *Growing Up in New Guinea* was published, I was asked by a large national magazine to write an article discussing our educational system. In the draft of the article I contrasted the atmosphere of American college campuses and their freedom from economic pressure and permission to get "C's," their emphasis on friendship and ease, with the demands of the business world which those same young men would enter. I permitted myself to wonder whether such a halcyon youth was the best preparation for a contented adulthood. The article was indignantly rejected as "calling into question all the values for which the United States stands." So much for the response to the questions raised by the discontinuity in the education of the children of Old Peri.

But there was an even deeper question, and one which is more relevant to what has happened in Manus and what is happening all over the world today. How was the experience of childhood—the kind of character which children of one society developed as they grew as compared with the kind of character which children in a different kind of society developed—related to their capacity to change? Because we had no adequate theory of character formation, this question reduced itself to the simplicity of asking: Can you change the social system by changing the way in which children are reared? This question seemed to turn on the strength of the culture, the strength of tradition. Here in old Manus there seemed to be a fine historical experiment—apparently children were reared to value one kind of human relations and were powerless against adults who determinedly lived in terms of a quite different kind. It was very clear that the adult world did win. This I documented in detail, and set against it the emptiness of the hopes of those who thought they would change the competitiveness of American life by creating a few protected little enclaves where children learned co-operation, or thought they could raise the low standards of the arts in America by permitting children to draw and paint spontaneously, without models.

Pursuing this argument I described how the happy, carefree children in old Manus were transformed into sulky, inwardly rebellious but outwardly conforming young adults who would some day demand from other adolescents what had once been demanded from them.

It was the young men just returning from their work experience who provided the models for Kilipak and Pomat, Loponiu* and Kapeli,† (Plate I) themselves eager for the adventure of going

---

*Loponiu, called Johanis Lokus in 1953.
†Kapeli, called Stefan Posanget in 1953.

away to distant islands to work for the white man. So I described that work-boy world through the eyes of those who had described it to me:

It is a world where the boy is often lonely and homesick, overworked, hungry, sulky, shrinking and afraid; where he is as often well fed, gay, absorbed in new friendships and strange experiences. It is a world which has nothing in common with the life which he will lead on his return to the village; it is usually no better a preparation for it than were the old days of war and rape. Furthermore, the leaders in the village, the substantial older men who have the greatest economic power and therefore the greatest social power, did not go away to work. Their tales are of war, not of the white man's world. In deference to them, all Pidgin English must be discarded except the few terms which even the women understand, like "work," "Sunday," "Christmas," "flash," "rice," "grease." In the world of the white man there was much evil magic afoot but at least his own Manus spirits were not concerned with his sex offences. He has suddenly returned to a world of which he has a fundamental dread, the details of which he never knew or has forgotten. The spirits whose oppressive chaperonage he has escaped for three years are found to take a lively interest in his surreptitious gift of tobacco to young Komatal who has grown so tall and desirable in his absence.

His return is celebrated by a ceremony which combines a family blessing and incantation with a feast of return. The blessing is called *tchani*, for the whole ceremony there is only the hybrid term, "*kan* (feast)—he—finished—time." Food is prepared and sent to other families, who have made similar feasts in the past, and the boy is ceremonially fed taro by his paternal grandfather or grandmother or aunt, while the following incantation is recited over him:

"Eat thou my taro.
Let the mouth be turned towards dog's teeth,
The mouth turn towards shell money.

The shell money is not plentiful.
Let the taro turn the mouth towards it,
Towards plentifulness,
Towards greatness.
The mouth be turned towards the little transactions,
Towards the giving of food.
Let it become the making of great economic transactions.
Let him overhaul and outstrip the others,
The brothers whom he is amongst;
Let him eat my taro,
May he become rich in dog's teeth,
Attaining many,
Towards the attainment of much shell money."

He feeds him taro, a lump so large that the boy can hardly hold it in his mouth. Then, rolling another handful in his hand, he says, calling the names of the clan ancestors:

"Powaseu!
Saleyao!
Potik!
Tcholai!
Come you hither!
On top of the taro, yours and mine,
I bestow upon the son of Polou,
Upon the son of Ngamel.
He will monopolize the riches
Amongst all of his clan.
Let Manuwai become rich,
Let him walk within the house, virtuously.
He must not walk upon the centre board of the house floor,*
He must walk on the creaking slats,

---

*Traditional phrase, i.e., he may not enter the house in a stealthy fashion, seeking to surreptitiously possess one of the women inmates. This is symbolic of any underhand dealings.

He must wait below on the lower house platform,
He must call out for an invitation (to enter),
He must call out announcing his arrival to women
That they may stand up to receive him.
Afterwards he may climb up into the house.
Let him eat my taro.
He must do no evil.
May he grow to my stature!
I endow the taro with the power of war!
And I now fight no more.
I give this taro to my grandson!
Let him eat the taro.
I am the elder, thy father is the younger.
It passes to this boy.
I give him the taro for eating,
I give thee power.
He may go to war,
He shall not be afraid.
There may be twenty of them,
There may be thirty of them,
He shall terrify all of them.
He shall remain steadfast.
He shall stand erect.
They will behold him,
They will drop their spears,
They will drop their stone axes on the ground;
They will flee away.
Let him eat my taro.
I give him my taro and he eats of it.
Let him live, let him live long. . . .
Let him grow towards a ripe old age."

This incantation blesses him, as the parallel incantation
blesses the adolescent girl, and gives him power to conform to
the ethical code of his elders, industry leading to wealth, open
and impeccable sex conduct, courage in war, health.

There are no tabus associated with this feast, nor are there important economic obligations. It is a family ceremony of blessing. The youth goes about as before, still unmarried, still free of economic or social duties, but with the shadow of his approaching marriage hanging over him.

This old chant dramatized vividly the second source of discontinuity in Manus life; the old men were still adapted to a world of warfare, still asked the blessing of their Sir Ghosts on their sons, that "he may go to war." The young men whom they were blessing had not been to war, and could not go. Instead of years with opportunities for reckless bravery on long sea voyages, in raids on the enemy, they had had instead the years away at work, which had increased their sense of independence. Even though the Manus had not been a very warlike people and the interest in trade had been primary, the old outlet for youthful exuberance had been more congenial than the new. In 1928, the adult world was making demands which were even more humdrum, more unromantic, than the past had made. On the whole, the older men had accepted the end of warfare gladly, but the images of warfare remained on their lips as they faced these young men who had to be turned into hard-working, tractable junior workers in the village.

And I described it, speaking of "the culture" and what "it" did, in a shorthand that we would not use today, when we would be more careful to stress that a culture "acts" only through its members.

. . . The little boy who slapped his mother in the face, demanded pepper leaf from his father and angrily threw it back when his father gave him only half, who refused to rescue the dog's teeth for his mother, who stuck out his tongue when he was told to stay at home and swam away under water, has grown to manhood with these traits of insubordination, uncooperativeness, lack of responsibility unmodified. He has spent all his years in an unreal world, a world organized by industries which he has not learned, held together by a fabric of eco-

nomic relations of which he knows nothing, ruled by spirits
whom he has ignored. Yet if this world is to continue, the
young man must learn to take his part in it, to play the role
which his ancestors have played. The adult world is confronted
by an unassimilated group, a group which speaks its language
with a vocabulary for play, which knows its gods but gives
them slight honour, which has a jolly contempt for wealth-
getting activities.

Manus society does not meet this situation consciously or
through group action. None the less subtle is the unconscious
offensive which the culture has devised. To subject the young
man it uses the sense of shame, well developed in the three-
year-old, and only slightly elaborated since. The small children
have been made ashamed of their bodies, ashamed of excretion, ashamed of their sex organs. The adult has been shocked,
embarrassed, revolted, and the child has responded. Similar
response to failure to keep the tabus of betrothal has grafted
the later, more artificial convention on the former. The small
boy also learns that he must not eat in the presence of his
married sister's husband, or his older brother's fiancée. The
onlooker, the brother-in-law, the sister-in-law to be, gives the
same signs of confusion, uneasiness, embarrassment which his
parents gave when he micturated in public. The act of eating
before certain relatives joins the category of those things which
are shameful. His embarrassment over his future marriage is
also intense. A boy of fourteen will flee from the house, like a
virgin surprised in her bath, if one attempts to show him a picture of his sister-in-law. He will scuttle away if he sees the conversation is even turning upon his fiancée's village. All of these
things are of course equally true of girls. To the boys' tabus
they add the ubiquitous tabu cloak and the shamed concealment of menstruation. But with the girls there is no pause—
the girl is ever more restricted, more self-conscious, more
ashamed. It is a steady progression from the first day she wears
a scrap of cloth over her head to the day she is married and sits

in the bridal canoe, inert and heavily ornamented, with her head drooping almost to her knees.

But with the boys there is an interval. By thirteen or fourteen all these early lessons are learned and they are given no new ignominies to get by heart. As in the old days of war and rape, so in the more recent adventure of working for the white man, the standards of adult life are not pressed more firmly upon them. But the old embarrassments are there, grown almost automatic through the years.

Now comes the time when the young man must marry. The payments are ready. The father or brother, uncle or cousin, who is assuming the principal economic responsibility for his marriage is ready to make the final payment, ten thousand dog's teeth, and some hundred fathoms of shell money. And in no way is the bridegroom ready. He has no house, no canoe, no fishing tackle. He has no money and no furniture. He knows nothing of the devious ways in which all these things are obtained. Yet he is to be presented with a wife. Not against his will, for he knows the lesser fate of those who marry late. He has been told for years that he is lucky to have a wife already arranged for. He knows that wives are scarce, that even on the spirit level there is a most undignified scramble for wives and the spirit of a dead woman is snapped up almost before it has left her body. He knows that men without wives are men without prestige, without houses of their own, without important parts in the gift exchange. He does not rebel at the idea of marriage, he cannot rebel in advance against his fiancée for he has never seen her. He knows there will be less fun after marriage. Wives are exacting, married men have to work and scarcely ever come to the boys' house; still—one must marry.

But as plan follows plan, he gets more nervous. So Manoi, the husband of Ngalen [Plate XI]; listened to the plans made by his two uncles, his mother's brother and his mother's sister's husband. He preferred the latter's house; here he had always chosen to sleep when he didn't sleep in the boys' house.

From his babyhood he has slept where he liked and screamed with rage if his preferences were opposed. But suddenly a new factor enters in. Says Ndrosal, the uncle whom he doesn't like, "You will live in the back of my house and fish for me. I am busy; your other uncle has already a nephew who fishes for him. You will bring your wife, the granddaughter of Kea, and you two will sleep in the back of the house." Embarrassment fills Manoi—never before have his future relations with his wife been referred to. He accepts the arrangement in sullen silence. After the wedding he finds his whole manner of life is altered. Not only must he feed his new wife, but also be at the beck and call of the uncles who have paid for her. He has done nothing to pay for his privileges. They have found him a woman—shameful thought—he must fish for them, go journeys for them, go to market for them. He must lower his voice when he talks to them. On the other hand his uncles have not completed the marriage payment. So he must go ashamedly before all his wife's male relatives. Not even to her father does he show his face. His wife's family are making a big exchange. He is expected to help them, but he cannot punt his canoe in the procession for his father-in-law is there.

On all sides he must go humbly. He is poor, he has no home; he is an ignoramus. His young wife who submits so frigidly to his clumsy embrace knows more than he, but she is sullen and uncooperative. He enters an era of social eclipse. He cannot raise his voice in a quarrel, he who as a small boy has told the oldest men in the village to hold their noise. Then he was a gay and privileged child, now he is the least and most despised of adults.

All about him he sees two types of older men, those who have mastered the economic system, become independent of their financial backers, gone into the gift exchange for themselves, and those who have slumped and who are still dependent nonentities, tyrannized over by their younger brothers, forced to fish nightly to keep their families in food. Those who have succeeded have done so by hard dealing, close-fisted methods, stinginess, saving, ruthlessness. If he would be like them, he must give up the good-natured ways of his boyhood. Sharing with one's friends does not go with being a financial success. So as the independence of his youth goes down before the shame of poverty, the generous habits of his youth are suppressed in order that his independence may some day be regained.

Only the stupid and the lazy fail to make some bid for independence and these can no longer be friendly or generous because they are too poor and despised.

The village scene is accordingly strangely stratified—through the all-powerful, obstreperous babies, the noisy, self-sufficient, insubordinate crowd of children, the cowed young girls and the unregenerate undisciplined young men roistering their disregarding way through life. Above this group comes the group of young married people—meek, abashed, sulky, skulking about the back doors of their rich relations' houses. Not one young married man in the village had a home of his own. Only one had a canoe which it was safe to take out to sea. Their scornful impertinence is stilled, their ribald parodies of their culture stifled in anxious attempts to master it; their manner hushed and subdued.

Above the thirty-five-year-olds comes a divided group—the failures still weak and dependent, and the successes who dare again to indulge in the violence of childhood, who stamp and scream at their debtors, and give way to uncontrolled hysterical rage whenever crossed.

As they emerge from obscurity their wives emerge with them and join their furious invective to the clatter of tongues which troubles the waters daily. They have learned neither real control nor respect for others during their enforced retirement from vociferous social relations. They have learned only that riches are power and that it is purgatory not to be able to curse whom one pleases. They are as like their forbears as peas to peas. The jolly comradeship, the co-operation, the cheerful following a leader, the delight in group games, the easy inter-

change between the sexes—all the traits which make the children's group stand out so vividly from the adults'—are gone. If
that childhood had never been, if every father had set about
making his newborn son into a sober, anxious, calculating, bad
tempered little businessman, he could hardly have succeeded
more perfectly.

The society has won. It may have reared its children in a
world of happy freedom, but it has stripped its young men
even of self-respect. Had it begun earlier, its methods need
have been less abrupt. The girl's subjection is more gradual,
less painful. She is earlier mistress of her cultural tradition. But
as young people, both she and her husband must lead submerged lives, galling to their pride. When men and women
emerge from this cultural obscurity of early married life, they
have lost all trace of their happy childhood attitudes, except a
certain skepticism which makes them mildly pragmatic in
their religious lives. This one good trait remains, the others
have vanished because the society has no use for them, no
institutionalized paths for their expression.

Nothing that I found in 1953 calls this general description
into question. The records of the pre–World War II years show
that in spite of their childhood those young men had been helpless against their elders, powerless to challenge tradition. But this
is the sort of conclusion which I drew from the material—writing
in 1929. I have underlined the interpretations which would no
longer be made, either in terms of modern theory or of what
actually did happen in Manus.

We have followed the Manus baby through its formative
years to adulthood, seen its indifference towards adult life turn
into attentive participation, its idle scoffing at the supernatural
change into an anxious sounding of the wishes of the spirits, its
easy-going generous communism* turn into grasping individ-

---

*Used here in the sense of possessions held in common.

ualistic acquisitiveness. The process of education is complete. The Manus baby, born into the world without motor habits, without speech, without any definite forms of behaviour, with neither beliefs nor enthusiasms, has become the Manus adult in every particular. No cultural item has slipped out of the stream of tradition which the elders transmit in this irregular unorganized fashion to their children, transmit by a method which seems to us so haphazard, so unpremeditated, so often definitely hostile to its ultimate ends.

And what is true of Manus education in this respect is true of education in any untouched, homogeneous society. Whatever the method adopted, whether the young are disciplined, lectured, consciously taught, permitted to run wild or ever antagonized by the adult world—the result is the same. The little Manus becomes the big Manus, the little Indian, the big Indian. When it is a question of passing on the sum total of a simple tradition, the only conclusion which it is possible to draw from the diverse primitive material is that *any method will do.* The forces of imitation are so much more potent than any adult technique for exploiting them; the child's receptivity to its surroundings is so much more important than any methods of stimulation, that as long as every adult with whom he comes in contact is saturated with the tradition, he cannot escape a similar saturation.

Although this applies, of course, in its entirety, only to a homogeneous culture, it has nevertheless far-reaching consequences in educational theory, especially in the modification of the characteristic American faith in education as the universal panacea. *All the pleasant optimism of those who believe that hope lies in the future, that the failures of one generation can be recouped in the next, is given the lie.* The father who has not learned to read or write may send his son to school and see his son master this knowledge which his father lacked. A technique which is missing in one member of a generation but present in others, may be taught, of course, to the deficient one's son. Once a technique becomes part of the cultural tradition the proportions to which it is common property may vary from generation to generation.

But the spectacular fashion in which sons of illiterate fathers have become literate, has been taken as the type of the whole educational process. (The theorists forget the thousands of years before the invention of writing.) Actually it is only the type of possibilities of transmitting known techniques—the type of education discussed in courses in the "teaching of Elementary Arithmetic," or "Electrical Engineering." When education of this special and formal sort is considered, there are no analogies to be drawn from primitive society. Even if, as sometimes happens, a new technique may be imported into a tribe by a war captive or a foreign woman, and a whole generation learn from one individual, this process is of little comparative interest to us. The clumsy methods and minute rules of thumb by which such knowledge is imparted, has little in common with our self-conscious, highly specialized teaching methods.

It must be clearly understood that when I speak of education I speak only of that process by which the growing individual is inducted into his cultural inheritance, not of those specific ways in which the complex techniques of modern life are imparted to children arranged in serried ranks within the schoolroom. As the schoolroom is one, and an important, general educational agency, it is involved in this discussion; as it teaches one method of penmanship in preference to a more fatiguing one, it is not. This strictly professionalized education is a modern development, the end result of the invention of writing and the division of labour, a problem in quantitative cultural transmission rather than of qualitative. The striking contrast between the small number of things which the primitive child must learn compared with the necessary educational attainments of the American child only serves, however, to point the moral that whereas there is such a great quantitative difference, the process is qualitatively very similar.

After all, the little American must learn to become the big American, just as the little Manus becomes the big Manus. The continuity of our cultural life depends upon the way in which children in any event receive the *indelible imprint* of their social tradition. Whether they are cuddled or beaten, bribed or wheedled into adult life—they have little choice except to become adults like their parents. But ours is not a homogeneous society. One community differs from another, one social class from another, the values of one occupational group are not the values of those who follow some different calling. Religious bodies with outlooks as profoundly different as Roman Catholicism and Christian Science, claim large numbers of adherents always ready to induct their own and other people's children into the special traditions of their particular group. The four children of common parents may take such divergent courses that at the age of fifty their premises may be mutually unintelligible and antagonistic. Does not the comparison between primitive and civilized society break down? Does not education cease to be an automatic process and become a vital question of what method is to be pursued?

Undoubtedly this objection is a just one. Within the general tradition there are numerous groups striving for precedence, striving to maintain or extend their proportionate allegiances in the next generation. Among these groups, methods of education do count, *but only in relation to each other*. Take a small town where there are three religious denominations. It would not matter whether Sunday School was a compulsory matter, with a whipping from father if one didn't learn one's lesson or squandered a penny of the collection money, or whether Sunday School was a delightful spot where rewards were handed out lavishly and refreshments served by each young teacher to the admiring scholars. It would not matter, as long as all three Sunday Schools used the same methods. Only when one Sunday School depends upon parental intimidation, a second uses rewards and a third employs co-educational parties as its bait, does the question of *method* become important. . . .

The rapid assimilation of thousands of immigrants' children through the medium of the public schools, has given to Americans a peculiar faith in education, a faith which a less hybrid society would hardly have developed. Because we have turned the children of Germans, Italians, Russians, Greeks, into Americans, we argue that we can turn our children into anything we wish. Also because we have seen one cult after another sweep through the country, we argue that anything can be accomplished by the right method, that with the right method, education can solve any difficulty, supply any deficiency, train inhabitants for any non-existing Utopia. Upon closer scrutiny we see that our faith in method is derived from our assimilation of immigrants, from the successful teaching of more and more complicated techniques to more and more people, or from the successful despoiling of one group's roll of adherents by some other group of astute evangelists. In both of these departments method counts and counts hard. Efficient teaching can shorten the learning time and increase the proficiency of children in arithmetic or bookkeeping. . . . The parent who rigorously atones for his own bad grammar by tirelessly correcting his son may rear a son who speaks correctly. But he will speak no more correctly than those who have never heard poor English. By method it is possible to speed up the course of mastering existing techniques or increase the number of adherents of an existing faith. But both of these changes are quantitative not qualitative; *they are essentially non-creative in character. Nor is the achievement of making Americans out of the children of foreign parents creating something new; we are simply passing on a developed tradition to them. . . .*

Those who would save the world by education rely a great deal upon the belief that there are many tendencies, latent capacities, present in childhood which have disappeared in the finished adult. Children's natural "love of art," "love of music," "generosity," "inventiveness" are invoked by the advocates of this path of salvation in working out educational schemes through which these child virtues may be elaborated and stabilized, as parts of the adult personality. There is a certain kind of truth in this assertion, but it is a negative not a positive truth. For instance, children's "love of music," with the probable exception of those rare cases which we helplessly label "geniuses," is more likely simply an unspoiled capacity to be taught music. . . .

So that if by "natural to children" we mean that a child will learn easily what an adult, culturally defined, and in many ways limited, will not learn except with the greatest difficulty, it is true that any capability upon which the society does not set a premium, will seem easier to teach to a child than to an adult. So our children seem more imaginative than adults because we put a premium upon practical behaviour which is strictly oriented to the world of sense experience. Manus children, on the other hand, seem more practical, more matter-of-fact than do the Manus adults who live in a world where unseen spirits direct many of their activities. An educational enthusiast working among Manus children would be struck with their "scientific potentialities" just as the enthusiast among ourselves is struck with our children's "imaginative potentialities." The observations in both cases would be true in relation to the adult culture. In the case of our children their imaginative tendencies nourished upon a rich language and varied and diverse literary tradition will be discounted in adult life, attenuated, suppressed, distorted by the demands for practical adjustment; while the Manus children's frank skepticism and preoccupation with what they can see and touch and hear will be overlaid by the canons of Manus supernaturalism. But the educator who expected that these potentialities which are not in accordance with the adult tradition could be made to flower and bear fruit in the face of a completely alien adult world, would be reckoning without the strength of tradition—tradition which will assert its rights in the face of the most cunning methodological assault in the world. . . .

When we look about us among different civilizations and observe the vastly different styles of life to which the individual has been made to conform, to the development of which he has been made to contribute, we take new hope for humanity and its potentialities. But these potentialities *are passive not active, helpless without a cultural milieu in which to grow.* So Manus children are given opportunity to develop generous social feeling; they are given a chance to exercise it in their play world. But these generous communal* sentiments can not maintain themselves in the adult world which sets the price of survival at an individualistic selfish acquisitiveness. Men who as boys shared their only cigarette and halved their only *laplap*, will dun each other for a pot or a string of dog's teeth.

So those who think they can make our society less militantly acquisitive by bringing children up in a world of share and share alike, bargain without their hosts. They can create such a world among a few children who are absolutely under their control, but they will have built up an attitude which will find no institutionalized path for adult expression. The child so trained might become a morbid misfit or an iconoclast, but he cannot make terms with his society without relinquishing the childhood attitudes for which his society has no use.

The spectacular experiment in Russia had first to be stabilized among adults before it could be taught to children. No child is equipped to create the necessary bridge between a perfectly alien point of view, and his society. Such bridges can only be built slowly, patiently, by the exceptionally gifted. The cultivation in children of traits, attitudes, habits foreign to their cultures is not the way to make over the world. Every new religion, every new political doctrine, has had first to make its adult converts, to create a small nuclear culture within whose guiding walls its children will flourish. . . .

. . . Those who wish to alter our traditions and cherish the Utopian but perhaps not impossible hope that they can consciously do so, must first muster a large enough body of adults who with them wish to make the slight rearrangements

---

*In the original this word read "communistic," where I used the term in its old descriptive sense before it had become politically loaded.

of our traditional attitudes which present themselves to our culturally saturated minds. This is equally true of those who wish to import part of the developed tradition of other societies. They must, that is, create a coherent adult culture in miniature before they can hope to bring up children in the new tradition. . . . Such changes in adult attitudes come slowly, are more dependent upon specially gifted or wise individuals than upon wholesale educational schemes.

Besides encouraging a most unfounded optimism, this over-valuation of the educational process and *under-valuation of the iron strength of the cultural walls* within which any individual can operate, produces one other unfortunate result. It dooms every child born into American culture to victimization by a hundred self-conscious evangelists who will not pause long enough to build a distinctive culture in which the growing child may develop coherently. One such group negates the efforts of another and the modern child is subjected to miseries which the Manus child never knows, reared as it is with unselfconscious finality into a Manus adult. Not until we realize that a poor culture will never become rich, though it be filtered through the expert methods of unnumbered pedagogues, and that a rich culture with no system of education at all will leave its children better off than a poor culture with the best system in the world, will we begin to solve our educational problems. Once we lose faith in the blanket formula of education, in the magic fashion in which education, using the passive capacities of children, is to create something out of nothing, we can turn our attention to the vital matter of developing individuals, who, as adults, can gradually mould our old patterns into new and richer forms.

When I look at this discussion now, the unasked question stares me so directly in the face that I wonder how I could have failed to ask it then. I state so definitely that to have real change, the sort of change that can be reliably transmitted to children, one must have *changed* adults. But the recognition of the success with which a society with a homogeneous culture could pass that culture on to the next generation, regardless of the educational methods used, completely obscured the other issue: What sort of childhood experience laid the basis for change in an adult, even the fairly simple change of preferring the standards of another society, or another class, or another religious group? I recognized that different methods of education might subserve the interests of propaganda, but not that they would produce different kinds of adults, with intrinsically different capacities for change.

Nowhere did I ask, nor did any of the reviews which I have seen ask, how the adults who must be responsible for changing children—and so changing society—were themselves initially to be changed. We were clear about the great differences among children, about the existence in every society of the especially endowed and the especially sensitive, whom we called the "gifted," and the child who temperamentally did not fit the culture, the "deviant." But we lacked enough understanding of the mechanism of character formation to see how a society could, in effect, produce a "latent deviance" in its members which was not due to an innate deviance of temperament, but was the by-product of the whole system of character formation. We had to learn to focus on the way in which character was formed, to see that *any method wouldn't do* the same things.

Early in the thirties, under the impetus of the interdisciplinary work with real cross-fertilization of theory among anthropologists, sociologists, political scientists, psychoanalysts, and psychologists, we began to attack the problem of how character and political and social structure fitted together, what were the social, economic, and political requirements of societies within which individuals might be expected to be co-operative or competitive, or individualistically rivalrous or mutually helpful. Students interested in the relationship between social forms and character spent much of their energy showing that categorizing individuals by social class corresponded to, was indeed a form of, categorizing them by character structure. It was not, however, until World War II, until we became by necessity applied anthropologists in international relations—applying all we knew to the problems of co-operation among the Allies, to outguessing and defeating the enemy, and to keeping the morale of our own people high—that we began asking the right questions. How did it happen that Japanese in America sloughed off their loyalty to Japan in just the way they did? If there were such striking differences between English, Scots, Germans, French, Roumanians, Greeks, Dutch, as we began to find as soon as we looked at them, how did it happen that the descendants of all of them, in America, became Americans? Somewhere, somehow, something happened in adults which could be transmitted to their children as part of their cultural inheritance.

All through the war we wrestled with this problem in various forms. What was there in the character of the Japanese which, when they were captured, made them willing to broadcast at once for us and express hurt that we did not put them immediately into our uniform? How had the descendants of Europeans, where the father's role was dominant, become Americans who expected their children to exhibit their independence to spectator fathers? Then, after the war, when we attacked the question of Soviet Russia, what had there been in old Russian character out of which Bolsheviks had developed who, in turn, after capturing the governmental apparatus, began to mould the character of all the children in the Soviet Union? And as China fell to Communism, it became clear that what we needed to know was what were the elements in a cultural character which made change possible, welcomed new ideas and welcomed them in particular ways, or responded—positively or negatively—to a new environment or a new invention.

Ruth Benedict[1] had laid one kind of groundwork in her studies of how some Indian groups refused the ecstasies promised by the peyote cult or the secular excesses of "firewater." Hans Sachs,[2] who was a historian before he was a psychoanalyst, had laid another important stone in the structure when he asked: Why did the Greeks not go on to invent machines? Geoffrey Gorer[3] had asked why did not the Lepchas, a primitive people of Sikkim, develop a higher civilization when more advanced forms were available to them? All of these questions, essential precursors of our present questions, had one thing in common, however. They assumed one procedure as normal, and then asked why it didn't happen. Questions about why something didn't or doesn't occur are intrinsically very difficult to answer. When we stated the question the other way, we made it more answerable.

So stated, the question becomes: What is the relationship of the character formation of any people to possibilities of change? Looking at the Manus in 1928–29, one could note the discrepancy between childhood and adulthood, the sullen misery of the young people as they met the demands of the adult world, and one could note the counterparts in the United States in young people who found the world to be something different from what they had been led to expect. Myrdal,[4] writing in the middle of World War II, could state the contradiction between what American children were taught of Christian and democratic ideals and the reality of American race relations as "the American dilemma," rather than recognizing that in the tension between high ideals and practice which must fall behind those ideals, lies the dynamic of American democracy—that the whole point of hitching one's wagon to a star lies in the tension on the rope.[5] "The trouble with Americans," remarked a Chinese student visitor at a student conference, "is that they are still unhappy when they are so rich. If a Chinese had even a fraction as much, he would be perfectly happy." But a Chinese girl student objected violently: "I think that the good thing about Americans is that they aren't happy *even* when they have a lot of material things."

The form of Manus childhood, in 1928, contributed to the unlovely and unlovable aspects of Manus adult character, to their sense of being driven and oppressed by a tyrannical system, to their lack of pleasure in the lives they led. The great avidity with which they seized on new situations, their great adaptability to the new inventions which came with European contact, was partly rooted in this driving discontent with things as they were. But the contrapuntal experiences of childhood were not sufficient in themselves to enable Manus adults to throw over the institutions of their culture, no matter how much they chafed beneath them.

There is no reason to believe that without a change in the external world—without the coming of Europeans and later of the Americans—the Manus character, which included a childhood experience that life could be lived without the anger and tension which suffused all adult dealings with one another, would ever have produced any fundamental change in their culture, beyond giving them a restless receptivity to small inventions, a driving capacity to organize their social structure so that it was responsive to the will and ambition of the more vigorous-minded. The outer events could not have been predicted, but had we known what we have learned in the last fifteen years, it would have been possible to make better predictions as to how the Manus would have responded to such situations as those which confronted them in World War II.

When I decided to go back to Manus, I already had this new theoretical orientation at my disposal, so that in the preface to the Mentor edition of *Growing Up in New Guinea*,[6] written before I set out for Manus in 1953, I wrote:

> As I read over the original introduction and conclusion, I am conscious of one emphasis that may prove misleading today. I laid great stress on the need of an adult tradition within which children could grow up, on the inability of adults with a poor tradition to rear children who would show the characteristics of having been reared within a rich and rewarding one. In pressing this point home—in counterpoint to the over-enthusiasm for progressive education in the late twenties—I stressed far less than I would stress today, than I did stress in the sections on the Manus in *Male and Female* (1949), how closely the Manus children's character structure could be related to *the way in which they learned their culture*. We knew very little—in 1930—about differences in upbringing among different peoples; we knew still less about the importance of character formation or how to phrase what I call here "being a Manus adult," or in a more generalized way "Manus culture," in terms of the precise learning experiences of the infant and

young child. Were I writing this book today, I would emphasize the mechanisms by which the Manus child's attention is
turned toward the outer world, motor behaviour insisted upon
at the expense of passive dreaming, moral and physical necessity so joined that the Manus are almost unique in the way
they handle the mechanical devices of modern civilization. I
would have stressed how the form of their educational experience gave them potentialities for change which would be lacking in people differently educated. As it is, the reader, grown
wiser now than I could be then, by sharing in the developing
climate of opinion to which anthropology, psychoanalysis, and
child development studies have contributed, will have to make
these interpretations for himself.

I was puzzled by the accounts I had heard that the Manus
had invented a new religion, compound of American bulldozers
and their old ghosts, and, turning their backs on Christianity,
had become complete mystics. This didn't fit. I couldn't imagine
the Manus whom I had known treating machines mystically
unless, just conceivably, after years of happily coping with every
engine they met, they had been intimidated—as so many more
sophisticated people have been—by electronics.*
I expected the way in which the children had been brought
up to be determinative of the way in which they were responding
now. But for the anthropologist, new theory grows from new
field work, and I went back to Manus not to demonstrate, not to
test, developed theory, but to find out something new, something about the actual process of change which can occur in one
generation.

---

*Actually they took to our new machines with great enthusiasm, using a dif-
ferent grammatical form to differentiate essentially static machines—even
including moving picture cameras which made a sound after being wound—
from the tape recorders, which were grouped with objects having a life of
their own. During the latter part of his stay, Ted Schwartz could leave part of
his linguistic work to Lokus (Plate I), simply saying, "Start the generator, and
take this informant and make a bilingual text on the Magnecorder."

part two

7

the unforeseeable: the coming of the American army

With an increasing scientific knowledge of culture and character, we should then be able to say that, given such and such conditions, the people of a given country will behave in one of a number of ways and *not* behave in one of a number of other ways. This can be done for very large nations, or for very small ones. But for small groups of people like the Manus, while it is easier to describe their culturally regular characters, it is harder to foretell any of the conditions they will have to meet as a group.

At a given period of history the relative position of a major historical power—like the United States or Great Britain, or the Soviet Union today—can be somewhat reliably foretold. Even a major invention like the atomic bomb only changes the main plot to a relative degree. But the fate of two thousand people on the South Coast in an obscure archipelago in the Pacific is subject to such a large number of conditions totally beyond the initiative or the control of the people themselves that prophecy becomes much less reliable.

Granted that each generation of Manus children developed an unrealized desire for a different kind of relationship to one another and to the world, what were the chances that this would ever be actualized? And even if we could have predicted that some

day they would be able to witness the great technological achievements of a modern army—Japanese, Australian, American—and
that these would mobilize their imaginations, so ready to probe
the mysteries of a Diesel engine and to recognize the value of
writing and law, there was still another factor which would have
been a great deal harder to foretell.

If we consider the total population of the Admiralty
Islands—between thirteen and fourteen thousand people—
would there be, at the moment when the external conditions
were right, an individual alive and of the right age and position,
with sufficient genius to take advantage of them? For this is a
mystery of history which is of quite a different order from the
possibilities inherent in the struggle for world power in the
related positions of the major nations of the mid-twentieth century. Given the idea of nationalism, given increasing communication and increasing power to integrate large groups of people, it
was reasonably clear that as we approached a state of larger and
larger powers, and surer and surer devices of communication and
control, there would have to be a struggle between a few powers
before the world either fell into a pattern of relatively unified
organization or so destroyed itself in the process that the big powers fell apart into little ones, and the world went back into a
period of disorganization, and possibly, once thermonuclear
weapons were discovered, would witness the disappearance of civilization altogether, or even the extinction of the human race.

These large-scale patterns of history were in fact far easier to
foresee than the involvement of a tiny island people through the
course taken by the armies in World War II. At the beginning of
the war, the Territory of New Guinea seemed to many of those
working there to be a backwater which the war would never
reach. It was unfortified, ungarrisoned, and quite indefensible
against attack. When the Japanese thrust southward, evacuation
of the Australian civil government was the only possible course.
But it had its disadvantages vis-à-vis the natives, who saw the
authorities whom they had been forced to obey displaced without
a shot, just as these same authorities had formerly, in the still

green memories of the old men, displaced their former German rulers. Many of the former Australian officials and some civilians joined the Coast Watchers, but these individually dedicated men, hidden in parties of two or three, only served to emphasize how few the Australians were against the hordes of invading Japanese.

In many parts of the Territory it was the Australian army which drove out the Japanese and presented to the natives a new picture of a great mechanized co-operative effort. From the contrast between the Japanese and the Australians, on the one hand, and between the Australian soldier—friendly and egalitarian—and the typical "master," used to keeping natives in their place, natives in many other parts of New Guinea drew many of the same conclusions that the Manus drew from their contacts with the Americans. Everywhere contact with these great armies stirred the imagination of the people, setting off little flurries of political demands and mystical "cargo cults"—the two sometimes separate, sometimes joined together. Far up in the highlands of New Guinea, people who had never seen Europeans built "radio towers" to communicate with the supernatural.[1] All over the Pacific there were echoes and reverberations, villages were redesigned like army camps, cargo was promised from Australia or America.

And whether it was an Australian or an American army which was primarily involved in the reconquest of any given island was entirely fortuitous from the standpoint of the people on the island. But in Manus it was the Americans who provided the stimulus, and the exact nature of the response to that stimulus is of special interest to Americans. In Manus there was a group of people peculiarly adapted to appreciate and welcome certain aspects of American culture, especially the delight in machines. And in Manus, in addition to a "cargo-cult" outbreak—familiar all over the Territory—there was a leader, Paliau, who was able to use the "cargo cult" as part of a real political movement. As the Manus had been Catholics, the break with the Mission had a world-wide significance, greater than it would have had if the break had been with a smaller, less internationally ramified mission. Finally, this movement can be, and has been, studied as no other movement of this sort has been, because we have the earlier anthropological work, which this later work of 1953 built upon.

The uniqueness of the Paliau Movement in Manus is therefore an accident of history, but an accident which it has been my intention to take full advantage of, as I try to extract every drop of significance from the details of what actually happened in Peri.* I propose to present the history of the war years as the Peri people relate it now, for it is in this form that it has become part of the living texture of their lives.

The Australian administration was evacuated from Patusi. The Japanese took over the islands without bloodshed, and the Manus experienced a long, relatively uneventful period of Japanese rule. There is a striking discrepancy between the lack of any detailed stories of this period and the intense fear and horror which the people express. They say that the Japanese conscripted labour, fed them with what they regarded as inadequate rations of rice, paid them in cigarettes. For the system of carefully enforced law to which they had become accustomed, they had to face a military power whose officers were primarily interested in getting as much, in the form of labour, and giving as little, in the form of supplies, as possible. They picture the Japanese as permitting a good deal of licence to those natives who co-operated—there are tales of police boys who went about among the land people pouring out years' supplies of feast oil as a way of abolishing the old native economic system. They describe one public torture which a large audience of natives were "invited" to witness, where a native who had been insubordinate was tied to a tree for several days and then killed brutally by having red-hot stones thrust into his mouth. "If the Japanese had stayed, they would have exterminated us, every one of us," the people say.

The Japanese established a small outpost in Peri village itself, and one day fighter planes appeared in the sky, which, the Peri

---

*Ted Schwartz in *The Paliau Movement of the Admiralties, 1946–1954*, will deal with the history of the movement in the whole council area.

people say, the Japanese maintained were Japanese planes until the planes came so near that people were machine-gunned from the sky. The Peri people, in a panic, hid under their houses. When the planes went away the people fled into the mangrove swamps. "The Japanese lied to us," they say unforgivingly, and hold no grudge at all against the Americans whose planes bombed them.

They describe the arrival of the Americans on the South Coast in this way: "The Americans had an Australian with them who could speak to us, and he came and got us, he gathered us all together on Ndropwa (the little island about four miles away) and then those of us who were men helped the Americans rout the Japanese out of the mangrove swamps." There is a version of the battle of Los Negros, generally current among Australians in the Territory, in which the Manus say that "Master McCarthy," now a District Commissioner, "took" Manus, and *then* the Americans came in—a tribute to "Master McCarthy's" single-handed skill as advance guard in the invasion.

Then came the years when American forces occupied one of the largest American bases between Pearl Harbor and Guam. It is claimed that over a million men poured through the Admiralty Islands, a million Americans representing all the services, all the major races of mankind, every sort and kind of American male, fighting a war with the most highly developed technical equipment the world had ever seen. Some fourteen thousand Admiralty Island people were exposed, quite capriciously, if one looks at the matter from the standpoint of their career lines, to this tremendous spectacle, as miles and miles were packed with barracks, built on the spot from wood sawed in saw mills set up in the bush. The Americans knocked down mountains, blasted channels, smoothed islands for airstrips, tore up miles of bush—all with their marvellous "engines."

"The Americans treated us like individuals, like brothers," which meant that the Americans took no responsibility for the preservation of the caste relationships which existed between

Europeans and natives. Like other Anglo-Saxon peoples, they were remarkably friendly and sympathetic to "other people's natives." They didn't have to worry about the labour situation after the war, how a plantation could run if natives got used to being paid a dollar to launder a shirt, or how the extraordinary lavishness with which American GI's disposed of Uncle Sam's property would compare with the enforced and orderly penny-pinching of a post-war civilian administration, with meagre funds which must be strictly accounted for. Rather than set up a special commissary for the small number of native labourers, it was easier to hand them a regulation food tray and put them in the food line, where they shared the ham and ice cream of the men in the service. The people of the Admiralties might have got a taste for ice cream out of this experience. But what they definitely did get was a passionate realization of what it meant to be treated—by civilized men, by white men—as people, people with individual names like anyone else.

There is no reason to suppose that the Americans, the some million Americans, who went through Manus represented in any way a specially selected, better mannered, or more idealistic section of the United States than any other such cross-section. Yet the Manus experienced them as a people whose relationships to each other were casteless and classless, where each man treated each other man as a human being.

Here again we are faced with the difference between our inability to foresee particular events and external conditions, and the type of prediction which is based on a knowledge of culture. It has been assumed in most discussions of race relations in the United States that negative attitudes toward non-white races and an inherent insistence on the superiority of the white race is the basic American attitude, tempered and confused by Christian and democratic ideals and by grudgingly expressed FEPC legislation, anti-lynch laws, and Supreme Court decisions. This assumption has been enormously aggravated in the last twenty-five years by rising European and Asian criticism of American race relations—where the criticism has been either a reaction against American criticism of European imperialism or a consciously calculated left-wing political weapon. The Nazi persecution of the Jews further aggravated the situation as the words *Jews* and *Negroes* began to be substituted for the words *Catholics, Protestants,* and *Jews,* thus typing the very difficult problem of integrating a people who were physically different from the dominant group together with a quite different problem related to the historical effort of a white minority group to keep their cultural identity by treating it as racial. Yet a careful look at American attitudes demonstrates clearly that the core of the race problem in the United States is the question of visibility: to be an American is *to choose* to be like other Americans, and this is more difficult the less you look like the other Americans.[2] It was harder, twenty-five years ago, for an Italian to look like an American than it is today. Toward "Americans" who cannot be classified as "Americans," Americans show a kind of irrational rejection, which is very much reduced in dealing with peoples of other countries where this aspect of the question of racial difference is no longer relevant. Thus the immediate shift of behaviour in an American restaurant when a coloured customer speaks French or Spanish, or conversely, when someone who "looked like a Mexican" speaks cultivated English, is explicable.

The basic American preference, overlaid as it is with the residue of slavery and the scars of immigration of peoples with markedly lower standards of living—and this includes migrations of "Oakies" and "Arkies" to other parts of the United States, where they were often treated as if they were not full citizens—is to treat every man on his merits, regardless of creed or colour or national origin, and regardless also of where he came from, what his possessions, or who his kin are. Furthermore, social differences are maintained in the United States by women, while men on the whole react against them—except where women are concerned. The million Americans who went through Manus were men, and men denied access to Manus women, by the fortunate circumstance that the Admiralty Islands have a lot of little islands where the women could be placed, so that I did not see or hear of a single half-caste Manus child.* The scene was laid for the Manus to realize acutely the difference between being treated as a "native," as belonging to another category of creature, and being treated as an individual.

To many of the American troops, these active, alert, curious people, with their intense interest in and great aptitude for handling machines, with their physical skill and zest, which made them delightful guides for an off day of fishing, were not "niggers," but "Joes," "good Joes," a term which the Manus still repeat with affection. The note of authority with which every European in the Territory addressed a strange native as "boy," implying a subordinate status, was absent by default. This it seemed to the Manus was the "brotherhood of man" about which the Mission had told them, but of which they had never seen such vivid illustration before. One of the things these wonderful Americans understood was how to make the "brotherhood of man" a reality. And this they thought of as something that "Americans" had, as a group, as members of a society with a particular kind of culture. All of them had experienced from individual Australians or Germans year-long sensitive care and kindness far exceeding that which any American, busy fighting a war, could possibly give them. But when this happened, when a European master sat up all night by a sick servant's bedside, when a missionary spent years in patiently teaching a whole village, when a patrol officer gave his only blanket to a police boy who had lost his—all these countless acts of kindness and of love were regarded as the behaviour of individual good men or men who were good on occasion. The Manus—in fact all the New Guinea peoples with whom I have had contact—are adept at distinguishing individual character traits and capable of great loyalty to individuals. The annals of World War II are filled with the deeds of exceptional devotion of individual natives to individual Australians,

---

*I am told by Dr. Ian Hogbin that this absence of half-castes, either Japanese, American, or Australian, is striking throughout the New Guinea area.

the Coast Watchers hidden in the bush, who survived through incredible hardships to radio out messages which saved hundreds of thousands of Allied lives. Behind each heroic saga of men, racked by fever, walking long journeys in the jungle without food, keeping their radios going another day knowing that capture by the Japanese was only a matter of hours, back of each such saga is the story, so generously told by men like Eric Feldt in *Coastwatchers*,[3] of natives who risked their lives for these same men. But it was as individuals, devoted loyal natives being faithful to brave, considerate individual Australians.

Behind the Japanese lines, there were devoted natives who were faithful to individual missionaries who told them the Japanese had come to stay. They were not loyal to a country—they had no country, no idea even of the Mandated Territory as a political unit. The Territory itself; as a mandate, could make no such demand on loyalty as was possible in Papua, which was a full Australian territory, so that natives could be given some idea of loyalty to Australia and the Crown. In the Mandated Territory they had seen first the Germans, then the Australians, administer a system within which certain individuals had treated them kindly, others had treated them badly. There were "good masters" and "bad masters," that was all.

And there were so few of them. In Manus, in 1928, there were half a dozen missionaries, a district officer, a chief clerk, a medical assistant, a patrol officer, half a dozen European traders and planters, a Japanese trader, perhaps two more Japanese, a Chinese or so, possibly five foreign women at the most. Even as the number might be swollen to a hundred foreigners, they stood out with highly individualized roles against the mass of the population. Even a Manus who went to work in Rabaul saw only a few hundred Europeans, a smaller number of Chinese. There were two thousand Manus, fourteen thousand Admiralty Islanders in all. The spectacle of a million Americans, bent upon their own business, with a system of interrelationships which was related not to governing, converting, recruiting, or trading with New

Guinea natives but to goals of their own, was something quite new. Those Manus men who were away at work, behind Australian or Japanese lines, also experienced, in less concentrated form, something of the same thing—a sense of the strength and power of a large modern society, organized so as to get the smooth co-operation of a great number of people.

Before the war, "fashion belong white man" included modern machines, money, Christianity, law, but these were seen as expressions of modern civilization vis-à-vis the native. There was little enough opportunity to observe, in the tense and highly particularized relationships among a handful of isolated and often disgruntled Europeans, *a European way of life.* But watching the Americans in Manus, and watching the various invading armies elsewhere, the Manus grasped the idea that there was a total civilized way of life, not an unrelated assemblage of detailed superior weapons, gadgets, and religious beliefs, etc., about which the civilized man knew and they did not. The Americans had got hold of something—a total form of social organization, of culture—which made it possible for them to be so many, to produce and keep together as one people such wonderful human beings who believed in getting on with each other without continual recourse to outbursts of righteous anger, who treated each other and the Manus with whom they came in contact "like brothers," who, in fact, treated their neighbour as themselves in the fundamental sense that he was assumed to be the same kind of a guy. This American attitude, which underlies American hatred of officers who claim special privileges as well as American angry uneasiness about racial differences, shone through the interrelationships among Americans and between Americans and natives on Manus.

There were American Negro troops on Manus. It was reported by a missionary commentator that at one time American Negro troops on Manus, or in some installation, outnumbered white troops seven to five, and that Marines had to be brought in to quiet a disturbance which followed a belief among the American Negro soldiers that they were being discriminated against in date of returning home. The Manus did not recognize a status difference in the treatment of the American Negro troops, because here were "black" men like themselves, who were dressed like every other American, who spoke and acted like the rest of the Americans. Today a Manus will comment on a bridge that was built by the "black" Americans, saying this with the pride of racial identification and without any sense that treating Negroes as labour battalions was demeaning. From their point of view, the Americans had helped the Negroes—originally primitive people like themselves—to be "all right," and this contrasted with their own status in the Territory, compound as it was of special legal provisions for natives who were both "black" and primitive.

And the Manus watched, fascinated. They seem to have got into every kind of installation, and I never knew when I would encounter either a superior toleration for my quite good field glasses because they didn't have a search light attached, or, as I squatted on the floor trying to stop a case of arterial bleeding, an account of the magnificent equipment of the operating room of an American military hospital. Manus were down in the engine rooms and up on the bridge; a people who enjoyed machinery as much as they did presented constant entertainment to the American troops. Out of this experience with the most lavish, the most intricate equipment of the modern world, they drew a second conclusion—the Americans believed it was better to let machines do the heavy work, that there was no advantage per se in human labour which tired the body and drove men into an early grave. The Americans did everything with "engines"; they had engines to cut down trees and engines to saw boards and engines to lift loads and engines to fire guns, and, so their American friends told them, at home they had engines to wash the dishes.

It is important to remember that this information was not given to a lazy, easy-going people who were reacting against the kind of work which the white man had imposed upon them. When the Manus went away to work in the late nineteen-twenties and -thirties, they came back with tales of how much *easier* it was to work on the European plan, with a bell to start and a bell to stop, and no work at all on weekends. This contrasted with the behaviour of their own elders, who drove themselves day and night, seven days a week, never stopping. But before World War II they had seen only a limited number of labour-saving devices, and these had not been presented to them as labour-saving so much as simply technically superior; with a steel axe one could cut down a tree faster than with a stone axe, supplemented with fire. An iron pot lasted many years where a clay pot cracked; a canvas sail stood up in a storm and could be patched again and again where a mat sail broke and had to be discarded. But the European technology of pre-World War II in New Guinea was a matter of make do and mend. Pinnaces and schooners left over from German times had their engines tinkered with and fastened together with wire. Even the airplanes of the period were subject to almost unbelievable improvisation. The only response of the Manus to the greater complexity of the American machines was the comment: "We can understand Australian engines [meaning pre–World War II models] easily because they are all open and you can see them work. The American engines are closed up in boxes, and all you can see is the button that starts it and the button that stops it. But," they add confidently, "if we could just see inside, we would understand how it works."

Native labour had been cheap enough so that, with the exception of mining machinery, very little heavy machinery had been brought in. The native worked by the sweat of his brow for the Europeans and for himself. To native eyes it was immediate power over people that made it possible for the European to escape heavy work also, just as at home in the village the big man could also send his economic dependents out fishing. The idea of machinery which would make heavy work by the poor and subordinate unnecessary was a new idea to the Manus, as it still is to most of the world. Industrious and driven as he was by his own economic system—and it must be remembered that in 1942 this system drove him as hard as ever; that the Mission in introducing Christianity had not altered the native economic system—the Manus saw this American idea as a life-giving one. "The Americans believe in having work done by machines so that men can live to old age instead of dying worn out while they are still young." The belief that it was good to reduce human labour, that there was no virtue in back-breaking activity which made men old before their time, they caught so completely that after the Americans went away and there were no machines left, no bulldozers to knock down mountains, it still persists. The Manus want to organize their society so that men no longer die of overwork.

From American hospitals they got the idea that the most complete, expensive underwriting of the health of individual human beings was an American ideal. Nothing, nothing, so it seemed to them, was too much to do to save a life, to heal a wound, to replace an amputated limb. Up to this point their relationship to the medicine of the European had been slight. Initially, with their spiritualistic theories of illness and death, they had been uninterested. After they were Christianized, men who had gone away to work, even on plantations in the Admiralties, had also brought back tales of sorcery from other areas as an explanation of disease. Their Christianity had taken the form of substituting one supernatural system for another; it was not that the God of the Christians and His Saints and Angels were the only supernaturals; ghosts were still ghosts, only their power was seen as very limited and unimportant. As a Christian it was possible to ignore the ghosts completely; their skulls were no longer treated with honour, and between a man and the ghosts of his ancestors there was no longer an on-going personal relationship. All the trappings of that relationship had been thrown away—the skulls, the broken rib bones worn in mourning, the divining bone from the forearm of an old woman who had trembled, the hair of the dead once woven into beaded pendants and pouches, and the rituals of herbs and paint and intercession which had accompanied dealings with the ghosts. Like a lapsed trade friendship between men who had not seen each other nor sent presents to each other for a generation, the relationship between men and ghosts no longer had any social reality. "We have thrown everything away." Only a few bits and pieces of charms of the black magic that killed babies were still hidden in the boxes of practitioners. In 1947, during the "cargo-cult" outbreak, the last of these bits was thrown into the sea, and Peri was left with only one owner of magic which could kill babies, and who knew the charms with which to protect babies against his own magic. He still possessed this magic because, being based on nothing but words, "There was no way in which he could throw it away." The Manus felt they had substituted one religious system—a better and stronger one which contained many more "truths" about matters on which they had been ignorant—for a local, inferior, limited religious system of their own, very much as a people might accept a more complex form of government or a more universal system of currency for a local one, or even as many American communities have done, relinquished a small local telephone system and taken on the nation-wide Bell system.

But the Mission, while always generous in the care of the individually sick or hurt, had not interfered with the primary connection between illness and death and sin, so that missionization had brought no increased sense of the importance of Western medicine, but rather an increased sense of the power of God over matters of health and illness. No ideas of public health or preventive medicine seem to have been introduced, either by the Mission or by the Administration's medical services. Those administration rules which were public health rules—like rules about burial—were seen by the natives as laws unrelated to matters of health. The pre–World War II government hospital was a very primitive affair to which natives with bad sores might be sent from their villages, and to which they violently objected going.

Then came World War II, and the wonders of American medical care which contrasted dazzlingly with anything that had ever been seen anywhere in the peacetime administration of the Territory. As the Manus report it today, the Americans believed that every human being's life and health was of inestimable value, something for which no amount of property, time, and effort was too much to sacrifice. This idea was completely congenial to the Manus, who under the old culture had laboured early and late under the sanction of threats to the health and life of themselves and their relatives. But the Manus felt that they had only kept a small proportion of their people alive by abstaining from sexual and economic sins and slaving their lives away in a backbreaking economic system. The Americans were so able to reduce labour and treat property carelessly that human health and welfare could really come first. "From the Americans we learned that human beings are irreplaceable and unexpendable, while all material things are replaceable and so expendable." When Raphael Manuwai was scolded by the trader and plantation manager, for whom he was drying copra, for deserting the copra at the time of the volcano—there might have been a two-hundred-dollar loss— his comment was that his children were more important: what was two hundred dollars' worth of copra! And this comment he would also now apply to his own property. "From the Americans we learned that it is *only* human beings that are important."

Perhaps this response to American culture will seem the most confusing and unexpected of all, not only to Europeans and Asians, but to those Americans who have incorporated negative criticisms of our culture. Non-materialism, valuing human life above property—how can any people have learned such a thing from the American Army, with its endless gadgets, its tendency to raise the standard of living inside a tank to the point where the tank won't function, its orange juice and seven kinds of hats? When the American Army was in Britain, people joked, they had to dismantle an airplane factory to wash the underwear of just one American division. Thus the European comment! Non-materialistic—when what an American household wastes in a day would sustain a Chinese or Indian family for a week! Thus the Asian! Valuing human life—when Mexican wetbacks are exploited as if they were cattle, when the beet fields or the railroads or the tunnels of the United States have been built with exploited labour of European immigrants and coloured peoples! This from the left wing outside, or the aroused conscience inside.

Here again it is important to realize what the Manus saw. They saw the American forces and no one who was not in a uniform, no one who was not protected from exploitation by others by an elaborate system of safeguards built by a people who believed only in a civilian army, who distrusted power over persons and all those who want power over persons. And further-more, the Manus, out of the closeness of their potential under-standing of American values, shrewdly recognized that the American willingness to sacrifice things for people came from having plenty of things, so many that there were always more, so many because of the apparently inexhaustible productivity of a machine economy, built not upon the limited strength of human beings but upon the unlimited potentialities of machines. "The Americans had so many possessions they did not have to quarrel and care about particular ones," they said. "The Americans were willing to give anything away."

When a unit pulled out, popular natives who had either been servants or guides or brought fresh fruit to favourite American officers were heaped with valuable objects. GI's who worked in machine shops gave their favourites fine tools to take home to their villages.

Meanwhile, the Australian military administration unit, ANGAU, striving to preserve some continuity between prewar and wartime administration, suffered from terrific difficulties within the Australian administration itself, and from endless harassments in trying to carry out the task of keeping the natives steady, keeping native life policed, preserving some semblance of the sort of relationship between Europeans as employers and natives as employees which would ensure a labour force after the war. Here they were doubly hampered, by the Japanese past occu-pation and the American present. The Japanese had shown little concern for the maintenance of law and order among the natives; they had concentrated on disciplining and frightening the natives by demonstrations of ruthlessness, and on intimidating them into working for them, not helping the Allies, and providing the Japanese with food. Whether the native population among them-selves committed murder, rape, or adultery, gambled, drank, or stole, was not a matter with which they took time to deal. They attempted to consolidate their power with native police and native officials and leaders, and thus, perhaps inadvertently, backed up unscrupulous power-seeking natives who wished to indulge in every sort of crime against their fellow villagers or their age-old enemies in the next village. This is, of course, a common-place of any occupation situation, where the occupying army is not ready to consolidate a civil government position. The favour and help of powerful local people is bought by permitting them a great deal of licence, especially against those who are not co-operative with the new power.

The Coast Watchers, individual Australians, alone or in small groups entrenched in the mountains throughout the Territory, especially when they were ex-government officials, were continu-ally faced with the difficulties of maintaining discipline and refus-ing to approve or endorse illegal behaviour from one native to another, not alienating the very natives on whom the continua-tion of their mission—maintaining radio contact with our troops and preventing surprise air attacks on our ships and installa-tions—depended, and convincing the natives that, despite what the Japanese and what many missionaries said, "We, the Aus-tralians, are coming back." It was enough of a touch-and-go game for an ex-district officer to hide, with a week's supply of food and a radio that might break at any moment, somewhere in the Solomons, with all the coastal villages well intimidated by the Japanese occupation and the German missionary on the next hill telling the natives that the Japanese had come to stay. Should he also try to maintain the standards of lawful behaviour which the Administration had set up in peacetime? And, if not, what of the day when the Australians were back, and he, as magistrate, might face these same natives from the bench?

But this situation, dreadfully dangerous as it was, was part of the calculated risks the Coast Watchers took. After all, if an ex-district officer, barefoot, virtually without food, with a few rounds of ammunition, would still take the trouble to maintain pre-war Australian-type discipline, he must be *sure* he was com-ing back. And, upon the native belief that the Allies would win and come back rested, except in the case of the individual devoted police boy or personal servant, any hope of getting continued co-operation from them.

All this was difficult enough, difficult for the isolated Coast Watchers behind the Japanese lines, difficult for the Australians who had to restore customary civil law after the Japanese troops were driven out. But it was a simple and clear situation compared to the one which faced the Australian administration wherever there were large concentrations of American troops, their Allies, yes, but Allies who were profoundly uninterested in what happened to the natives after the war. With the fine indifference of the obstetrician whose only concern is to get the baby out of the hospital with the mother doing well, and who can then turn all the problems which result from those first few days over to the pediatrician and let *him* worry, the Americans were concerned with the present, with good relations with the natives in the immediate present, so that they would have them as carriers, guides, dock labour, so that they would provide fresh food or thatch, now.

Perhaps the best story occurred not in Manus but in Guadal-canal, where an ANGAU officer, finding that native gardens were being permanently injured because all of the natives were working for the Americans, ordered them back to the village to repair the damage so that their wives and children would have something to eat. A nearby American commander, furious at having his local labour arrangements interfered with, piped electric power into the village and installed large laundry equipment so that the natives could go on working for his unit.

On Manus the situation was made all the more difficult because it was an American base. The seasoned Australian officials realized the complications which were bound to result after the war if the people of New Guinea became accustomed to obtaining money and goods so easily. They naturally disapproved of the Americans giving Manus natives, surreptitiously, United States Army property. (This disapproval had the added background that Australians felt more keenly the difference between individualistic extra-legal behaviour on and off duty than the American soldiers did, and so, although they shared a generally similar set of attitudes toward military service, disapproved of such behaviour by Americans on duty.) So the ANGAU officials tried to prevent the Manus from benefitting from illegal American generosity with Uncle Sam's property. From situations of this sort, the Manus built up a picture of Americans who were infinitely generous and of Australians who took away the things that the Americans had given them. To this day "Angau" is the nickname for a man who takes things away from other people by excessive winnings at cards, or over-use of his privilege of begging from his relatives.

With the coming of the Japanese, the people of the Mandated Territory for the first time really became conscious of what differences in national cultures might mean, separate from race, separate from individuality. In the past there had been a tendency to lump all English-speaking white people together. Now the Americans were clearly distinguished from the Australians, and this meant that where the army had been American rather than Australian the inevitable contrast between the lavishness of a military installation and the relative poverty of a civilian administration was expressed in invidious national terms. So on Manus immediately after the war, when native idealization of the American occupation was pitted against the hard experience of having to set up an Australian civilian government again, the contrast seemed to many Australians as distinctly and cross-nationally unfair. The time when the Americans were there is now called by the Manus "the time without taboos"—it was at this period that gambling (said to have been introduced into the Territory by the Chinese but kept rather rigidly in check before the war) became widespread on Manus.

During the Japanese occupation, the Mission had been completely disorganized by the Japanese, and the sisters and several of the priests had perished. The religious experiences under the Americans were discontinuous, and there are still tales which are hard to trace of various evangelists whose connections with the

American occupation remain ambiguous. After the war, the Mission, like the Australian civilian officer, had the task of calming the people down, settling them back into a routine of devout observance.

Traders also faced the task of building up plantations which had been trampled and run down, of setting up shops again in which the cheap trade goods which had seemed so desirable before the war now had to compete with the presence of American Army equipment of far higher quality. Goods were not only poor but hard to come by and, after the pleasant, easy days of give-aways, seemed exorbitantly expensive, while the wages the planter was prepared to pay seemed chicken feed, hardly worth working for.

The Australian administration set up a system of redeeming the American money within the shifting exchange rates throughout the subsequent years. The combination of this system and the imperfect explanations of the difficulties of changing money between sterling and dollar areas has confused a people who had just prepared to embrace the superiority of a currency that was good all over the world!

The native labour system was in confusion;[4] three-year indenture was not resumed, and the cost of recruiting labour at a distance and transportation by plane instead of boat was so great that even the most exacting high officials living in Port Moresby, which was now the capital, were forced to engage casual labour, at the door. The head-tax was also not reinstated; the local medical services were disorganized, attempts were being made to train native medical assistants, but initially it was decided that the literacy of the Admiralty Islanders was too low to qualify any Manus for entering this training. (This is the most vivid factual evidence of the extent to which the education which the Manus had expected to obtain through missionization had failed to meet their hopes.) The price of copra was at an all-time high, which meant that planters exploited their existing plantations— holdovers from German times—rather than planting new ones.

This, then, was the wider historical situation within which the fate of the Manus of the South Coast was decided. A war, which might hardly have affected them at all, had come straight to their doors. They had been exposed to an extraordinarily massive impact of American culture—one million to fourteen thousand. And this on top of the general exposure in New Guinea—in which many Manus natives shared—to modern warfare, modern technology and organization, and the experience of great armies where before they had encountered only a few individuals.[5] History had staged for them a unique experience, and for us—voluntarily or of necessity students of change—a unique experiment. A handful of people on a remote little cluster of islets in an archipelago whose very name was unfamiliar in the great chancelleries of Europe and the Americas, became important enough in history so that their attempts to set up a culture like ours have involved them with the highest authorities in the world: they have been discussed in the United Nations, and the high councils of the Roman Catholic Church know the history of the heresy which has lost to the Church, perhaps only temporarily, an ardent group of converts.

This new role, a place, as a people, in the history of the twentieth century, is in strong contrast to the kind of involuntary impact which the same group of people have had on modern thought because their names and characters were recorded twenty-five years ago in an anthropological record which has become part of the intellectual stock-in-trade of modern anthropology and the sciences which draw on anthropology. *Growing Up in New Guinea*[6] has been reprinted in the dollar edition of the early thirties, in *From the South Seas*, twice as a Penguin edition, in 1942 and 1954, and as a Mentor Book. The life of the Manus formed one chapter in *Co-operation and Competition*, provided the cover design for an issue of *Story Parade* and an issue of *Natural History*, slides to develop a course in homemaking for high school girls in Newark, New Jersey, the basis for an article on the relation between culture and neurosis in a German psychoanalytical journal, the background of the concept of continuities and discontinuities in character formation, part of the stock of examples for lectures in sociology and social psychology as well as anthropology. Several hundreds of thousands of people (students, specialists, lay readers) have used the materials, smiled at the small boys in tiny canoes on the jacket of the first edition, pitied the joyless bride dressed in dog's teeth and shell money, envied the children in their water world, and marvelled at the financial acumen which made the Manus so clear headed about the inflationary effects of the importation of real dog's teeth, or questioned how a secret could be kept so well that the men were ignorant that women menstruated more than once before marriage. The portrait of "The Five Retainers" grinned in perpetuity from the pages of the original edition of *Growing Up in New Guinea*, and then all the photographs disappeared from the later editions, and the Manus became either pleasant fiction, hardly to be distinguished from the natives in *Orphan Island*[7] or *White Shadows in the South Seas*,[8] or else counters in the discussions of social science, where people had long since lost track of whether one Manus was a Manus or a Manu.

This strange existence which one small group of people lived in the pages of books they had never really seen—although a visitor once showed them a copy in the early thirties—in the minds and discussions of people of whom they had no comprehension, was, as far as they were concerned, quite meaningless. To them, two white people had come and lived in their village and gone away forever, and were probably dead. That the conditions of their lives could be used to illuminate discussions on economics or mechanical ability, that groups of MIT students who were to be the engineers of the next great step in technology might hear about the way in which their curious upbringing gave them great facility with machines—of this they knew nothing, and would not have comprehended had they known. They were actors on the stage of the modern world, unconscious that it was a stage, unconscious that they played a special role. If it had not been for two events, both unpredictable, so they would have remained.

1. Stefan Posanget   2. Petrus Pomat   3. John Kilipak   4. Johanis Lokus
   (Kapeli)   (Loponiu)

1953
FROM SAVAGE BOYHOOD
TO MODERN MATURITY
1928

(PLATE I)

5. Kapeli   Pomat   Yesa*   Kilipak   Loponiu
                                                *Yesa has moved away.

1. New Peri, 1953

THE OLD VILLAGE AND THE NEW
(PLATE II)

2. Site of Old Peri, 1953

3. Old Peri, 1928

1. The volcano as seen from the Patusi hilltop

2. Petrus Pomat addressing the meeting

THE MEETING ABOUT THE VOLCANO, JUNE 30, 1953
(PLATE III)

3. The meeting: people listening

1. Josef Bopau, 1953

2. Teresa Ngalowen and the "starved baby," 1953

THEN AND NOW

(PLATE IV)

3. Bopau, 1928

4. Ngalowen and Ponkob, 1928

1. Canoe Races, 1953

SAILING CANOES HAVE NOT CHANGED
(PLATE V)

2. Carrying a corpse, 1920

1. Fishing with two-man nets, 1953

THEY STILL FISH IN THE OLD WAY
(PLATE VI)

2. Fishing with two-man nets, 1929

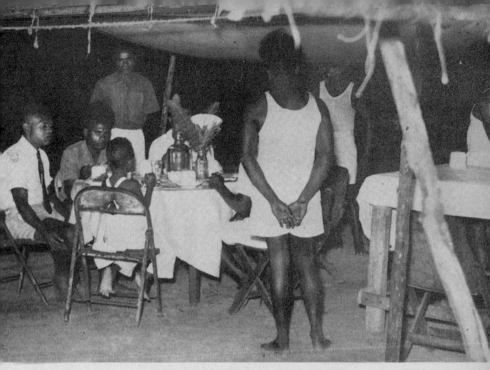

1. Feast given for Paliau in Bunai, August, 1953

EUROPEAN MODELS, DISTANT AND ATTAINED
(PLATE VII)

2. Peri child sailing model of
a European pinnace, 1928

3. Ponkob playing at being
a European, 1929

1. Paliau shaking hands in Peri, July, 1953

(PLATE VIII)   PALIAU AND A TRANSFORMED ELDER

2. Pokanau and his son,
   Matawai, 1929

3. Pokanau making a speech,
   December, 1953

1. Stefan and his three-year-old son, 1953

CHILDREN
STILL LORD IT
OVER
PARENTS
(PLATE IX)

2. Demanding betel nut, 1928

PARENTS AND CHILDREN
(PLATE X)

Above: 1. Raphael Manuwai and his five-month-old daughter, 1953

Above left: 2. Raphael Manuwai and his daughter at three months old, 1953

Left: 3. Mother and adopting mother teaching eight-month-old child to walk, 1953

Below left: 4. Luwil Bomboi and his daughter, Piwen, 1929

Below: 5. Mother of Teresa Ngalowen and her baby, 1929

MARRIAGE: OLD, MIDDLE, AND NEW
(PLATE XI)

Above: 1. Marriage in a church, June, 1954

Above right: 2. Karol Matawai and his second wife, 1953

Right: 3. Pokanau and his fourth wife, 1953

Below left: 4. Tawi bride dressed in money, January, 1929

Below center: 5. Taliye wearing headcovering of betrothed girl, 1929

Below right: 6. Catholic marriage of Johanis Lokus and Pipiana Lomot, 1946

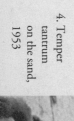

CHILDREN COME ASHORE

(PLATE XII)

1. Children in play canoes, 1929

2. Children playing on land, 1953

3. Temper tantrum at low tide, 1929

4. Temper tantrum on the sand, 1953

1. Michael Nauna and his
father, Ngamel, 1929

2. Michael Nuana and his son,
Pwochelau, 1953

THREE GENERATIONS
(PLATE XIII)

3.   Pwochelau and his mother, 1953   4.

KAROL MANOI: MAN OF MOODS
(PLATE XIV)

Above left: 1. Gaily relaxed

Above: 2. Withdrawn

Left: 3. Rigid and anxious

Below left: 4. Baleful and angry

Below: 5. Posing for farewell picture

LOCAL FINANCES
(PLATE XV)

1. Village economics, old style, 1929

2. Beginning of economic self-government, 1954

3. Women citizens with their money ready, 1954

Above: 1. Small girls making fish baskets

Below: 2. New tools for old

Above: 3. Peranis Cholai, gay

Below: 4. Peranis Cholai, stiff

CHILDREN OF THE NEW WAY,

1953

(PLATE XVI)

Above: 5. Peranis Cholai, the teacher, preparing a lesson in the church

Below: 6. School children with teacher

Sophisticated observers of the twentieth century might comment on these strange relationships between people who have never heard of those who read the most intimate details of their lives, who know their names, their foibles, their indiscretions, and their ambitions "better," as a friend of mine wrote me when the book came out, "than I ever expect to know the people in this London boardinghouse where I have lived for six months," and readers who could not have located, without a long search, the tiny island about which they had been reading. The links between the two worlds were so slender, so unexpected—no one would have known if they had snapped altogether, so that there was no one alive who could recognize and be recognized by this cluster of people.

But when we arrived in 1953, the leaders of the little communities were prepared to come into communication with the modern world. So Samol, the leader of Bunai, held a meeting and exhorted the citizens: ". . . that for these months everything you do will be recorded, filmed, put on tape . . . and *all America* will know whether we are succeeding in our new way of life." Two accidents of history—that we had chosen them and not some other tribe to study in 1928, and that Manus had been a major American staging area—were involved in this strange emergence of a group of erstwhile savages twice upon the world stage, once unconscious of their role, now fully aware of it.

But there was the third accident of another order without which all this still might not have happened. It was the sort of accident which has puzzled historians for centuries—the presence of a leader with the political genius to take advantage of a situation created by history. Many parts of New Guinea were unsettled after the war; many parts of New Guinea during the last sixty years have had at times some sort of religious movement combining elements of new and old. What made the Manus situation develop into something which could catch the attention of a busy world while hundreds of thousands of other natives, after brief flurries of "cargo cult," refusals to work, attempts at behaving in some new way, sank back into obscurity?*

The cultural character of the Manus people, their peculiar position in the Admiralties, and the events of World War II were the necessary but not sufficient conditions for the development of the New Manus Way. The presence of a leader, Paliau of Baluan, turned widespread predictable unrest into a socio-political movement.

---

*There have been several other movements in the Pacific with gifted native leaders on which we do not have as much information and which cannot be properly compared with the Manus movement, although it is known that the personality of the leaders was of great importance.

8

Paliau: the man who met the hour

Paliau,* whose imagination turned unfocused post-war discontent into a full-fledged political movement, came, like so many political leaders of history, from outside, from Baluan. He was not a member of the group on whom his strength rested—the Manus. Most of the people of Peri had only seen him a few times in their lives, when he had visited Peri or when all the Peri canoes had gone to his capital on Baluan for Christmas celebrations. A few individuals had worked with him more closely. Peranis Cholai had spent several months on Baluan being trained as a lay preacher; others of the leaders had made short trips to take their financial contributions to the treasury that he had built up. Karol Manoi had known him as a police boy behind the Japanese lines in Rabaul.

But Paliau was essentially—as far as Peri was concerned—a leader at a great distance. Inevitably I saw him through their eyes, also as distant. I actually only saw him on three occasions: in court in Lorengau before we had met, when he visited Bunai and Peri two weeks later, and on my visit to Baluan in August. He had a

---

*Ted Schwartz is publishing a long study of Paliau.[1] I have presented Paliau here as the people of Peri and I myself encountered him at a distance.

wholly disarming frankness behind which his reserve never broke down, and, while he might seek counsel from those around him, or attempt to enlist their help, he never seemed to lose his air of extreme loneliness. I can find no better way to present him than in the series of scenes through which I made his acquaintance.

Human imagination tends to treat any fit between a leader and a situation as a miracle. When there is a man of the hour, the drama of the match between man and history is so striking that we ignore all the hours without men, and all the men without hours. Our own historical tradition is steeped in messianic and nativity stories, in which the emphasis is laid upon the birth of the hero who will set his people free or lead them toward the light. The log cabin where Lincoln lived, the house where Jefferson was born, take on some of the luminosity of the Manger. Even the determinedly antireligious Soviets in the early days of the revolution approved readaptations of Georgian folklore, which spoke of Stalin's birth with all the imagery of a Nativity, and also contained the traditional hope of an apocalyptic leader who would right all wrongs. People after people who have experienced the presence of a gifted leader have tended to regard his entrance into the world as the great miracle, forgetting that had he died before he came to leadership, or had the time not been ripe, such a gifted man or woman would have gone "mute and inglorious" to an unsung death.

In October, 1946, Paliau returned to the Admiralties from Rabaul, and the Manus date the beginning of their new era from 1946. For earlier periods they know that Europeans have calendars and dates, names of great men, names of battles and discoveries and events, but for the Manus history begins in 1946. Each detail of Paliau's past history and experience became fateful because of what happened after he returned. If he had died on shipboard between Rabaul and Manus not only would he never have been heard of but the aspiring discontent in Manus, the desire to work out a new social system, would probably have fizzled out and died.

It was Paliau who had a program for action, an organized picture of change which involved genuine ethical ideals—he wanted all the people of the Admiralties to become one people, eschewing the narrow rivalries and hatreds between the different tribes and different villages, pooling their specialized skills and possessions. The land people should share their gardens with the sea people, the sea people should invite the land people to fish on their carefully guarded reefs. By banding together as one people, there would be many of them, enough, if they used their resources wisely, to get good European goods and to live the way of life of the Western world.

I had first seen Paliau while attending a court case in Lorengau, where a man who had nearly killed Paliau with an axe was being tried. There I had been able to watch him from behind the barrier of my temporary identification with government, as the District Commissioner's lady and I chatted with the judge during the recess of the court. I had seen something of the way he was regarded on the government station as he stood, dressed very correctly in white ducks, with two or three other men, similarly dressed, at a distance from the court building.

During the court proceedings, when the judge instructed him to attend and listen to the evidence, I had watched Paliau stand, very still, and spiritually withdraw from the whole proceeding. His stocky frame was drawn in to the smallest possible compass, his face expressionless and non-participating. The trial of the man who had nearly killed him was conspicuously ignored, even while he stood and answered questions respectfully, coolly indicating it had, quite clearly, nothing to do with him. This behaviour contrasted sharply with the picture which many of the officials involved still held about him. He was called, even in the notes made by officers on patrol, "The Emperor," whose attempted reign was illuminated with comparisons to Hitler and the Emperor of Japan. He was said to have maps of his proposed empire which would include not only the Admiralty Islands but New Ireland and New Britain as well! He was said to maintain a huge harem of women, to have his food served to him by a line of kneeling servitors—Japanese style—to have established a totalitarian régime which flouted every canon of free government and used such loathsome devices as drilling, bells, curfews, passes. He was said to have claimed to have been given a key to Heaven by God, with whom he was in personal communication, so that he could "drop in any time."

These stories I had encountered in various forms from the time I decided to go back to the Admiralties. It had been much easier to get accounts of a conspicuous wicked totalitarian, traitorous, native leader who had broken the ties of a whole people with the Roman Catholic Church than it was to find out what had happened in Manus villages, which, according to some reports, were engaged in the flourishing and contented worship of a new pagan cult with the Blessed Virgin as the Goddess of Fertility, and, according to others, were living idle sinful lives in a "slum" which stank to high heaven, running illegal hospitals where miraculous cures were attempted for broken bones, and "schools" which taught "Hitlerian race hatred."

I knew also that the top administration of the Territory had followed the most advanced post-war thinking and attempted to encourage Paliau, as a native leader, by having him taken to Port Moresby, the capital of the two territories, where he had been shown infant welfare clinics and functioning co-operatives and had explained to him the new system of local councils, modelled on British systems in Africa, whereby natives could, with careful guidance, manage their own affairs. I knew that local officials had felt that all of this attention had spoiled him even more, and that plans to wreck the whole Administration's attempt to regularize his movement by transforming it into a council had been promoted locally. Paliau had been given a council—for *half* of his unified group—with the capital in his native village on Baluan, thus splitting apart the whole Manus group who formed the backbone of his movement and leaving the South Coast Manus and the Usiai, who had come down to live with the Manus on the South Coast, without a council and with only a vague promise that one would materialize some day. This nineteenth-century divide-and-rule policy, combined with the subtler and more ethi-

cally complex plan of electing Paliau president of the truncated
council, was said by some to have clipped his wings.

In between 1946 and the present, Paliau had been to Port
Moresby and had been put in prison as responsible for the
excesses of the 1947 "cargo cult." He had also been put in prison
for adultery—the principal legal recourse against unpopular local
leaders, rather like income-tax accusations in the United States.
His adherents claimed it was an unfair frame-up while local white
opponents claimed it was only a partly successful frame-up and
government officials stressed it as totally non-political in nature.*

As I watched him, there in the courtroom in Lorengau that
July day, I thought of how his name had emerged in the conversation of the people of Peri, during the previous three weeks, as
they came to trust me and the caution introduced by the Patusi
immigrants faded away. Upon my arrival I had not mentioned
Paliau's name, for I wanted to see how the Peri people would
present their changed life to me. And no one mentioned his
name. When I asked, referring to all the dramatic changes with
which they presented me, "Who thought of this?" they said, "We
all thought of it together. It is *our* idea." "And who were *all of
you?*" The names of all the leaders in the village would be mentioned. Once, during the second week, while we were all up on
the mountaintop waiting for the volcano to subside, I heard a
rapid conversation behind me, mentioning Paliau's name and an
order about making graveyards, which had an entirely different
ring to it—the blind, uncompromising note of bigotry which distinguishes a cult from a great religion, a minority or subversive
political movement from a seasoned and tolerant political system.
Once I had caught the same blindly fanatical note in the voice of
Peranis Cholai, the Peri teacher and clerk, when I had suggested
that they were using the words *council* and *committee* incorrectly
to designate single individuals. "*We* call them *committee*," he said

---

*In 1954, Paliau was again imprisoned for beating his adopted daughter for
marrying against his will and while he was out of the village a clerk from
another tribe.

with a stubbornness so close to the closed delusional system of the insane that I began to wonder whether there was only a difference of degree between the rigidity of the psychotic and the fanaticism of any human being trapped in a cult.

Finally, after I had been in the village a little over three weeks, the young preacher, Tomas, a Patusi immigrant, decided to trust me, and, after a long speech about trusting no European, told me the *Great Secret*, that Paliau had planned the New Movement *before* 1946 while he was still in Rabaul. The air of import with which he told me this seemed wholly disproportionate to the content.

It took a long time to sort out the complicated story of how Paliau had developed a coherent plan, returned with it to his people, been received with very moderate enthusiasm by the Baluans, with greater enthusiasm by the Manus of Mouk. A conspiratorial atmosphere had developed because the young men who wished to follow his modernization plan feared the opposition, not of the government, but of their elders. But it was not until the intensity of the atmosphere within which Paliau advanced his new view of the Manus place in the sun—combining as it did mystical elements of reinterpretation of the Old and the New Testament, town planning, sumptuary laws, and an economic design for building a treasury—had generated a prophet, Wapi on Rambutjon, who started a "cargo-cult" movement in which people destroyed all their possessions in expectation of a millennium, that Paliau's movement took hold. The mystical phase died down; a period of sober planning had followed, but in the minds of the leaders of the New Way there was still a profound recognition that Paliau had been responsible for it all. In secular terms, this was phrased as his having had the complete idea *before* 1946, before he went to Port Moresby and was taught about the organization of partially autonomous local councils. In religious terms, it meant that God had especially chosen Paliau to bring His Truth to New Guinea after having watched first the failure of Captain Cook and then of the English, the Germans, the Australians, the Japanese, the Americans, to bring about the transformation of Manus man into the kind of society which Christ, by paying for men's sins, had made it possible for men to build.

Both Paliau's vision of a world transformed—a world which would have the outward and visible forms of European life and the inward spiritual grace of Christian ideals, brotherhood of man, shared resources, peace—and the reactions of the Europeans in the Territory stemmed from the same source. While Paliau saw himself as establishing the order which had been meant for all men, government official, planter, and missionary saw him in terms of Western history, within the messianic tradition and the tradition of apocalyptic Christian cults which have tried to establish a state of ideal Christian community on earth or preached the immediate end of the world against the wreckage of spent and unspent Führerships in European countries. Thus some of the drama of the defecting sects of the Reformation was being re-enacted in this out-of-the-way corner of the world.

As in the Reformation, where the right of the common man to have access to God's Truth and Word and the sense that men had fallen from the way of life preached by the early Church had combined with other social movements of the time, so had the native sense of the darkness and benightedness of their low cultural state, plus their desire to be treated as dignified human beings, combined to produce a religious-political doctrine in which there was some of God in every man. "Every he in England is as good as every other he," as the Levellers had said. The parallel between Paliau and a variety of historic figures had already been pressed upon the natives, and those who remained faithful to the Catholic Church spoke of Paliau as being like an evil character called "Henry Wesley," a combination of John Wesley and Henry the Eighth.

All of these things, about which I knew so little, were in Tomas' voice as he confided in me. In his intensity lay the knowledge that somehow Paliau and his reforms met with far more resistance if they insisted that Paliau had acted with direct revelation from God than if they phrased their movement as something "we all thought of together." When later I commented to Paliau on how people had first insisted on multiple origins of the plan to protect him from attack, he added with quiet bitterness, "And to protect themselves also."

In the pot-pourri of fear and reproach and historical model, within which everyone was trapped, there was only one missing element. No one had accused Paliau of being a Communist, or ever having heard of Communism. What collectivist and possibly "socialist" taints his movement was believed to have were referred to the evils of German National Socialism, of Fascism, of Japanese imperialism. In other parts of the Territory there were dark tales of a mysterious stranger who had appeared off the coast of New Guinea in a double outrigger canoe, claiming to have sailed from the Philippines, and who sailed up and down the Territory talking to natives, telling them that later they would be kings. Some Europeans believed he had been dropped off a German steamer, canoe and all. He was said to have made his exit, by way of jail, as an impostor in the British Solomons. This story provided a kind of gross secular version of designation by a sacred leader—a man appearing as mysteriously as an angel—to choose his followers.

Australians emphasized that Paliau had been inside the Japanese lines, had not been on "our side," and people hinted at the disgraceful tales which could be told if the secret security files (which, however, seemed never actually to have seen the light as evidence in the war-criminal trials held in Rabaul) had actually been produced. These commentators knew that when the Australian government had been forced to evacuate at the beginning of the war, the native police and labourers from other islands had been instructed to obey the Japanese for their own safety, as the Japanese would be the government. There was also the question of what constituted treason for a native of a mandated territory.

But still there was an understandable prejudice in favour of the native police who had been in a position to risk their lives for the Allied side. In fact, this preference was probably strengthened by the very circumstance that there had been a discussion, just before the Japanese advance (which resulted in the capture of Rabaul, the old capital of the Mandated Territory), as to whether or not the ethical thing to do was to send every one of the hundreds of native labourers imported from other islands home to his own village where he could confront a ruthless occupying power on his own ground, with his own kin, his gardens, and his knowledge of the bush behind him. But the counsels of those who objected to this policy had won, and this seemed to have exacerbated the advocates of evacuation who, in turn, had taken extra care to exhort the faithful native police and the helpless hundreds of work boys that they "must" obey the Japanese. It was believed by Paliau's adherents, I learned later, that it was only because Paliau had understood this position and argued it effectively that he and the other police who had worked *for* the Japanese were not punished as "war criminals."

It seemed curious to watch this quiet man, that day in the court in Lorengau, dramatic only in his fixed intention to be non-dramatic, standing unostentatiously in the little local courtroom where the whole majesty and tradition of British law was concentrated—abstracted from the pageantry of the Old Bailey to come to life again in a little frame building on a South Sea island—and to know that he appeared to the Australians against the backdrop of their whole complexity of historical tradition, which carried the memories of long-dead kings and religious reformers, the lecheries of the Tudor Court, the combination of a revolt of the working man and non-conformity which was Wesleyanism in eighteenth-century England, the excesses of Hitler's totalitarian régime and of the partly modernized Japanese Empire. I wondered how Paliau saw himself, how much of this weight of history lay on his mind. One confusing little bit from a report I had read teased my mind. A government official who had had some dealings with Paliau had reported that the latter "asked me what he should do next." This little bit suggested some ground for the accusation that he was a man without a program, hungry for power. I had yet to learn that his peculiar genius consisted of drawing inspiration from every opportunity that came his way.

It took us a long time to understand the way in which he had to combine a sense of divine mission sufficient to focus the enthusiasm of the mystical fanatics in his following, an ability to make rational and effective plans that would keep the loyalty and admiration of responsible, intelligent lieutenants, and finally, an ability to work with government officials as someone anxious to learn all he could about their methods and goals. All three—the sense of mission of one who was fulfilling a divine plan, the sense of great intellectual competence in which he towered over his followers in statesmanship and planning, and a genuine puzzlement as to how his program was to be related to the wider world—all three of these threads in his complex personality were real and integrated. He, indeed, adeptly and responsibly, tried to be "all things to all men."

So Paliau stood in court and heard the man who had just failed to kill him given only nine months more in jail. Did he know, I wondered, that there were plenty of people who said the man ought to be executed for having *failed* to kill such a pestilential character, and that even those not so deeply involved would see the case as a kind of test between Paliau's assumed "power" and the constituted authorities, so that the lighter the sentence for the vaguely smiling psychotic assailant who stood in the dock and explained that he had not really meant to kill Paliau, only to hurt him badly—when he fell upon him unexpectedly and cut his chest open with an axe—the greater the victory for law and order.

Although Paliau's marriage (after a divorce) to the attempted murderer's wife had occurred much earlier, it made the attack into just the sort of mixture between the consequences of "imperial" debauchery and political subversion which served to confuse the whole issue of how the New Way was to be regarded. A light sentence was passed, the issue of the defendant's insanity taken into account in the only way open at present as an amelioration of the severity of the sentence, as there was in any case no place but the jail for the violent insane. And afterwards I was entertained with stories of what a theatrical act Paliau, carried into the station half-dead from a chest wound, had put on in the hospital. Meanwhile, Paliau never approached me, nor I him. He had been a boy when I went away, and a member of another tribe, inhabitants of an island on which I had had only a brief visit.

Then I returned to the South Coast and Paliau came to Bunai. After being prepared for the enormous importance of the occasion by private conferences with Samol, the "councillor" of Bunai, Ted and Lenore Schwartz first met Paliau accidentally, when he was wearing only a pair of shorts. He greeted them, they felt, with presence of mind, but some slight embarrassment. Ted Schwartz asked him if he would make a tape recording, and this was agreed upon. That afternoon, the whole group, Paliau, Samol, and four of Paliau's Baluan associates— variously designated by European critics as his "court," his "lieutenants," and his "cabinet"—accompanied by Ted Schwartz, appeared without warning in Peri, where they were greeted by our Peri high officials without fanfare, in a state of informal undress. Only old Pokanau, "the old lawyer man within the village," went and got dressed up and hastened to prepare some food.

Paliau and his group entered my house and accepted cigarettes, Paliau at ease, charming, completely master of the situation. It was not until a little later, as I watched him speak to other people or help me out of a canoe with just the very slightest extra flourish, that I realized what there was about him that had perhaps contributed more than any political aspect of his New Way to the title of "Emperor." For his manner was definitely "vice-regal," not "regal," merely vice-regal, but a native who can play a vice-regal role skilfully—with style, without any subservience, wearing his higher allegiance like an accolade—carries an air of aristocracy about him which is especially detestable to many Australians in the prevailing egalitarian climate. I remembered again that Paliau was from Baluan, the island on which I had found twenty-five years ago many of the qualities of Polynesia, in the grace and open sex appeal of the women and the style and pride of the ceremonial. The Baluan people had been regarded by the Manus of twenty-five years ago as light weights, slighter, fairer, gay and idle, and given to light living. Our little house boys had brought back just one phrase in the Baluan language: "Come out in the bush and make love."

After the first interchange of amenities as we established the fact that we had never met and that Paliau had been away at work as a young boy when I had been on Baluan, on an impulse I went and got an article[2] in which I had published a picture of a small naked Peri child, Ponkob, sitting on a chair trying to write with pencil and paper. "This is the picture," I told Paliau, "with which I have ended my talks about Manus. I showed it to Americans and said, 'The way the Manus meet the white man is to try to *do* what the white man *does*.' " His eyes lit up, and I saw for the first time the strength of the imagination which had conceived the possibility of skipping five thousand years of history. "You *knew!*" he exclaimed. "You knew, you understood twenty-five years ago what we only knew much later." I was to learn that he was especially caught by the appearance of foreknowledge of any kind. The first precursor of an idea carried a heavy weight of emotion for him, one which it would be easy for devoted followers, imbued with partly understood Biblical doctrine, to endow with even greater force of prophecy. There were many tales told of the way he had foretold Japanese attacks in the early days of the war.

The next day, when he made the tape recording giving an account of his life, I was to see, as he concluded his speech, another facet of his particular kind of genius, his ability to seize the materials offered by a situation and use them: "We—my generation—were born too late—what we want is to *make a good chair for our children to sit down on*," and the photograph of small sturdy Ponkob, his toes tensed as he imitated the writing of the

white man, was transmuted into a symbolic statement.* No wonder the Manus, themselves possessed of a language shorn of habits
of spontaneous metaphor, had found Paliau's oratory irresistible,
when he spoke, to use the new words which have been coined to
describe such speech, in "talk picture."

During the week that he stayed in Bunai, held fast by an
unfavourable wind which made the crossing to Baluan impossible, among us we saw several other sides of his multifaceted personality—the gracious guest at a feast, the angry leader betrayed
and humiliating his followers for putting on a dance in which
men still "dressed as women," wearing *laplaps* [loincloths] instead
of trousers (for Paliau understood that to wrest treatment as
equals from trousered white men, trousers worn properly would
in the end be essential), the "experienced" parliamentarian
drilling the council members who were novices, the law-giver
upbraiding a habitually adulterous woman who attempted to
misrepresent her case. The ambiguity between what he demanded
and that which his followers insisted on according him remained.
Yet when he left I was surprised to see that, while small groups of

---

*Paliau's treatment of projective tests showed the same ability to bend
everything presented to him to his own purposes. When presented with a
Rorschach (inkblot test) card, he would use each card plausibly, competently, would relate it as a whole to an ethical theme, not concerning himself with the small details but assigning top and bottom, middle and sides
to roles in the drama between good and evil. On the Thematic Apperception Test he also turned each incipient theme into an item in social reconstruction. On one card a group of young native girls were represented
standing outside a hut, while one stood at a distance. After identifying himself with the lonely one, he then went on to give a lecture on the education
of women. He skilfully avoided personal content, and worked with the
experimenter through the gross morphology of card structure. He worked
two days on his Mosaic test, building a beautifully balanced statement of
good and evil. In each case he performed in such a way that his percept,
although part of his own system, could not be challenged by the structure
of the card.

people brought farewell presents as they did to any departing canoe in Peri, he left with little fanfare but with a gesture of grace and an intrinsic sense of himself as a person. Much of his charismatic power to lead was said to have dimmed by this time. James Landman described the tense excitement which used to grip the group when he spoke, standing high on a lighted stage. Now, sobered by the small, humdrum tasks of local administration, he less often reached such heights.

Soon after, on his own island of Baluan, I was to have one more series of encounters with Paliau with which to help me fill in a preliminary picture of what his role had been in the New Way. Osmar White, the Australian journalist who has devoted years of effort to understanding New Guinea, stopped in at Peri, and I went with him to Baluan, thinking it would be fascinating to listen to him—representing Australian readers' queries about what was going on in the islands—talk with Paliau. He had heard all the tales about Paliau's "empire," and he had come after seeing various other sorts of native leaders in New Guinea and Papua.

Osmar White and I were both keenly aware of what we were going to see, what kind of solutions had developed for a governmental plan, a plan brewed from the discussions of the post-war years, from the particular aspirations of a Labour Government in Australia, from the efforts of local anthropologists—who had been advisers to government before the war—to develop standards for self-government, from the considered careful recommendations of experienced administrators. We both knew how fateful for New Guinea, and therefore for Australia, and therefore for the whole free world, were tiny social experiments like this one which was slowly coming into focus as our clumsy work boat lunged and plunged in the southeast and a schizophrenic Baluan woman who thought she was the Virgin Mary sang out of tune against the roaring of the wind.

Had New Guinea a political future? Could its hundreds of tribesmen be transformed fast enough into members of a modern society to which they could feel and show loyalty? And how was it to be done? By the Australian government, whose annual appropriation for the Territory of over four million pounds was still hopelessly inadequate to the need?* By private industry, by opening up lumbering in the rain forests, or by some new type of agriculture, all activities which would draw on the people of New Guinea as a whole as a labour force? Were the thousands of tiny communities, at present tied to their land by government policy, to be replaced by a territory-wide labour force? This Osmar White felt was a crucial question. He saw as a test case what was happening on Baluan, where the government had met the native demand for local schools by giving them a school—through which the worried officials said the people would educate themselves out of existence. Would the people, once educated, be willing to go away and work? What did they think of the Administration, of Australia? What did they want? What did they understand of all that was happening? Both of us knew that some of the answers which we might find on Baluan were terribly important to the free world, from the fate of the people of New Guinea to the fate of the inhabitants of Sydney, London, Oslo, New York, Paris, Djakarta, New Delhi.

Yet it seemed hard to believe that this pretty little island, framed against landscaped roads and flowering trees, with the young Native Affairs' Officer, James Landman, standing on the dock, could possibly provide answers to problems of such import. I wondered, as the smiling, lean young officer extended a hand to help me clamber ashore, saying that I must be me, how he saw himself in this world drama about which Osmar White was going to write and about which I was bent on collecting as rich and vivid detail as possible. The well-worn joke, "It must have been two other fellows," is only one side of an unbearably momentous coin—"Who was there?" at any given moment in history. Paliau's presence had transformed a movement which without him would have flickered out into discontent into something with enough vitality to echo up the political stairways of the world, to the very top. What had James Landman's presence done to make a success

*See Appendix II for details.

of the attempt to regularize a movement that might have been a disastrous rebellion into a steady social experiment, without robbing it of its dynamic impetus for change?

Everywhere in the world where those who had power met those who were asking for it, where those who were the inheritors of thousands of years of literacy came face to face with men who were just learning what writing meant, where men in various kinds of suits were face to face with people of another ilk, whether they were of another race or class or region, who were demanding "the suit you are wearing," the problem that James Landman faced was agitating the councils of the great. How to do it, how to give to movements which owed their strength to the very fact that they were spontaneous the kind of help, the kind of pattern of contact within which they could actually realize what they so passionately desired? It was so easy to promise such fulfillment, so terribly hard to deliver any part of it. And a democratic state cannot liquidate those who realize how slowly a dream comes true.

When the council—the careful legal structure designed to give groups of natives local self-government and power for co-operative economic ventures—had been set up, James Landman happened to be an officer with the right degree of experience for the new type of appointment, and he happened to have been in Manus earlier. These two conditions had determined the choice, out of all the available personnel of the government service, of a man who was peculiarly able to respond enthusiastically to the challenge of guiding people who had been branded by every epithet—subversive, heretical, totalitarian—who, merely as a political move to keep them quiet, were being given a council long before they were "ready for it," at least to the minds of experts on councils. And it can be regarded only as a historical accident that Jim Landman had a wife who had been trained as a teacher and who responded with equal enthusiasm, with freshness and delight, to the task of making a group of small native children literate. There were other felicitous factors. Jim was an enthusiastic builder and landscaper, so that the little capital, with its new

council school and store and council house, was built of a beautiful combination of native materials and good design and was set
in a re-landscaped station on the prettiest island in the Admiralties. And the Landmans had a baby boy, who was a focus of affection and concern of all the people of Baluan.

So we came ashore and talked far into the night, after Jim and
I had both commended Osmar White to Paliau and the other
officials and he had been given a chance to walk about and talk
with them alone. The next morning there was a full-dress discussion in the council house, designed to give physical expression to
democracy, with a long, curved, half-moon table, around which
the councilmen, Paliau as president, Osmar White, Jim Landman, and I sat. This in itself represented a tremendous achievement: a government official who lost no jot of the dignity and
responsibility of his office had created a situation, and embodied
it in a form of architecture, in which he and the natives sat down
together on the same level and talked.

Here I saw in miniature, as we sat around that table, the
problem that faces the world. Here sat Paliau, with a mind as
gifted as that of men who have led millions and changed the face
of the earth, able to speak no wider language than the local lingua
franca—Neo-Melanesian—supported by a group of Baluan
elected officials with their narrowness and stubbornness and
alternation between blind devotion to him as a great leader and
petty factional hatred against him because he wanted some minor
change or had looked at or been looked at by a local woman.
Here sat Paliau, a prophet in his own country, his wings very
surely clipped. There sat Osmar White, with his self-chosen task
of interpreting the position of countries like New Guinea and
interpreting it to Australia, with his self-elected concern with the
political fate of his part of the world, which included the little
group of officials around the table. And there was Jim Landman,
who by the accident of war had first encountered a native people
as a guerrilla fighter in Timor, where he was dependent for his
very life on the personal loyalty of people whose lives he endangered. He brought to his task a kind of understanding unusual for

the service in which men customarily began their bureaucratic careers with extraordinary authority for which they were unprepared. And there I sat, alternating with Jim in translating for Osmar, who understood but preferred not to speak in Neo-Melanesian, matching the new against the old, bridging a quarter-century of the most dramatic change that my world, and Paliau's world, had ever known.

It became clear what Jim meant when he expressed boredom and annoyance with Paliau—a thorn in the flesh locally because of his confinement to the too-local scene—at the same time recognizing that Paliau's organizing talents were not being used as they might be. But it was also clear that he was grateful for Paliau's often wearisome gift of explaining over and over again a point that Jim had made and only Paliau had understood. It was clear too that the short-sighted policy of divide and rule—which had cut over half of the Manus people out of the council and so given the Baluans a disproportionate role which they were in no sense up to—had been almost fatal.

Jim had been given the task of establishing a council on Baluan. The Manus people of Mouk, whom Paliau had invited ashore on Baluan to live on his ancestral grounds, were still strangers to Baluan, fellow citizens only in name and in Paliau's dream. Instead of being caught up in the impetus of the Manus picture of an organized world, the dream which Paliau had energized, Jim had been held down in a backwater of the Baluan community, itself split among Paliau followers, Seventh Day Adventists, who only most reluctantly participated, and disgruntled local factionists hoping for the return of the Catholic Mission. A failure on the part of the Administration to take even the most rudimentary cognizance of the ethnological situation on Manus had established a situation which made Jim's task many times as difficult and many times as unlikely to succeed. But meanwhile, Jim's patient demonstration of how democratic government worked, how men sat around a table, how a subject was threshed out, was a daily model for Paliau, who, in turn, could carry it over and explain it to others as I had seen him do in Bunai. And, while we sat and talked, we could hear the school children's voices from the hilltop where Marjorie Landman had created a school. She made her own adaptations of Neo-Melanesian—which the education department was not permitted to make—and day by day, in the steaming heat, with no siesta, taught the first group of children the literacy which made it possible for the fifth grade to write compositions in English. It was all part of the patient demonstration hour after hour, day after day, of what lay behind the way of life of the Western world which the Manus had glimpsed, but the mechanics of which they did not yet understand.

"Ask them," said Osmar White, "whether they think if there were no government here, could they go on with this new way of life they have started." The uncomfortable local officials squirmed a little, as might the undistinguished dignitaries of any small town, unexpectedly called into the high councils of the land. And Paliau answered, "No! Behind the council there must be force, courts, police, armies. If such force did not exist we would fall apart and fight among ourselves." "Ask them," said Osmar White, "if they think they can do everything themselves." And one of the local Baluan officials answered, "No, we must cry to Australia to help us."

"Cry to Australia." It was the first time I had heard the phrase, and it struck me as most unpleasant, with its note of plaintive petition. These were indeed not Manus speaking. The men who had formed the strength of Paliau's system were too proud and too autonomous to cry to anyone. These were Baluans, men of a less definite pride and strength, who had in their confusion exhibited the kind of behaviour which is so frequent the world over among natives divided among themselves by echoes from the wider world.

And suddenly I realized something else. While taxation without representation is tyranny, and taxation designed only to force a native people to seek European employment is a form of slavery, *government without taxation is degradation.* Territory taxes had never been reinstated since the war. What else could the people of Baluan do except "cry" to Australia to take pity on them and help them? Untaxed, they had no right to ask such help as free men. In the little local council they did pay taxes, they ran their store, they could pass local ordinances. And Jim Landman sat with them, patiently explaining the relationship between taxation, voting, and community decision. The council system had been planned to fit in with a reinstatement of Territory-wide head-taxes, so that those who were far enough advanced could pay local taxes instead of territory taxes. The territory tax was not reinstated, and here in Baluan the weeks of patient labour, in which Paliau's original organizing vision and Jim Landman's patient exemplification were possibly being welded into a new model of democracy, were being endangered by action at a higher government level, action taken in an out-of-date climate of opinion about taxes and salt mines, because the people were not being required to make an explicit contribution to the government of the Territory.

The whole drama of the modern world was spread out around that table. Here we had the historical situation of the gifted native leader who had crystallized a movement of tremendous possibilities, who, if given scope, might have done much to give a sense of unity and purpose to the whole of the Bismarck archipelago—the Admiralties, New Britain, and New Ireland—the Territory officer in the whole group of officers available who, with his wife, was most fitted and willing to undertake the slow, thankless task of translating dream into day-by-day practice, the journalist who had taken the trouble to understand New Guinea and had the ability and the audience to write about it, the anthropologist who had had an opportunity to study this particular situation before and after.

And yet, were we among us, each of us as unlikely ingredients in the situation as a viable mutation in a colony of fruit flies, going to be able to make anything at all out of this curious historical accident which had brought us together? Paliau, seen as subversive, denounced by Church and State, could easily make the front pages of Australian newspapers or set the Secretariat of the United Nations a job of memo writing and explaining. He could be played up as the harem-keeping mogul of a "cargo cult" so that disordered female newspaper readers wrote letters to the "King Farouk of the Admiralties." The corridors of communication are open so that news of a political invention made anywhere in the world can echo from one end of the system to another, and men come out of tunnels dug under the flooded rice fields of Indo-China to set the world by the ears. But have we any technique adequate to ensure that a felicitous historical accident—the combination of a gifted man, a people set on fire by the vision of another way of life, a government plan which was at least halfway ready to receive them into a legal fold, and a government official especially suited to the task which needed a dedication far beyond the call of duty—will echo and re-echo around the world?

"I have come," Osmar White explained to them, "to find out the truth about you. I do not belong to the government. I can promise you nothing. I can only promise to tell the people of Australia what I learn, and they can tell their government what they would like to have it do." And I remembered a conversation we had had—several men of Peri and I—coming back from Bunai after Paliau's explanation of the latest details of democratic procedure. "Which is better," I asked, "to have disagreement, people on both sides of a question, or to have everyone agree?" They had learned the lesson—Paliau was a consummate expositor—and they answered, "Both sides must always be heard, both sides must speak, and those who are the fewer must agree to do what those who are more want. *But of course it would be better if everyone agreed to the right thing.*"

Democratic procedure, yes, it was part of the picture; it was the kind of political pattern which made the Americans so many, which kept their babies alive, which kept people from quarrelling, but wasn't there some way, some quick, forceful way of keeping things "straight"? Unequivocally devoted to our way of life, and convinced that they were learning the way we did things, they yearned nevertheless for a speedier and more decisive method. And what of the peoples of the world who have seen the West only in destructive terms, who have no faith in Western procedures dramatized by a million men, taught by a Jim Landman, and propagandized by a political genius like Paliau?

We will need many more detailed studies before the role of a personality like Paliau's will become clear to us. To be as intrinsically superior to his fellows as Paliau was to every other one of his people lays a burden of loneliness on any man. It establishes a sense of precipice, of distance, which may lead to madness and despair, or to greatness. Paliau sees his own childhood in such terms, fatherless, remote, and as a boy too young for the police work for which he succeeded in volunteering, sitting by himself, wondering. All his imaginative enthusiasm was for the Manus, to whom he proudly traced one ancestral line, and he saw them as a great and terribly unfortunate people because they had no land.

For his own people of Baluan he had far less enthusiasm. These were his grasping relatives who stripped him of his earnings each time he returned from work. When he went away to work, the people of his village had not yet been fully Christianized, and Paliau, alone of all the people he led, had never been christened and had no Christian name. He had been trained in no mission school, and during his work-boy days his experience of missions had been mixed and unsystematic. He had learned to write from no teacher who carried the authority of an orderly education, a right to teach the mysteries of education. Instead, with help from his fellow police boys, he had taught himself to write and had developed not the characterless printing or script of an unsuccessful schoolboy which most Manus use but a distinctive script and a real signature. Even his most ambitious organizing experience, providing for the homeless masses of native labourers left over in New Britain after the Japanese conquest of Rabaul, he saw as something which he had to work out himself.

PALIAU'S SIGNATURE

Perhaps most important of all, he had practically no contact with the Americans. The experience of the American way of life which had caught the imagination of the people whom he led was unreal to him. His vision of a world remade came from the same sources as the ideals which his countrymen sensed behind the imperfect actualities of American behaviour—human brotherhood, human consent to be governed, the dignity of each individual man, and the benefits of civilization, education, medicine, law. But his was a distillation from other sources, mediated by his long experience of working with Australians. So he brought his impassioned pleas for a New Way to a people whose imagination had been quickened by a different reality—the American occupation of Manus—in which he had no part.

Speaking in a language which was not his mother tongue, Manus rather than Baluan, he led a stranger people, whose spirit he admired extravagantly, whose landless fate he pitied far more than it deserved, whose vision of a world in which all children who were born lived, and no man's hand was turned against another, and no human being sacrificed for gain, he met with his vision of a unified people, whom he must lead toward a limited earthly paradise to be realistically attained only by hard work and controlled behaviour.

9

what happened, 1946–1953

It had taken just seven years to turn this little corner of the Admiralties from a remote, sleepy, forgotten bit of a territory, which had only itself been put on the map since the war, into a political actor on the world stage. We have seen how in the Manus cultural character we had found a point of leverage for change in the discontinuity between the happy comradeship of childhood and adolescence and the realities of adult life, but a point which without a changed external situation might never have been used. The Manus confronted with American culture could have been expected to respond positively. This was the one completely predictable element, given enough scientific knowledge of the way cultural characters are organized. Their delight in mechanical things, their sense of organization, their tendency to treat human beings both humanly and mechanically, their flexible here-and-now approach, their zest and optimism, their concern for children—all these were elements which would predispose them to appreciate American culture.

What we could not predict, except as a long-term possibility, was the course of World War II, and the choice of Manus as a battleground and a large base. What we could still less predict was that here, and not in one of many hundreds of other native tribes, there would be at the historical moment a man of Paliau's political genius. Even given the cultural character, the political situation, and the native leader, the actual course of events in Manus was still extraordinarily subject to day-by-day events: *which* patrol officer was sent on a patrol or relief, *who* was District Commissioner for Manus, *which* priests were in charge of the Mission, *which* plantation manager was placed on Pak or Ndropwa. There were so few people involved, they were so loosely tied up with one another, that it was important what religion each European confessed, or had once confessed, or how much trouble could be made in the Baluan and Manus communities by the personality of one woman or another. If they are seen as just one group of native villages out of thousands, one can think of the accidents and personal jealousies and misunderstandings of the illiterate or the unstable as simply one example of the way small face-to-face communities work, the world over. But seen, instead, in a world context, with the fate of the Paliau movement affecting the political future of all of New Guinea, and, by virtue of the significance of New Guinea to it, the political future of Australia and of the whole free world, it becomes more like the annals of those small significant groups of men who have made world changes, who were originally a handful gathered around a leader—in Siberia, in Vienna, in Gandhi's Wardha—too close to one another for detachment, struggling for the same office, or for the love of the same woman, or the favour of the same man, breaking with one another forever over some small point of belief or practice.

These were the men: Samol, the leader of Bunai; Lukas, the leader of Mouk and Paliau's principal rival in the Baluan Council; Banyalo of Peri, the sulky little clerk who, taken away from Peri to school in Rabaul in the early twenties, had lived all his life in Rabaul and who became a leader because he possessed the essential skills of keeping records and knowing what a school room was like; Napo, the leader of Mbuke, who had been a police boy so trustworthy that he had maintained an isolated police station on the mainland of New Guinea before the war; Kilipak, who had the brains and the authority to lead Peri, but was restless and restive, periodically throwing it all away, and counterpointing them, the unstable and the illiterate, the dreamers, preachers, and visionaries seeking the quick mystically rewarding solution.

It seems important to try to identify the point at which—given the culture, the wider historical situation, and the personality of the leader—the movement came into being so that, although its particular course hung by a thread and could be changed by a canoe going to one island instead of another, it was launched in such a way that all those who had participated in it would never be the same again but would carry deep shared traces which would become an element in their culture.

We may first consider what indications there are that some such movement would have resulted from contact between Manus and Europeans had there been no World War II and no Paliau. Here we have the whole history of European New Guinea contact to draw on. Since the beginning of contact there had been "cargo cults"[1]—mystical outbreaks in which a local prophet commanded the people to kill their pigs, destroy their property, and wait for the cargo. People having seizures had been an accompaniment of many of these mystical cults; one and all they had fizzled out, sometimes by the government interfering and sending the prophet to jail, sometimes by the people themselves turning against the leader of the cult. There had been other related semi-religious movements with local leaders who spread rumours that Jesus Christ was in Sydney, wanting to come to New Guinea but unable to do so because he had no clothes, so they took up a collection; earlier still, a local leader who had the secret of bulletproof shirts. The familiar ferment of half-abandoned old and half-understood new out of which religious cults spring was all there.

At the end of the thirties, somewhere in early 1939, Napo, a police boy from the village of Mbuke, went home to Manus on leave and began advocating modernization, elimination of the old marital-exchange system, and adoption of working only for money, buying European goods, and dressing like Europeans. He reports that he was preached against up and down the South Coast, denounced both by the Mission priests and by his own people. Here, then, was a moment at which something might have happened—a strong able man presented a program and the program was rejected and rejected dramatically enough to have possibly created an issue out of which a movement could be born. But none was. Napo's leave ended; he went back to New Guinea, returned to make a new effort at reform and was interrupted by the war.

Sometime before the war, the young Paliau, also a police boy who had begun his career as a very young police recruit in the famous inland campaign against the Kukakukas under Allan Roberts—who is now Director of District Services in the Territory—came home on leave. Annoyed with the practice by which men who could not pay their taxes were sent to jail, he used all his most recent accumulated earnings—which he had succeeded in keeping out of the grip of his relatives—to set up a revolving fund, administered by the administration-appointed village officials, to keep those in default of taxes out of jail. This was an extraordinary move for a young uneducated native coming from a tiny island like Baluan, which was divided into several politically disconnected groups. Had there been a historian present, these two moves made by Napo and Paliau might have seemed precursors of change: both were economic, and both focused on money and money relationships to Europeans. Paliau went back to Rabaul, where he was now a police sergeant, and spent the war back of the Japanese lines.

When the Japanese invasion came, it set up for the duration a line between those Admiralty Island age-mates who had returned to their villages—either on leave for a brief spell between periods of indenture or permanently—and those who were away at work. Events beyond the control of the Manus also determined who were back of the Japanese lines and who back of the Allied lines. Those who were in the Admiralties had the most massive exposure to the Americans only, but all were exposed to large military establishments. During the war, men back of the lines, some experiencing—like Johanis Matawai—first the Japanese and then the Americans, some the Australians only, some never seeing an Allied or a Japanese unit, talked about what would happen after the war. Two themes seem to have been important: economic reform, especially the abolition of the old system of marriage exchanges which was seen as exploitive of the young men and limiting because native currency bought only native goods, and racial equality. No one has yet traced where the emphasis on all men being brothers, regardless of race, came from. The general idea of the Brotherhood of Man certainly came from the Mission, but the phraseology used, the frequency of the word *blackman* rather than *kanaka*, and phrases like "all men of whatever colour, black, white, red, or green," suggests some point of propaganda, possibly an underground Chinese anti-Japanese weapon, or possibly an offshoot of some Japanese attempt to get the natives to identify with them against the *white* man, rather than something they learned from the Allies. The official view of the war as presented to the people of New Guinea by the Australian military administration officers, who could talk easily with them in Neo-Melanesian—in which communication is really symmetrical and mutually comprehensible—was simply, "The Japanese are bad. Kill them." This was reinforced with "We are coming back." There seems to have been no ideological presentation more complicated.[2]

The Manus theory of why the war was fought is remarkably simple and is based, as is their version of the War of the Angels— a story they prefer to the Fall of Man—on their conceptions of how human groups are motivated. Originally the English—in the person of Captain Cook—discovered New Guinea, and the English were going to come and occupy it. But they didn't hurry enough, and the Germans got there ahead of them. The English, however, kept on remembering that they were the ones who had discovered New Guinea, so in World War I they came and took New Guinea back. But then the Germans couldn't forget that they had had New Guinea, so in World War II they put the Japanese up to trying to get New Guinea back for them. Because there were so many Japanese, the Americans came to help the Australians drive them out. The Americans, however, didn't want New Guinea; they only wanted to straighten things out. The religious version of this same history, as presented by Paliau to his followers and as elaborated by them, represented God as sending one people after another to make the people of New Guinea "all right," that is, like the rest of the civilized world. And each in turn failed. The Germans treated the natives like "trucks," the Australians like "oxen," and when the Americans came, the Australians blocked their way. Finally God chose Paliau to accomplish this mission.

The period just before the war is spoken of as "When it was still a good time," but it is a period about which people can remember very little as a period, although they can relate long episodes in which they were personally involved. The transition culture of the thirties, with its combination of resident catechist, visiting priest, christening, confession, and communion, seems to have been in some ways very fluid—so that an event which impressed one person did not impress another—as compared to what is remembered of the period before 1930, in which a sure sense of what would have been done comes to the rescue of any one memory of what was done.*

Concerning old events before the native religious system was rejected, they could remember everything except the actual number of objects exchanged. They would refuse to falsify these facts, so that it was impossible to get them to improvise a speech about a feast in the past. In the late thirties, the culture, for all its appearance of good adjustment, was evidently moving toward the state in which each man's view of what was happening was more dependent upon his own individual experience, lacking the stabilizing effect of a great body of traditional behaviour. Carefully as they had attempted to master the rudiments of Catholic doctrine

---

*So Kilipak could tell the story of his long attack of dysentery, of who nursed him, in what house he stayed, several times without mentioning a ceremony called by the old name, *kano*, which was a Christian version of the old expiatory payment to an injured ghost, in which property was offered in the church and then given to the orphans of the village, Pomat, who had nursed Kilipak through the illness, could not remember this ceremony at all, even when his memory was jogged by details.

and practice, each selected different details from a body of religious knowledge so little understood and so much more complicated than anything they had ever known. Small rapprochements between the new faith and the old, complicated in turn by beliefs brought by returned work boys, by the former religious beliefs of catechists who were after all only very recently trained, all were present to individualize what they knew.

The fate of the institution of godparents is an example of this fluidity. In the early days when the adult population were being admitted to the Church, the first godparents came from among those already baptized on adjacent islands. In the early church records[3] for Peri, the name of the catechist appears frequently as a godfather. Only gradually, as Peri men and women became church members do they appear as the godparents of their friends and of their kinsmen's children. I was initially puzzled when I talked over the godparent position with the resident priest on the main island because he emphasized that the Manus made the relationship "too materialistic," and hadn't understood its spiritual side, and he expressed in general a feeling that too much emphasis on godparents was not wise. Yet in Latin America the godparent relationship and the co-godparent relationship have been principal bits of social structure in giving coherence to a Christian society built on an older base. Today, Peri people know who were their godparents and who were their godchildren, and when these were faraway people the relationships have been transmuted into trade friends, people with whom you exchange favours on a very long-time basis, while within the village having been some child's godparent is a reinforcement of kinship and co-operation. It was the distant godparent, introduced apparently during the period of converting adults, whom the Manus assimilated to their trading pattern, the only pattern they had for handling relationships to non-kin members of other villages or other language groups. It was the trade-friend pattern which the missionary had objected to as "materialistic." This is the same pattern to which friendships formed between work boys from different parts of Manus are referred, so that nowadays a Peri man may have three separate categories of trade friends: those he inherited from his father, those he made while working, and either a godfather or a godchild. In 1928, everyone knew the locations, and usually the names, of everyone else's hereditary trade friends, and these are still known today. But the other two categories—godchild or godfather, and work friend—are only known to a man himself, and possibly one or two close kin. Here again the new pattern was too unstable, too unformalized, to be easily grasped and remembered. Another generation of church membership might have brought about a coherent pattern in which those who had been godchildren became in turn the god-parents of their godparents' children, or cross-cousins or father's sisters and mother's brothers might have assumed the role. Then people would have found it easier to remember about others as well as themselves—who was whose godparent or godchild.

Evidence of this sort suggests that the late thirties was a period when each man was getting more out of step with each other man, when the grounds on which decisions were taken, as between the traditional economic arrangements of the culture and a series of casual or purposive attempts to alter them, were becoming more confused. Money had been substituted in many marriage payments, in huge sums which were regarded as very irksome. The European understanding of affinal exchanges as "bride price" had permeated the native idea of exchange valida-tion more and more. People were responding to the comment of "You people pay too much for women," which came up in court cases when there was a quarrel. As only the "hard side" of these marriage transactions came up in court—"I paid 10,000 dog's teeth for this girl to be my son's wife; she has run away with someone else, and I want my money back"—it was natural enough that a government official should comment on the high cost of women. The return transactions in food, grass skirts, etc., tended to fall out of the picture, although they were actually the part of the exchange which was essential to the Manus economy, because it was the food and other manufactured articles which the bride's family gave in return for the heavy payments that were made for the bride and for each child she bore which kept the people well fed.

This changing picture of the economy was also affected by the ideal of a Christian marriage advanced by the Church, with a disallowance of divorce, not because it upset financial arrangements, which had been the old Manus objection, but because marriage was a sacrament. The old taboos which made it necessary for women to go about hidden under pandanus rain mats or muffled in long calico robes seem to have been discouraged by the Mission, although photographs of them are recognized for what they were by people who were young children just before the war.

It is impossible to reconstruct a clear picture of this very unclear period, but it seems certain that people were getting out of step with each other, and this sense of being out of step became an important component in what happened later. During the war work boys behind the lines clung together, not only under the impact of the war and the protracted absence from their villages, but also in shared plans of how they were going to change things after the war, modernize them and bring them up to date.

The war period had one other effect; it kept many men away at work longer than would otherwise have been the case, and so slowed down the transfer of financial power from one generation to the next. By 1945, under the old conditions, Kilipak, Pomat, Manuwai, Lokus, would all have paid for their own marriages and have become heavily involved in preliminary payments for the marriages of other men. It was the payment for the marriages of younger men which gave the enterprising man a "vested interest" in the system. In terms of social change he was the man who would lose by change and would resist it, as the landowner, the retired pensioner, the people with fixed incomes from investments are motivated to resist change in more complicated societies. It was involvement in financial transactions which turned the companionship of childhood and adolescence into the quarrelsome rivalry of "big men," and, in the ordinary course of events, this would have already occurred for the younger men of the village, the men who would be the leaders after the war.

How important it was that in 1946 there were a group of strong, vigorous young men with the character and will to inaugurate activity, and who were still free of the entanglements of financial investment in younger marriages, is sharply illustrated by cases where this was not so. Just after the war, Kilipak paid for the wife of his younger brother, Karol Matawai, and paid for her in the old style, with lavish exchanges of the old kind of valuables. This marriage fared badly from the start. Kilipak expected from Karol the kind of service which a young man owes his backer. Karol, once the new way of life, with freedom of choice in marriage, was introduced, resented having been married without his own desires being consulted. When the marriage was finally terminated by a civil divorce, the child given to the mother to care for, and Karol Matawai remarried, it continued to vex the village in 1953 in a way quite different from the complications of quarrels and divorces under the new system. Kilipak's expectations from Karol were greater than the new system could stand. Kilipak continued to treat the divorced wife as if she were still a member of the family, and even let her walk in the funeral procession for the burial of her ex-husband's mother. Quarrels over who was to have the child were recurrent. But Kilipak, still feeling himself involved, also became disproportionately furious at Manoi—then an elected official—when he opposed Karol's second marriage. Karol got himself a jail sentence for disregarding a waiting time, imposed by the government officer, for the second marriage. The wife's remarriage was particularly unstable, and even bringing Paliau's authority into it failed to stabilize it.

Meanwhile, Karol remained strangely dependent for his age and temperament, caught in a phantom relationship to his older brother which neither of them knew how to liquidate. He couldn't repay his brother for a backing which no longer existed. His brother was forbidden by the rules of the New Way to invest money in a new marriage, and invested an undue amount of

quarrelsome emotion instead. He couldn't establish his independence either by paying his brother back or by taking over the
financing of his own marriage, for his new marriage didn't
require that sort of financing. There the two were stuck, getting
into meaningless quarrels; Kilipak, proud, arrogant, adoring
Karol yet "getting ashamed" in his presence, and Karol, sulky,
violent, and confused, giving expensive elaborate little feasts for
relatives whose exact genealogical relationship to himself he
didn't even know, feeling somehow that he was still a part of an
older system which he, however, didn't understand at all. And,
significantly enough, he did not know how to read. The village
could not have stood many such situations; the ghost of this single marriage stretched tempers and co-operation to the limit.

A second instance of conflict between an older man who had
invested heavily for his son and the son existed between Pokanau
and Arnold Bopau, Pokanau's brother's son whom he had reared,
and between Pokanau and Matawai, his own son. Between
Pokanau and Bopau there was a complete break, given vividness
by the hangover of childhood resentment. Pokanau's son, Johanis
Matawai, had stayed away at work for ten years and then had
come back to the village with a chronic illness which he had been
told by European doctors was "something belong ground," interpreted to mean "due to sorcery from a native of another island."
Pokanau was the strongest and most intelligent old man in Peri,
the man who could talk, the man who had survived into a sturdy
old age, whereas men who had played a more decisive economic
role than he had all died. Although he embraced the new order
enthusiastically his habits belonged to the old. To have subjected
his son, who had been away so long, by buying a wife for him,
very expensively, in the old way, would have been a possible
course of action for him. But both Pokanau and Matawai were
ardent exponents of the New Way. Pokanau took another course;
he maintained his sense of his own position by remaining the sole
support of Matawai and his wife and child, thus reducing them
all to the position of complete dependents, rationalized by
Matawai's recurrent illnesses and the fragility of his two children,

each of whom had been born prematurely and had demanded continual care from the rather passive young mother. So Pokanau remained a strongly youthful man, fishing and trading in all weathers, asserting loudly that he was equal to maintaining the entire household. The habit of subjecting the son was still there. Bopau had dealt with it by complete withdrawal. Matawai, despairing of ever getting well, and fluctuating between gambling and paying large sums to native magicians from other areas who claimed they could cure his illness, became a compliant partner in his father's plan.

The old system had been grounded deep in the character structure of the Manus, in the bullying small child and in the way the devoted father acceded to the child's bullying requests. When Karol was a baby, his tall arrogant father had left a ceremony to come and beg a balloon from us because his baby had cried for it. When Matawai was a small boy, his father took him everywhere, indulged his every whim. But back of this parent-child picture there lay the expectation that some day the father, living or dead, would get his own back, some day the child whose every whim he had indulged would work for him, would pay back this love and attention, while the child, accustomed in babyhood to autonomy and autocratic command, was prepared to chafe under and get out from this temporary dependence on his father. This was the deep-laid pattern which developed while one was a lordly baby, intolerant of any frustration or delay, accustomed to having a father who was a devoted slave and who cuffed mother into a less pliant slavery, and learning to rage and stamp and shout and curse when crossed. The big men and the babies showed the same uncontrolled pattern of behaviour. The more equitable companionship between equals in childhood, which provided the model for the institutions of the New Way, was tenuous in comparison.

It was this system of transfer of power that Napo had challenged just before the war, and it was such systems which were to be challenged in village after village just after the war, by Samol in Bunai, by Manoi in Patusi, by Banyalo in Peri, by Kampo in Lahan, by Lukas of Mouk, by Napo of Mbuke, and by Lungat of

Nriol. In each Manus village there were returned work boys, old
for their junior status, and men who had been too occupied during the war with working for the Japanese or the Americans to get
deep into native financial arrangements, who were now ripe for
some kind of entrepreneurship.

In Peri, the first thing they did after the war was to rebuild
the old village in a new realignment of the pile houses on the
lagoon. Kilipak built a house for Karol and started plans to
finance the marriage of another youth of Peri. Raphael Manuwai
decided to reassert his membership in Tchokanai (a clan which
everyone had regarded as extinct in 1928 because Manuwai's
father, the only male survivor, had been adopted into the clan of
Peri) and went back to build his house in the bit of lagoon which
had belonged to Tchokanai.

Mateus Banyalo came home, retired for age. Banyalo was a
peculiar representative of historical accident, of the fruition of
plans made long ago. In the early twenties a school had been
established in Rabaul and pressure had been put on local district
officers to find boys to send away to school. The Manus people of
the South Coast, who were just coming into real and unenthusi-
astic contact with Europeans, did not wish to send their cherished
small boys away to school. Only the orphaned, the very stupid,
the unstable were selected to go—either to a school where they
were to become literate in English, to be trained as clerks, and
schoolteachers, or to a trade school to learn carpentry, etc. We
analyzed the group who were away at school in 1928—four of
the ten had histories of instability. Banyalo had been at school for
six years when Dr. Fortune and I arrived in Rabaul in 1928. We
were planning to go to Manus and spoke neither Manus nor the
lingua franca, then called Pidgin English. Mr. Chinnery, government anthropologist at the time, suggested that he could get us a
schoolboy to take back to Manus as an interpreter. Banyalo was
produced: a short, stocky, unresponsive boy, conspicuously
dressed in white shirt and shorts, a costume which set him completely apart from the bulk of the native population. He was told
that he was to go with us back to Manus, and I doubt if any effort

was made to involve him in the decision. I know that he did not want to go: he liked school, he was resentful of being taken away from his schoolmates, and only the promise that he would be returned to Rabaul when we left mollified him at all. It is significant how very little I remember about Banyalo, who was our interpreter and original liaison with the village during the whole time we were in Peri. Although he spoke practically no English, he understood just enough to tide us over the first weeks of learning Pidgin, so that when we asked a question about Manus in English, he could answer slowly, with experience of the ways in which Europeans spoke Pidgin English. He worked with Dr. Fortune on texts, with both of us on language. As we became more proficient we dispensed with him more and more, as he understood very little about the culture and treated every task with grudging lack of enthusiasm. He wore khaki shorts and sulked.

In discussing the position of fatherless boys, I wrote:

> The loneliest children in the village were the boys whose fathers were dead. Banyalo was one of these. His father had died when he was seven. He had passed into the care of his father's sister, an old widow living alone. No new man took his father's place. His mother went to live with her brother and later married again. When the recruiting officer came through looking for school children, Banyalo was given to him. Fatherless, there was no one to object to his going. When he returned to the village after six years in Rabaul, he came home as a stranger. His mother he hardly knew. His mother's brother extended a formal welcome to him. He might of course sleep in his house, but he did not feel himself as having a real part in his household. After wandering about from place to place, he finally settled down in the home of his mother's younger sister's husband Lalinge,* who took upon himself the duty of paying for Banyalo's wife. To the constraint and embarrass-

---

*Called Paliau in earlier publications.

ments which belonged to the brother-in-law relationship was added the invidious dependence of the wifeless upon him who bought his wife. Banyalo turned finally to a warm friendship with a younger boy and so staved off his loneliness for a little.

Banyalo seemed to us, watching him through six months of close association, intellectually dull, sullen, lazy, uninterested in anything except getting away from the village in which he felt so alien. He was being trained to be a schoolteacher; he wanted to be a clerk. Characteristically, the only group in the village with whom he associated were the young, just adolescent boys. His particular friend, Kutan, was the one adolescent who did not work for us. (Kutan was a steady, warm, reliable person, who by 1953 had become the father of seven children, Kilipak's best friend, and an able all-round member of the community.) In 1928, we saw Banyalo as an example of wasted money due to the system of recruiting, which meant that a government officer recruiting for the schools got only those boys whom no one valued. Even seen with the perspective of experience with many other interpreters, with their minds confused or bemused by complications of partial European contact, I would still classify Banyalo as the stupidest, least inspired, and least likable of any interpreter I have ever worked with.* When we left Manus we

------

*He contrasts particularly strongly with I Made Kaler, the Balinese secretary without whom the Balinese work would have been very different. I Made Kaler also showed an initial dislike for working among people of his culture. After having accepted what he thought was a job with a visiting anthropologist in Java, where he had gone to get a position as a clerk, he was hauled back (not against his will exactly, only because he had found no job in Java) to work among a group of simple mountain peasants whom he, as an urban, culture-contacted Balinese, looked down upon. At first he made great difficulties, refused to obey "old-fashioned" rules of caste etiquette, but gradually his beautiful mind became fascinated with the task itself, and on our first Christmas in Bali he wrote us a letter saying how much he appreciated the opportunity to become integrated in his own culture.

fulfilled our promise and returned Banyalo to Rabaul, never expecting to hear from him again, and we never did. His English was inadequate for letter-writing, and he was being taught not in English but in Neo-Melanesian, which would have been a language in which he could have written fluently. He seemed to represent a dead end, as if the expense and effort of educating him would result in one more slow, inept, badly educated clerk in a government office, to discourage both the Europeans and the natives with whom he came in contact about the possibilities of a native ever functioning like a European.

I have stressed all of these negative aspects of Banyalo in such detail because Banyalo became the leader of Peri village, one of the important precursors of Paliau's leadership, and an essential link between the old way and the new. Where Paliau's role can be assigned to extraordinary ability, Banyalo's most definitely cannot. His intelligence was well down in the average of his group, and he showed neither extraordinary social nor moral capacity.

In 1953, my first encounter with Banyalo was his name in an educational report which I read in Port Moresby on my way to Manus, which described him as a Manus native with enough education to teach school, who was at present teaching on Rambutjon, but who had been retired from the government service "for age," and was "a sick old man." (At a conservative guess Banyalo was at least ten years younger than I, which made him about forty-two.) In Peri I soon began getting echoes of what Banyalo had done. Banyalo had started the vital statistics records, and had been severely reprimanded by a patrol for "tampering with the village book." Banyalo had tried to start a school for the whole South Coast, on Ndropwa, with the help of an interested wife of the European plantation manager in charge. He had been the first village *besman* [village head in the Paliau movement]. His young cousins, Lokus and Josef Bopau, both spoke of him with great affection and warmth, although they also reported that he had been a terrific gambler right after the war and had gambled away most of his wards' inheritance. Pressed more closely, beneath the warmth one could find violent criticism, as of the time that Banyalo had warned Bopau's wife of an affair which Bopau had religiously kept secret. It was clear that in Rabaul he had become a sort of dean of Manus boys, having lived there longer, and as clerk in a government storehouse in charge of rations, he was in a position to be helpful in many ways.

Later more details came out. It seemed that Banyalo had come home after the war and attempted to organize an economic revolt against the older men. He had gathered the younger men together in secret conclaves, out in canoes, sharing with them a turtle, plotting. All the younger men of any consequence, all of them younger than he, had been interested in his plan, which had centred around making money, getting rid of the old economic system, and working for wages in order to acquire more European goods. But the older men were adamant. Pokanau claimed later this was not because of the plan itself but because of the way Banyalo had proceeded without consulting them. How much the later secret night meetings of the Paliau movement, which had such a subversive aura, are to be attributed to this early style set by Banyalo, it is now impossible to say. But it is important that in this case he was plotting against neither Church nor Administration, but simply against his elders, whose authority he had never accepted. Seven years later, in describing Banyalo's role, the others told me that then, despairing of getting the support of the older men, or of doing anything without them, and after a showdown discussion in which Banyalo laid out betel nut and only ten men took his betel nut, Banyalo got his closest adherents to come with him to work at the Ndropwa plantation. This group also included Karol Manoi, a returned police boy from Patusi who had been in Rabaul with Paliau during the war and who had become the economic agitator of Patusi village at the same time. On Ndropwa, it was said, he planned to set up a school for the whole South Coast.

Meanwhile, before Banyalo took the group away to Ndropwa, Paliau had returned to Baluan and begun his agitation for a reform which was both economic and ideological. Paliau came to Peri to talk with Banyalo about how his reform attempts were proceeding, and all accounts agree that Banyalo told Paliau Peri was a hopeless place in which to accomplish anything, and that he had better go back and work on his own people instead. Paliau returned to Baluan without making any public speech in Peri, and Banyalo took his group off to Ndropwa to work.

This was in effect what I could gather about Banyalo's role from the people of Peri, and from Jim Landman, who commented on his present functioning on Rambutjon by saying that whatever one might think of him, he had by now taught some hundred people to read and write.

The months of my return visit wore on. Plans were made for me to go to Rambutjon, for the government work boat on patrol to bring Banyalo and two other men I wanted to see from Rambutjon to Baluan. Each plan was negated by some new, and characteristic New Guinea, turn of events. The work boat broke down, something would happen to the captain or crew, someone was sick, the volcano erupted again. So it was only a few hours before I left Peri that at last I saw Banyalo, who had finally taken a long dangerous canoe voyage in a bad sea to get to Peri to see me. We picked up contact with the story of the last time he had seen me, in that post office in Rabaul. He told me the story of his life since then, and I had an opportunity to question him about his role in the New Way. He was definitely both senile and ill, his face distorted by some sort of stroke. Beneath it all there showed two qualities: shrewdness—when I asked him what he would like, he asked for a sewing machine, which would benefit him *and* the community—and a kind of mischievous joviality which I think must have been the personality trait which, combined with his technical superiority and working literacy in Neo-Melanesian, had given him his leadership.

His own account gave an even less idealistic picture of his reform movement than had been given by his friends. The idea of a school on Ndropwa, he said, was just a blind. What he had really meant to do was to lure all the younger men away from Peri so that the older men, deprived of any young men or adolescents to help them, would have been forced to give in. This, again, is interesting in the light of later accusations against the schools set up by the New Way, and the accusation that the leaders of the New Way wanted to keep Manus young men from going away to work. Here was a shrewd, but unintelligent, over-educated and malicious man, who because of his education was able to assume leadership among his people. All the way through the development of the movement, we will find evidences of the dangers which inhered in the unintelligent, the uncontrolled, and the uneducated. It seems reasonably certain that the Administration would have been well-advised years ago to take greater pains to select boys for their intelligence and chance of reintegration with their own society.

I have described this slow piecing together of Banyalo's role in such detail to give some idea of how slowly the picture of the years between 1946 and 1953 emerged. The people of Peri were partly secretive and partly ashamed of the early excesses of the movement, particularly of The Noise—the "cargo-cult" aspect in which they had thrown much of their old property into the sea. They had made strong efforts to separate the cult and religious phases of the movement. In Peri, leaders of the council and preachers in the church were regarded as separate officers, only crossing over in terms of Peranis Cholai, who had been nominated by Peri and trained by Paliau to be one of the new lay preachers. Later, after Banyalo left Peri, Peranis had been selected as schoolteacher, and later still, he was elected clerk of the whole South Coast.

Furthermore, the Peri people recognized the extent to which I could sympathize with and share their secular ambitions to become part of the modern world, and shrewdly judged that I would be less sympathetic toward survivals of a belief in a mystical arrival of cargo, a belief that actually still did survive even among the most level-headed. So it was only very slowly that the history of what happened in Peri could be related to the history of the Paliau movement and integrated with the research that Ted Schwartz did after I left Manus. The account that follows is the result of a mass of comparing, reinterviewing, matching accounts given by natives and Europeans, and is a very much abbreviated statement of the details which Ted Schwartz will publish.

After the war, then, in each Manus village, although more highly developed in some than in others, there was discontent on the part of one or more of the returned work boys, and in various ways they attempted to persuade or coerce their elders into accepting a change. Samol of Bunai, a catechist and heir to the old paramount *luluai* on the South Coast, actually led a group of young men away and founded a new settlement. Banyalo took a group to work on Ndropwa. Each village had a slightly different emphasis on which aspects of the old system they wished to destroy—the expensive marital exchanges, the dominance of the older men in economic arrangements, the use of native money, traditional clothing, and kinship observances, etc. No one of these programs was inclusive, and all of them were more negative than positive.

Into this ferment came Paliau, who began a series of meetings on Baluan, with the Mouk Manus as his strongest supporters. Paliau began to outline his plans for the future—a complete repudiation of the old way and a break with the mixed unsatisfactory culture-contact way of life. The mixture of half-explained Christianity and the old dark ways of the pre-contact world were described by Paliau as a "poison" that was killing the people. Instead of becoming more long-lived and healthy, they were, he felt, actually dying out.* Abortion was spreading, children were dying, the people were becoming fewer. In the Baluan meetings Paliau harangued the people day after day, establishing a religious sanction by giving them his version of Christian Truth, which had been hidden from the New Guinea people by an order of some remote Western government which forced the missions to hide truth in many ways, bury it in metaphor, divide it up among several sects each of whom brought only a part. In these meetings Paliau also laid down the main design of his movement. All the people of the Admiralties were to work together; the old lines

---

*No evidence on this point.

between one people and another were to be broken down; the sea-dwelling Manus were to move ashore; a treasury was to be built up for future economic improvement; the old bad culture was to be completely abandoned for a new culture which would bring the natives into world culture.

From the accounts of this period it seems evident that Paliau had no very clear idea of how all this was to be brought about beyond the need for saving money, organizing, invoking a wider unity, resisting the mission teachings of conservatism. He was picturing a future in which the people of Manus would act, think, dress, and live as Europeans, but the means were vague. After his experiences in Rabaul after the war, he seems to have had little hope of help from white men of any sort.

The people of Mouk began preparing for the new way of life, wearing European clothes and going about looking very secretive and important. Gradually the word spread and people from Ndriol, a Manus settlement on Rambutjon Island, and from Mbuke Island, Tawi, Loitja, Bunai, and Pak went to Mouk to find out what was happening. But no one from Peri or Patusi was involved in this early period. Most of the more intelligent young men were away from these villages, working with Banyalo on Ndropwa or with Samol, who had led a similar insurgent group off to form a splinter village from Bunai.

Then, out of the intensity of a plan which contained as yet no apocalyptic elements beyond the sanction given by a reinterpretation of Christianity, there developed on the periphery, in faraway Rambutjon, a full-blown mystical movement complete with prophet, seizures, promises of immediate delivery of cargo sent by God and the spirits of the ancestors, and a demand that all present property be destroyed. The Rambutjon prophet, Wapi, was killed by his brothers after no cargo materialized, but not before the main outlines of his revelation and the manifestations of violent trembling ending in loss of control, visual and auditory hallucinations of planes and ships arriving with a cargo of European goods, and the excited pitching of property into the sea, had been communicated to the people of Baluan, Mouk, Mbuke, and a little later Tawi and Loitja. On Baluan, Paliau himself was caught up in the mystical phase of the cult for several days before he pronounced against it.

Peri had been prepared for something portentous to happen by vague rumours of what was going on in Baluan, although no one went to Baluan, as those who might have gone were still absorbed in Banyalo's scheme. News of the cult outbreak on Rambutjon and Mouk reached Peri after a Tawi native had visited Patusi, which was after the full-scale cult outbreak in Tawi, and after a Bunai woman, Piloan, returning from Tawi, spread the word that the cargo *had already arrived* in Mouk, Rambutjon, and Tawi, and was imminent for them.

This woman, Piloan of Bunai, arrived first in Patusi, only a half-hour's canoe trip from Peri, and claimed that she herself had seen the ships anchored at Tawi and unloading their cargo as she left. In Patusi, she told them that they must go ahead and destroy their property as their cargo was due to arrive at any moment. So the people of Patusi threw their property into the sea. That night footsteps and voices of the dead were heard on the little artificial island in Patusi lagoon, and the next morning after church the seizures began. The second man seized, speaking as from the spirits of the dead, told them that everything they had thrown away would be replaced with money, and while he was still speaking his wife shouted from the house that a pound note had miraculously appeared on the table. Later an additional ten-shilling note materialized.

Some Peri men who were in the village saw this money and brought back the account of it. The news spread to the young men of Patusi working on Ndropwa, who went first to Tawi to investigate the ship story, initially very skeptical but convinced by the money. They returned to Ndropwa and reported to the plantation manager that they had seen no ships, and then left his employ to return to Patusi, where such exciting things were going on. But Karol Manoi, whom I have called the New Manus man and who was councillor-elect of Peri while I was there, did not go along. He was still skeptical, and he threatened to report them to the government. That night an old-style séance, in which the ghosts communicated with the living through whistling, which was then interpreted by the medium, was held with Popei, the dead brother of Karol Manoi, speaking to them. Manoi acted as interpreter and grew exceedingly angry when the ghost indicated by whistling that the plan to become modern by working and saving their money was wrong, and that no government officer would come in response to his complaint. Manoi left that night, angry, and he took back to Ndropwa the miraculously sent one pound, ten, as it was said to have come from his brother!

The Peri men who had followed Banyalo were still on Ndropwa. In Peri were only the older and more conservative men who had refused to go along with Banyalo's plans of modernization. It was these men who had totally resisted who were bowled over by the promise of the cargo, illustrating a familiar occurrence in hypnosis whereby the strongly resistant is often the easiest subject to work with.

Piloan brought her message to Peri. The message was simply that the cargo had arrived; they had only to make the necessary preparations:

"There is a ship with many 'black men' of Tawi on board. It is very big. It has already anchored. Tawi village is completely filled with cargo. We saw all of this when we left. When we were near Loitja we saw many more ships running beyond the reef. There is one ship for each village. Our ships are on the way. Tawi's had already arrived. These ships are bringing the cargo and everything that belongs to you. Listen, people of Peri, many big ships are coming. All of our people who have died are now coming to us. The cargo has already been landed in Tawi. Why haven't the ships come in here? We are blocked by all the things of the past that we own. All of these things of ours are like a reef keeping out the ship. The ships cannot come inside. If you throw away everything, then the ships will come with your cargo. When the ships unload the cargo your village will be so full that you will have no room to walk. Your houses will be full."

So The Noise came to Peri, not as a mystical religious seizure in which people felt themselves shaken by an unseen power, but as the practical preparation for the certain arrival of a wonderful cargo of European goods, which had already arrived elsewhere—ships, planes, machines, food—a cargo sent by God especially to the people of Manus.

Once before, within the easy memory of all the adults, the people of Peri had prepared—for the coming of Christianity—by throwing out the skulls of their dead. Only by making a clean sweep of the paraphernalia of the old religion were they ready to receive the new. This was comparable. They were to become full participants in the physical blessings of the white man's world, and, in preparation, pitched the old into the sea. Shell money, dog's teeth, mourning costumes, grass skirts, dancing spears, baskets covered with rubber gum, pottery, all went crashing into the lagoon—and four miles away some of it drifted ashore on Ndropwa. In mounting excitement—very much like the excitement that accompanies a fire, when, in order to prevent it from spreading, volunteer firemen are given axes to hack down the adjacent houses—they pitched into the sea both the traditional and a certain amount of the new, which they had obtained from the Americans—tables, chairs, beds, clothes. Some people went in for the destruction more wholeheartedly than others, and the accounts suggest it was partly the inclusion of the new valuable objects, along with the remnants of the old way of life, which sobered the people up.

The first intimation the missionary priest had of this tide of destruction was the sight of the lagoon choked with property. When he remonstrated he was met by a blind stubborn insistence that this was an affair of their own—the same stubbornness which makes subject peoples with newly won nationality turn their backs on world languages and insist on their *own* language being reinstated. The people said the Mission had hidden the true word of Jesus from them, and now they knew the Truth.

But as the pleasure of pitching all the trappings of the old life into the sea abated, people began fishing a certain amount of the new American property out again. (That not all the old valuables were thrown away is attested to by the account that some months later, after a good deal of surviving shell money and dog's teeth had been sold to the conservative Usiai, the "rest" was placed in two big oil drums and sent out to sea.)

Meanwhile, after the first excitement, people sat waiting in their emptied houses, or gathered in the church for long, feverish sessions of prayer, and the Mission Father discontinued Holy Communion. The days passed, no cargo came. A boat crew was picked to go to Mouk and find out what was happening, but they refused to go for fear of missing the cargo. (In other villages this sort of desperate possessiveness had been even more marked; people from other villages had been driven away in the frenzy of the seizures, for fear they would claim a share of the cargo that was not theirs. In Peri there were fewer waking visions, and many dreams. Mikael Nauna, one of the most stable and respected men in Peri, dreamed that he saw a warship and an airplane in the passage in the reef with the ancestors on board, and with Cholai, deceased *luluai* and father of the present clerk-schoolteacher, Peranis Cholai, on board. A white man stood at the mast. But the ships did not land; their entrance was blocked. A woman dreamed that she saw the cargo landed in a village which was no longer on the sea but on land.)

During this week of excitement, Christof Noan, a man whose arm and leg were shrunken and crippled, probably from polio, became dangerously insane. Throwing off his *laplap*, he went about the village shouting obscenities (the usual accompaniment of insanity in Manus) and threatening to fight the white men, declaring that the planes that flew overhead really belonged to the natives, and that the white men were withholding them. He refused to eat, saying that the native food was no good, claiming that he was being fed by God. Four men were assigned to guard him night and day, and a serious effort was made to find out what had driven him mad. His wife reported that the madness had seized him one morning after he had first thrown away a wooden bowl and then brought it back. They succeeded in calming him by throwing the bowl away and talking to him constantly about having good thoughts. Still the cargo did not come, and no one else in Peri had any seizures.

Meanwhile, a Bunai man had returned from Mouk with a full version of the cult. The cargo had come but had been blocked by the errors of Wapi and by men's failure to purify their thoughts. He set off new spasms of seizures in Bunai, and Johanis Pominis, a Peri man who was present, became involved in the quaking. He and one of the Bunai cult leaders returned to Peri, gathered everyone in Pokanau's house to hear the Long Story of God, in which God's true plans for the natives, which had been interfered with so long by generations of white men, were recited. Here a mild quaking seized four other people, one of whom, Lukas Banyalo, remained a mystic adherent to the cult aspects of the movement.

At this point Lukas Pokus, who had set off for Mouk by himself in a tiny canoe seldom used outside the reef, returned to Peri denouncing Pominis' version of the truth and insisting on his own revelation. He had arrived at midnight in Mouk and in an atmosphere of intense suspicion and excitement convinced the people there that he came not from the government or the Mission or for trade, but to hear the Truth. He was tutored in the Long Story of God, the new laws, taught the songs and the marching ritual of the new cult. On his way back, the wind was against him and he had to paddle for many of the twenty-five miles. The sea was dangerously rough; he prayed, and it became calm. He felt the weight of God around his head, and heard the sound of an airplane in one ear and a whistle, as if someone was calling him, in the other. He then heard God speaking to him, telling him again the Long Story of God, duplicating what he had been taught on Mouk but felt by him as a separate revelation. Lukas stopped at Ndropwa and told his story to the other Peri people there, most of whom he persuaded to return to the village—all except Mateus Banyalo, John Kilipak, and Karol Manoi, who stayed to finish the work they had contracted to do.

Lukas Pokus arrived back in Peri just as Johanis Pominis was gaining ascendancy through his seizures, and, aided by Lukas

Banyalo, he denounced Pominis' inspiration as false. He and
Lukas Banyalo succeeded in curing Pominis, and in a dream he
was instructed to turn over the leadership to Lukas Banyalo, in
whom religious experience continued to take a hysterical form.

A few days later, a government patrol on the way to Rambutjon to investigate the murder of Wapi arrived, and this patrol
took Paliau away with them and sent him to the capital, Port
Moresby, for indoctrination. Peri sent other emissaries to Mouk
and participated in the post-cult phase of the movement, with all
its paraphernalia of the new order, marching, "customs"—local
passports from village to village—and a big collection for Paliau's
treasury, which he was building up to help with modernization.
This money was later taken over by the government and held in
trust for the various villages. Through the intervening years it
gave them something to dream about, and was also a focusing
point of bitterness as the people wondered whether it would ever
be returned to them to buy land or a pinnace. Mateus Banyalo
was chosen as *besman*,[4] or head of Peri, and plans were made to
move the whole village of Peri ashore. There were alternative
plans, to move to Bunai—in which case Peri would have become
a hybrid village, since Bunai now included four Usiai hamlets settled by Usiai from the interior—or to move to the little island of
Shallalou, on which Peri people still owned a little land, though
the bulk of it belonged to the firm of Edgell and Whiteley and
was managed from the island of Ndropwa.

The early mystical leaders of the cult phase were demoted.
Mateus Banyalo, Kilipak, Pomat, Raphael, Lokus, and Nauna
became the leaders of the more sober modernization movement.
The breach with the Mission widened. The Mission headquarters
for the South Coast, which had once been in Peri, were moved
ashore, and two of the older men, Joseph Paliau Lalinge and
Alphonse Manuwai, kept their houses in the sea and remained
loyal to the Mission. All the rest of the lagoon village was
destroyed; only the old house-posts (Plate II) remained to mark
the site. The new village was built by communal methods: the
plan was laid out, house-posts were cut for every house, thatch

was made communally, and, with a great sweep of effort, the new village church with posts and altar was moved from Old Peri, and a great dock floored with American airfield stripping was built.

Paliau, meanwhile, had returned from Port Moresby to endorse the Administration's proposal of a council, but when it was finally set up in 1950, Baluan was the capital and the whole South Coast was excluded. The intervening years were years of waiting, of attempts to maintain morale, of boredom on the part of some of the younger men, who went off to work against the wishes of the leaders of the community. There were conflicts between the leaders, who were still government appointees and who could be tried for abuse of government-given authority, and the more restive and insubordinate. At one point, all of the principal Peri leaders were put in prison. The most educated men, Mateus Banyalo and Samol, a carpenter (not Samol of Bunai), were taken from Peri to the capital on Baluan. Peri maintained a school. They had long protracted town meetings to discuss village problems—the new law in which women were emancipated, kin quarrels forbidden, old expensive customs eliminated, and new forms of parental behaviour advanced. In 1952, an Administration-sponsored election was held in which Karol Manoi, formerly of Patusi but now, together with some dozen other men, an immigrant to Peri, and Petrus Pomat were elected. They had a locally selected lay preacher, Tomas Keyai, also from Patusi, who had succeeded the lay preacher, Peranis Cholai, originally trained by Paliau and who had now become the schoolteacher. Without an ordained clergy, every vestige of the sacraments had been swept away, and a highly simplified liturgy prevailed, with church services twice a day.

Everything was essentially in suspense in a period which they described as "wait-council." In a sense, the coming of the council had absorbed many of the hopes once so intensely aroused by the promise of the cargo. When the council came, then they would be able to put all their aspirations into practice.

The particular way in which Peri had experienced the various phases of the movement—Banyalo's pre-movement economic reform, the lack of participation in the early days of Paliau's planning, the delayed and fragmented experience of the "cargo cult," the fact that Peri had been the headquarters of the Mission and that the Mission Father was actually in the village when the destruction of property started, their particular relationships with the plantation managers on Ndropwa and the alternatingly sympathetic and unsympathetic patrol officers in the nearby Patusi patrol post—combined with the special composition of the Peri leadership group to give the New Way in New Peri a style of its own.

Peri had no single dominating leader; it had had no extreme religious experience, the destruction of property had been done after the people were told the cargo had arrived, not on any visionary promise from a prophet in their midst. Seen from Peri, the whole "cargo-cult" episode appears as much less important than it looks from Mouk, or even from Bunai, with Bunai's complement of Usiai immigrants who came down to the beach after their own later version of The Noise, which included ceremonial burning of village census books and government appointees' hats. Yet Peri was undoubtedly the beneficiary of the integrating effects of The Noise, which brought old men and young in line together within the community and mobilized external pressures from Administration and Mission, which served to unite them against these forces which they felt threatened their new-formed identity. After The Noise, the people of the South Coast were sufficiently in step with each other to be able to go forward in working out a new pattern, half derived from Paliau's original design, half from Administration plans for councils and co-operatives.

part three

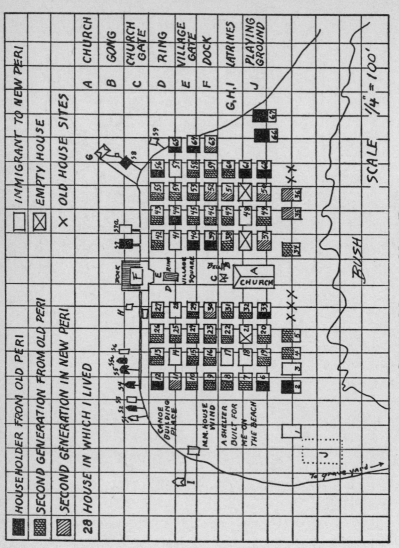

SKETCH MAP OF NEW PERI

10

New Peri

My first impression of New Peri came from a rough map, very thoughtfully made for me by a government officer, showing two blocks of houses arranged in rows, with roughly thirty houses on each side and a large rectangular space between them stretching from the church to the waterfront, where there was a huge, three-sided dock which was labelled "dock where no ships come." This diagram indicated a very crowded village. The comment about the dock was inexplicable to me then. It was in fact a joke about cargo which was to have been brought by magical ships. An accompanying note explained that there was no space for building a house for me, so it had been arranged that I was to have the "second house from the dock" on the right-hand side of the map, a native house with a plank floor which would be renovated for me. The people of the former village of Peri now lived, the note concluded, on the edge of the little flat island of Shallalou, which belonged to Messrs. Edgell and Whiteley of the Pak Plantation.

My second impression of New Peri came from an encounter, at one of the stops the airplane made, with a young patrol officer who said that it was a horrible, stinking, swampy slum; that people used to the sea had no idea of how to live on land. I received my third impression when District Commissioner English said I would need hip boots, and Mrs. English lent me a pair.

On June 21, 1953, it was after dark and pouring rain when

we arrived on the beach of New Peri, and, with a crowd of over two hundred people pressing around clamouring for remembrance or attention, I got no view of the village at all. A pressure lamp soon blazed from the verandah of the house which had been set aside for me and by its light I could just dimly see how well placed I was on the edge of a wide square, with a second row of gabled houses, identical with the row in which my house stood, facing me.

So my first real sight of the village came next morning at dawn, when a few sleepy children straggled across the square and a handful of men came or went about their fishing. But it is by moonlight, with people of all ages seated on the logs in the centre of the square, producing fantastic shadows as the light from my lamp streamed out over the square, that I remember the village best. Later in my stay, the children made shadow-play with my lamplight by holding up large squares of cloth and dancing behind them. In the daily life of Peri large lamps were too expensive to light except for major events or for fishing. But as fishing went on all through the night, the darkness and momentarily silent spaces would be punctuated with the roaring sound of lighting an old pressure lamp and a sudden blaze of light would shine in and out of the nearby twenty houses.

The village seemed very crowded with the houses all standing in straight lines, all built to an identical plan. The roofs were straight-gabled and thinly thatched in contrast to the old beehive-shaped roofs, which, thick-thatched, had resisted the raging gales of the northwest monsoon. The only variations visible from the outside were in the entrances to the verandahs, some at the side, some in the front, and in the faintly visible arrangement of the rooms, for in some houses the verandah stretched deep into the interior, in others most of the space was walled off into rooms. One was immediately conscious of a plan, not a plan to which the houses only partly conformed or which had to be laboriously teased out from irregular plots and accidentally planted trees which weren't where trees ought to be, but a plan that had all the over-articulateness of wartime housing, in which the builders had

had to work for speed, conserve materials, and cut out frills of all sorts.

The houses were in straight rows. They all had the same dimensions. The streets between the houses—I was to learn later that the idea of "streets" had never really taken and people thought instead in terms of "rows"—were all identical; wide ones running from the inland bush toward the curving waterfront, and very narrow alleys, across which one could almost reach with one's hand, running parallel with the sea. The houses were so close together that one could hear when someone in the house on either side stirred in his sleep, and the front verandahs of the houses to the right and left of my house constituted box seats from which one could watch anything taking place on my verandah. When there was a feast in the village, every verandah fronting on the square was filled with the watching faces of those not immediately involved.

At dawn and in the moonlight, the design of the village looked delightful, with the long immaculately swept square in the centre, the great wooden gateway which led out to the dock, the gabled houses on each side, the turnstile leading into the space in front of the church. The church itself, because it was built on the ground while all the other houses were on posts and because it was so much bigger, seemed like a huge mother hen, brooding over a set of identical neat little brown chicks. Just to the side of the church gate hung the iron gong in another smaller gate-shaped frame. Around the square there was a railing with occasional entrances against which people leaned to listen if they did not want to go out into the square and sit down on one of the logs which served as seats for those who attended a council. Or one could walk along the waterfront, and as evening fell, people stood gossiping, laughing, playing different melodies but always with the same rhythm on the homemade ukuleles. If one turned right, the shore curved away sharply, ending in two small houses built out over the sea, where the old and the sick lived. If one turned left, one passed another cluster of such little sea houses, in which the old were permitted to live again as they had once lived, close to the shifting tides which lapped steadily against the house-posts and flashed sunlight and moonlight up through the slatted floors. This was the "side toward the sea," and along the shore one would come upon an industrious man charring the hull of his canoe with a coconut-leaf torch, or a group of children squatting about a fire eating some special delicacy reserved for children, the candy-like new sago packed in bamboo tubes, or even meat of a captured turtle which had died before it could be sold to the land people for making a feast.

Old Pokanau's house stood at the corner of the village. Inside there was always talk going on and much coming and going, for he had many friends and kinfolk in far villages. Beyond his house, where the shore curved out more widely, there was a coconut grove where canoes were built, and where canoes were kept in all stages of construction: Big Manus, the biggest canoe in the village, ashore for recaulking, the crooked hull of a canoe which Nauna of Peri was making for a Usiai friend, and half a dozen other large and small ones scattered about. In Manus, work on canoes, like housekeeping, is never completed. One must always be refurbishing, recaulking, replacing an outrigger, and it is well to be always in the midst of making a new canoe. Beyond the canoe ground there was the long narrow bridge that led out to the men's latrine—the women's latrine was at the other end of the village. Walking farther along the shore, one came upon the playing ground where the young men played an odd version of soccer on Sunday afternoon. Beyond that still, there was a crowded little cemetery with a host of rough wooden crosses, planted with crotons. This shore faced the open sea, but there were no waves, for the reef broke a quarter of a mile out, and the wind blew cool and fresh. Here people could breathe at those times when the centre of the village, with the heat steaming up from the rain-soaked ground, was almost unbearable.

It was here in the place called "kompani," because this was ground which they used by courtesy of the company (Messrs. Edgell and Whiteley), that people gathered in the afternoon, women sitting together while their babies practised first steps on the cleared ground—but no mother ever relaxed for a moment her watchfulness of the coconuts hanging overhead. Each was swift to gather up her baby if it toddled beneath a dangerous tree. Here one found little groups of people, seemingly casually collected. But when one got to know each person, it was always possible to explain how and why each group had gathered, whose canoe was being worked on, who was helping, who was merely being a gossipy spectator ready to help when more manpower was needed for a brief heavy job, which women and which children had followed which men, and which children had joined the group from the side—as half a dozen ten-year-olds tagging along with the ten-year-old son of the canoe builder.

Occasionally, on a very hot day, when there was need for a daytime meeting, a council would be held with people seated in informal, seemingly inattentive clusters on the ground, as I had first seen them on the mountaintop during the volcanic eruption. And here on moonlit nights the adolescents used to gather and dance, the boys playing ukuleles, the smaller girls and boys dancing the new dances with a strange sort of loose-limbed grace, the older girls lumpish and self-conscious, but dancing with a dogged determination which reminded me of their mothers' dancing twenty-five years before, glum and joyless, behind the rows of men in the phallic shell dance.

There was just one house which stood out of line from all the others, the house of a Solomon Islander who spoke neither Manus nor Buka, his own tongue, but only Neo-Melanesian. He was one of the second generation of Solomon Islanders who had been imported to work on Ndropwa, the plantation island which could be seen four miles out to sea. Far away, on good days, one could see Baluan, the capital, and Lou, the island on which lived people who had been grim and hated cannibals and hard-fisted traders for the obsidian which they made into spears and daggers, and who were now determinedly "outside" the New Way.

Beyond the graveyard lay tangled bush and plantation. The bush was where children gathered berries, men foraged for bits of lightwood and dry old coconut leaves, and where occasional coconuts were stolen, each nut looming into a major point of friction with the company,* as the ethics of hunger, the rigours of the old sanctions against theft, and the new commercialism, which let individual natives work the copra on shares, collided. Into the bush fled women who had provoked their husbands' tempers to the danger point; and in this bush the people and I found the gifted young boy who had been driven to a suicide attempt, seated with a huge knife in his hand, and his kin, faces stiff with fright and repulsion, standing in a ring around him. It was said that a few particularly timorous people were unwilling to go near the part of the bush where two men had attempted to hang themselves, although neither had succeeded. Even on this flat little island, separated from the main island by a wide deep strait and by a winding river, the old dislike of the bush survived. It was still a place to be visited in broad daylight or in an extremity of self-destructive anger. (Suicide threats seemed to have become more frequent since warfare was forbidden.)

On the beach beside the place called "kompani," the children built houses in the sand. Here men and women came to bathe their babies and one had the unbelievable experience of seeing an adult head and a baby's head sticking out of the sea, a foot apart and apparently unconnected. It was a gay, busy spot, and only the remembering eye could recognize a certain shyness in the women, a certain slight awkwardness or assertiveness in the small girls, reminiscent of the days in Old Peri when women did not go on the canoe-building island. There was an odd over-emphasis on following—women followed their husbands, husbands their wives, children their parents, parents their children—along the easily trodden smooth sand path from the village, as if still celebrating the new freedom of living on land, where the smallest toddler could follow where he wished, and the new freedom from taboos which made it permissible for women to go where the

---

*Edgell and Whiteley of Ndropwa.

men went. I wondered, as I watched them, whether this sense of extra freedom, this intentional following, would survive as a permanent way of behaviour long after those who remembered what it was like to live in houses separated by water and to shrink from the voices of relatives-in-law were dead.

All along the curve of the beach, back along the village waterfront, in front of the houses facing the water, canoes were drawn up on the shore, set up on forked sticks, turned over for charring, occasionally fastened to the sides of the little water houses or drawn up beside the long log bridges that led out to them. Only once in the whole six months was there a quarrel over a where a canoe was placed. When old widow Kisai, who had been quite mad twenty-five years ago and was less mad but very bad-tempered now, railed at old Poli for putting his canoe where she usually put hers, he shouted back that all that was finished, no one owned special spots any more, each person could put his canoe where he liked. The dock itself was a huge three-sided structure, built of salvaged American metal airfield stripping—a kind of giant iron lace work—set on strong posts, but now falling into disrepair. Officially it was used only for spreading out fish. When there was an especially large catch a portion of the fish would be brought out on the dock and distributed among widows, the sick, and those who had not fished that day.

When it came to mapping the village, I found that there were actually many gaps in the design; some houses had fallen into disrepair and been pulled down, others were abandoned and falling to pieces. People did not live where they were said to live, but had traded one home for another, sometimes with complicated and little-understood arrangements about the exchange of floor planks, which were individually owned, or about the status of houses which had been built collectively for each man, but which were now partly owned, partly not. In designing the houses the Manus had attempted to follow a Western plan—verandah, the back of which served as a living and dining room, separate sleeping rooms, preferably with floors made of sheets of plywood, and a cook house attached to the back of the house.

This cook house was a miniature of the old lagoon dwelling house, with a fireplace of sand and ashes edged with logs, a shelf for drying fish suspended above it, and stones on which to support pots in the centre. These cook houses were the focal points both of freedom and of regression to the old way of life. They were small enough to be moved easily by a group of men. They were light enough to be set up on a type of house-post which was not too much trouble to sink in the water. Old people, accustomed all their lives to sleeping beside a fire, preferred them for sleeping. Although small fires for a new mother or a sick person could be made in the new style bedroom by putting a piece of sheet iron on the plywood floor, this wasn't really a proper fire but just a cigarette-lighting fire. So cook houses were moved out over the sea, either temporarily for a sick person who expected to be immobile only for a short time, or permanently for the old. And most sick people, before being moved to a house over the sea, were carried into the cook house and placed near the fire, near the old women who were more at home there.

Walking carefully through the village in order to map it, I found that some of the back streets were swampy and littered with the piles of palm leaves and other bits of old wood with which the people were attempting to fill in the swampy land. The people described to me with what enormous effort they had brought stones and sand and filled up the centre of the village so that it was now dry except at the end of November and in early December when the high winds of the northwest monsoon lashed the village, lifted the thatch, sometimes broke walls down, and the high tides came into the village forming clear pools on the swept sandy floor. But of the filth of which the young patrol officer had accused them, there was not a trace. The people had given up keeping pigs because they littered the land.* There were

---

*In 1954, after I left, they built a new pigsty in the old style, out over the lagoon.

only two dogs in the village, and although small children were permitted to defecate through the house floor at night, they were trained to report this to their parents so that everything could be carried away and thrown into the sea early in the morning. The anxiety to get the sick moved immediately into one of the little houses over the sea stemmed almost entirely from their concern with the problem of sanitation. And only once in the six months during which I watched the crowds of playing children did I see a child who was able to walk and talk, urinate on the ground. To their land life the people had brought their passionate prudish insistence on sanitation and on privacy during the act of excretion, just as they had also brought their habits of making land where there was no land before.

During the first days when I was learning the new village and its ways, I heard the sounds of uproarious laughter coming from groups who were gambling for fun, playing cards for betel-nut stakes by the light of a pin-sized flame, which barely flickered among them. I watched the comings and goings and the brighter lights of the serious gambling groups to which men would come from another village to challenge someone or to get their revenge. I got accustomed to the ubiquitousness of laughter and singing, where once there had been mainly quarrelling and haranguing. I could weave together the old sounds—the noisy lullaby which matched the baby's cry and rose with it in a crescendo, the keening for the dead which was set up whenever anyone lost consciousness, the tempestuous crying of the children when they were left behind or awoke in the night to find a father gone—and the new, the beating on the great iron gong for rising, for church, for "the line" (when the whole village was supposed to gather to hear announcements and get assignments for village work), the sound of the metal whistle being blown by an adolescent boy trying to collect enough players for a game of football, the strumming on the ukuleles which went on all day long. The village was far noisier than the old village had been, for before the houses had been farther apart and people shouted to each other only in emergency or anger, and spoke discreetly from a canoe under a house when their errands were friendly.

In the old village loss of consciousness had been signalled by the screaming of many women and the shouting of men, who, stirring wooden bowls with long sticks, shouted for the ghosts to return the soul stuff of the stricken. Where today the big iron gong, beaten by official beaters, announced important communal events, in the past wooden slit gongs rang out through the whole village only when announcing some major trouble or disaster; otherwise all communication was carried on from individual to individual with even an event such as a great feast being announced to those concerned on a man's own slit gong. Now people were always calling each other, the call picked up and repeated from house to house by anyone who heard it. This could not have happened in the old days, for names were never said aloud for fear a tabooed relative-in-law might be listening.

I noticed now that some names were called oftener than others, and I found that I could plot the interrelationships and the individual personalities of the children by listening to which children were called oftenest and for what. Only for an hour or so just before the "call of the first bird," would there sometimes be absolute silence in the village. And sometimes in the heat of the afternoon one could walk between the houses and hear only slight rustlings or stirrings from within. But for the rest, the sounds of night and day were very similar: children waking from sleep and screaming for food or absent parents, people calling to each other, outbursts of laughter, and the almost continuous strumming of ukuleles or sound of singing, punctuated rarely with a genuine temper tantrum by a small child, or by a sudden outburst of quarrelling—dreaded because anger was so contagious, so many people might be swept into an altercation no matter how casually it began.

Beneath the ease of manner with which people walked about, greeted each other, jested, and young boys sat or leaned on the railing of the square playing their ukuleles in the twilight, there was still some of the old watchfulness. People marked where others walked, wondered why they took that short cut between two houses or stopped to beg a bit of betel when they really still had betel tucked inside their belts. The new gaiety and ease sat a little strangely on people who had traditionally been so rigorously censorious and watchful of one another. But the sounds were far happier sounds. Anger and laughter had changed sides so that now there was much more laughter and much less anger; and gay music had almost completely replaced the occasional "crying" chants for the dead or absent.

The day began with a rising bell, beaten out on the gong. People then would begin to come out of their houses in little groups; the men and boys walking in one direction, the women and girls in the other, both groups wrapped in the large coloured squares of cloth which were fashionable village attire for young dandies and made good cotton blankets for both sexes. This early parade to the beach was distinguished by a display of conscious virtue. Formerly also the Manus had paraded the trip to the latrine. In Old Peri, fathers had sat proudly in canoes punted by prouder small boys forming a veritable procession moving toward the latrine island. The gong was described as the gong for washing and dressing, and an hour or so later there would be a gong for church, just as the morning light fell slantwise across the square and in at the open church doors.

Weekday morning congregations were small, sometimes not more than fifteen or twenty people, often silent, as they stood quietly (standing prayer was the rule because of the wet sand floor), after making the sign of the cross, until one of the more experienced intoned the words of the closing sign of the cross, "In the name of the Father and the Son and the Holy Ghost," after which they drifted out again. The most faithful daily church attendants were the older men who were obedient to the New Way, who came faithfully to "the line" to hear about work assignments, and who clung to the church as a memory of a more comprehensible past before so much of the old culture had been changed. These were the men who seldom wore full European dress. Other people came because they happened to live near the church or to be near the square at the moment. This was true of Makarita, wife of the one Peri police boy who had been fortunate enough to be inside the Allied lines, and so was to receive a medal for distinguished loyalty. Makarita, who had been born when I was in Peri before and had been named for me, always came to church bringing her fretful baby, who was so cantankerous that no one wanted to carry it. And Josepha Sain, widow of Lalinge— our old house builder who had been a leader in his day and had fought the New Way and remained faithful to the old priest until he was dying—Josepha who called me "sister-in-law" because her husband had said, in 1928, that I might treat him as a brother in case my husband beat me, came often. Sometimes Lukas Banyalo, the preacher who preached from emotional conviction rather than because he had been chosen by the village to be the preacher, would preach, and his voice, very loud and strained, like a voice which seeks to compensate for a loud-speaker that has gone suddenly dead, rang out implacable, repetitive phrases into the village.

Whether people went to church or not, they were conscious of church because when church was over "the line" was formed. Like the rising gong, "the line" was reminiscent of plantation dis- cipline when all the labourers formed into a line to be given assignments for the day, although in the minds of young Euro- pean critics of the New Way, who had never seen an old-style plantation, "the line" was regarded as part of the militaristic "totalitarian regimentation" introduced by Paliau. In form "the line" followed the style traditionally used in government or med- ical inspections. Nowadays a modern government officer did not line up the inhabitants of the village but let them come up one by one from groups that waited in the shade. But the morning line continued as one of the symbols of the New Way, and the size of "the line" was regarded as an indication of village morale. School- boys to the left, schoolgirls to the right. Behind them a line of men, young men to the left, old men to the right, and behind them the long line of women, undifferentiated by age, with babies in their arms, small children clinging to their legs. "The line" would form slowly. The churchgoers drifted out from church and went and stood in their places. A few women and babies would sit on the meeting logs in the middle of the square until the last possible minute. Or a sudden squall of rain might send people underneath the eaves of the houses along either side of the square. The village officials, two councillors, three committeemen, and old Pokanau, did not line, and other important men who had or would hold positions were likely to vacillate between lining very conspicuously, not listening at all, or standing casually at either end of "the line" where it merged with the corner houses, so that although it wasn't quite certain whether or not they were lining, it was clear that they were there to hear what was said.

Theoretically, "the line" was "made" by "committees," of which there were four in the village, one "committee" one week and another the next, but more often it was made by Pokanau, who, true to the former role of a big man, was likely to announce decisions that had not been cleared with anyone. The official who "made the line" stood under the gate facing the people, haranguing them on various points, announcing decisions that had been made in meetings, sometimes setting aside several days during which there would be no village work but people would be told to "go about and find food for themselves," to fish, trade, and work sago. At other times tasks would be assigned; groups of men would be told off to go and get caulking material for the hull of Big Manus, the village canoe, or cut bamboo in the bush for mending the thatch of all the houses. "The line" would break up very suddenly in contrast to its leisurely formation, and, if it was to be a school day, the school children would wait in a cluster until their teacher set them drilling and then marched them into the church, the back of which served as a school room. If it was a market day, people who had fished all night might already have left for market. Now others would hurry to get together the dozen small commissions which their neighbours and kin were giving them, and in canoes laden with chattering, shouting people would set off for market. On other days there might be much work going on at the canoe-building grounds. A few women would go out fishing with their cone-shaped hand fishing baskets; others might sit under their houses making a pandanus mat or sewing a little thatch. Houses needed to be broomed, garbage thrown away, late-rising children washed in the sea.

The régime called for a bell at twelve o'clock and a "bello-back" at one. When everybody was scattered or concerned about their own affairs, these bells were often irregular or forgotten altogether, but when there was village work on, then people remembered and watched the clock—my clock, which was more reliable than theirs—for time to stop work. Timing they regarded as one of the great inventions of the European world; especially that there was a signal stopping work. The canoe ground, where every tool was laid down when the bell rang at noontime, was to them a symbol of the orderly work world of the European. But the extent to which native life was a creature of the wind and rain and sea was recognized when on rainy mornings everyone lay abed. There might be no rising gong until after eight o'clock and no "line" at all and sometimes no church, while people stayed in bed as the only way to keep warm when the thermometer fell to the low eighties, so that teeth chattered with cold, and the village square lay flooded with rain water. As the rain began to lessen, the small children, who habitually wore no clothes, came out like mayflies to gambol in the shallow pools, loving the rain, while their elders, who owned too few clothes, had to scurry for shelter to keep dry. The people still wore their old native-manufactured rain capes, a rectangular mat sewn up on two adjacent sides and fitting over the head like a peaked hood. It was becoming the fashion, where people could afford it, to carry a brightly coloured plastic umbrella over a small baby if it had to be taken out in sun or rain, but most adults were still too practical to spend their money on umbrellas.

For everyday wear the men wore only *laplaps* (short sarongs). The women and girls after puberty wore *laplaps* and blouses about in the village, but inside the house, while cooking, or while in the bush or out fishing, they too went about naked to the waist. In church, at special meetings, when one had a court case, the important men dressed in complete European clothing, including shoes, socks, and ties. If they had to go to another village they carried their clothes wrapped in a waterproof mat, and put them on when they got there. The women's handling of clothes was more erratic. A woman might bring a good blouse to court and leisurely put it on while the court was convening. Except for the very narrow and rigid requirements of modesty, people dressed and undressed in each other's presence with complete lack of embarrassment.

At four o'clock in the afternoon the bell to end work sounded, a signal for men who were tired to lay down their canoe-building tools, for women to carry newly bathed babies out to the "kompani" grounds, and for a general relaxation to set in. This was the hour when the boys played ball, when the younger children played noisily in the square, when—during the thatching period—there were little clusters of people, young men, young men and women, young women, adolescents, children down to the two- and three-year-olds, playing darts with the slender shafts made from the ribs of the sago palm leaf and throwing them alternatively at targets of palm trunk set up in the ground. The men played with a flourish. An occasional young woman, near her own or her father's house, exhibitionistically joined the men's group. The small children stood over the targets and practically stuck their darts in, taking no chances. Everywhere thatch was being made; everywhere there were ribs which could be gathered into bundles of ready-made darts as the contagion to play spread with the lengthening rays of the sun. Suddenly it would be six o'clock, and time for church. The games would break up; someone would lead a whole group of boys, or a group of girls, to the front benches in church—the day would be over.

On many evenings there was a council meeting. The gong would be beaten for council right after evening church, and any official who had something he wanted to say could beat the gong or request an official gong beater—there were two of these—to beat it for him without necessarily telling anyone beforehand what he wanted to talk about. So any announcement of a council meeting meant a certain pleasant anticipation. It might not, after all, be about what you expected it was going to be. But people gathered at council meetings very slowly, coming to sit on the logs arranged in a rectangle in the centre of the village square—the so-called "ring"—with the rectangle of logs reminding one of the old-style fireplaces. Sometimes women sat with their husbands; sometimes with other women. The small children often arrived first and perched like rows of birds, in groups of eight or ten. A few older men, possibly an official, might come and sit down ostentatiously. The ones who came early had a right from there on in to be indignant with those who came later. People talked together quietly, and the low murmur of casual conversation would be interrupted from time to time by loud remarks, sometimes through the old American megaphone which was used for village-wide announcements, about how slow people were, how few had come, how few seemed to be coming, etc. No one ever knew when the person who had something to say would decide there were enough people to begin. Meanwhile there were *sotto-voce* remarks about its getting cold, or being windy, or about the number of mosquitoes, as people sat with big cotton squares half-wrapped around them looking, in the pale light of the stars or when silhouetted against the light from the lamp on my verandah, rather like a group which had been buried under sand. A few people brought chairs—iron chairs left over from the American occupation—and sat down a little outside the "ring"; others leaned on the railings which edged the square, close enough to hear, far enough away to fade out unnoticed. All this might go on for an hour before an official or occasionally someone else would stand and, looking at no one, throw his voice into space, addressing the meeting either in Manus or in Neo-Melanesian or in a mixture of both. There were long pauses between speeches, but usually the end was clear. Someone would say, "*Ki ne pwen*" ["It is finished"], and people would get up and go home. "Without a meeting every two or three days, things are not straight," remarked Pomat sagely.

This was in the style of other Manus events. The beginning was always slow. The very knowledge that there was soon to be action seemed to have a soothing effect upon people, whether it was the little group of women around the woman in labour, the fishermen in a canoe just before the watchman sighted the school of fish, or the people who were gathering for church, forming a line, or attending a meeting. No one hurried; there was no sense of scurry or tension or rush; almost a sleepy hush fell over their usually more taut movements, as if this was the most perfect mobilization for the decisive complete action to follow, whether it be focusing every muscle in holding the parturient woman, or leaping into the sea with nets to meet the onrush of the fish, or plunging into the intention of a meeting. Even little children already showed this tendency to make sudden transitions from extreme relaxation to activity, lying sound asleep one minute and then beginning to scream vigorously before they were even awake.

All village life was geared to the fish. Nights when it was good fishing and nights when it was poor all depended upon fine points of wind and rain, tide and moon. For the women's type of fishing, days when it rained steadily were best, and one could look into the flat lagoon with the rising dark hills of the mainland behind and see a dozen canoes each with a rain-hooded woman fishing. Then there were the crab traps to be baited and watched; two trips a day for bait, two to watch the traps and put in the bait. And there were the deepsea fish baskets on which men who worked alone depended. All of these operations were necessary to provide daily food for the market and food for the next meal. People cooked at all hours, when the fishermen came home, and it was a commonplace to hear a whole family rouse after midnight to cook a meal.

On evenings when there was no council meeting and no expectation of a big catch of fish of the kind which caused every man in the village to go out fishing, there were usually one or two small gambling games, often women's games at which the men stood about and gave advice. These games were played for stakes which the men considered too small to take seriously, but which were not completely negligible to the women, who had few sources of income in the New Peri. Sometimes the games were mixed and played for bits of tobacco or betel nut, and then there was only good humour and laughter as a crowd sat about a lamp made of a small can with a wick of twisted rag, set on the floor so that the faces were dimly lit from below as they used to be by the light of the old fireplaces. Those who walked about and wanted to see who was in any gambling group could stand in the doorway, their faces just above the house floor, and look straight into the centre of the circle. The presence of a gambling group of this sort was taken as an invitation to enter a house. In the long evenings, with the shortage of light when there was no moonlight, these little gambling groups, and perhaps one or two men poring over records written laboriously in old exercise books, a group of schoolboys sitting together chanting the half-understood English words of the Papuan reader, a group of women sitting up about a sick person or a new baby, were the only centres of activity in the village, for there was no meeting house where people could gather otherwise.

In the old villages, the men's house was a definite institution which stood separate from and opposed to the domesticity of the homes of married men, presided over by the Sir Ghosts so jealous of marital virtue. The men's house was for the young men; there they could tell uproariously funny pornographic stories—if no brothers-in-law were present. There cross-cousins could come into their own; there two boys could act out, with over-vivid detail, the rape of some woman of another tribe or village; there a captured woman of another tribe could be kept on as a prostitute. It was always a little dangerous, a secular spot, but still within the village. Sir Ghosts did move about with their wards, and could occasionally take umbrage if jesting went too far. Married men theoretically should have stopped going to the men's house, although it is said that wives were less angry if all the married men slept with a captured prostitute—then it became the "affair of all the men." But even then the women would sometimes go to the small rocky island near the men's house, where the children's long rattan swings were fastened, and swing in a group far out over the water, combining shrieks with jeering laughter designed to badger the men in their relationships with the prostitute. This unseemly laughter echoing through the village entered the houses, and, like the idle whispered salacious gossip of the women alone, was annoying to the ghosts. The men's house was periodically a brothel, a brothel for isolating that which no one believed could be completely controlled.

When the government had forbidden prostitution, the prohibition had coincided with the young men's going away to work, and the institution of the men's house fell into disuse. Periodically the men of New Peri talked about having a men's house, periodically some dwelling house became specialized as one where bachelors and widowers and divorced men slept, but the old incentives were gone. Young men tended to prefer to gather and to sleep in dwelling houses. But house owners, their old nervous fear of the ghosts transformed into a nervous apprehension of an illegal brawl developing, disliked having their houses used for any gambling in which money changed hands. As in the old days, when those who sinned in a dwelling house had to pay an expiatory offering to the Sir Ghost of that house, so today, those who quarrelled in a house had to pay the house owner for the quarrel. And this payment could be demanded in the local village "court."* As of old, when the fear had been that the greatest sin—forbidden types of sexual behaviour—would invade the home, so today there was a comparable fear that the greatest sin—anger and quarrelling—would invade and endanger the home. So the evenings in New Peri—if there was no deadly sickness or mourning for the dead—were lifeless. Sociability tended to be hushed and limited, people who walked about without visible purpose

---

*I have put the word *court* in quotes to distinguish the local institution, unrecognized by the Administration, but which the people feel is a real court in which officials of the New Way mete out local justice according to their ethical codes.

tended to be questioned in "court cases" later. Even the school children, dancing quietly on the beach in the moonlight, might be complained about by old Pokanau. It was dull but not silent, for the screams of waking children, the shouts of fishermen calling for their fishing partners, the roar of lamps being lit, went on through the night.

Wives, angry that their husbands were walking about instead of sitting quietly at home, would provoke their children into wailing for the absent fathers, and husbands whose wives tarried in some card game, however innocent, were likely to fly into rages. The fury of being left behind, so prominent in the attitude of small children toward their parents, who interspersed complete permissiveness—carrying the children for hours on their backs— with occasional sharp, decisive refusals to take their children with them, was carried over into married life, where in turn the husband or wife, re-experiencing the sense of childhood desertion, goaded on the new generation in their expressions of rage, so that the evening was punctuated with the fury of three- and four-year-olds shrieking for absent parents. Only in the rejection of any music and dancing that lasted late into the night did the one-time puritanism of the village show, a pessimism about pleasure which was shared by the Church. But, while the priest spoke of them as being a very "sensual people," it would be more accurate to say that they were a people almost unused to any form of joy, or of physical contact not based on sex-anger or on sorrow.

This uneven tenor of daily life was keyed to the Christian week, which the Manus had accepted with enthusiasm. Markets were now held on definite days of the week instead of at three- or four-day intervals which had to be continually reckoned ahead. Saturday was devoted to preparation for Sunday, and on Sunday work was forbidden. Saturday morning "the line" was told that "tomorrow is Sunday. Today is Saturday and you must all prepare for Sunday. Sweep your houses, wash your clothes, tidy the village in readiness for Sunday." Sleeping blankets were hung out to air on the railing around the square, together with bright *laplaps* drying for Sunday wear. The school children were set to brooming the church, the square, and the graveyard, working rhythmically in a long line. Women broomed the sand beneath their houses, gathering up the light litter—bits of leaf and coconut husk—in a piece of bark and carrying it down to the sea or throwing it into some swampy place in the back of the village.

There was never enough of this Saturday preparation to please the leaders of the village, and the morning was likely to be interrupted by short, angry, and loud tirades exhorting more people to do more washing and brooming and tidying up. No one fully understood that the difficulty arose from there not being enough to do. The houses which were broomed several times a day took very little additional cleaning. With so few clothes, people usually washed the clothes they wore each time they bathed. There was no starch, and irons were used exclusively for the men's best white clothes, worn only on important ceremonial occasions. No one had introduced the idea of a feast meal every Sunday on which the whole of Saturday could have been expended. So people did less than they were asked to do; many people went on working as usual, and there was a general uncertainty about the moral position of Saturday afternoon, the likelihood of a big illegal gambling game on Saturday night steadily looming larger. Here was a place where the Manus had been given only part of a pattern, the pattern for hours of work, the pattern for making Sunday a day set aside for godly cleanliness, but the element of joy associated with celebrating the Sabbath, of feasting and gaiety either on Saturday night or on Sunday, was missing.

Sunday morning meant church, and when the sun shone preparations for church happily occupied the early morning hours. People dressed in their best; children were decked out in smart little sunsuits—special ready-made children's clothes imported from Australia and bought in cellophane packets; women borrowed other people's extra children so as to have a child to take to church; new babies made their début, mother and child appearing for the first time together. Important men took their little boys with them to the men's side of the church, and sometimes a small son accompanied his father up to the altar step and stood beside him as he preached. Men carried very tiny babies, too, and as the service went on men stepped back and forth across the aisle, handing over hungry babies to their wives. The church was full, with each age and sex group sitting together, the women preoccupied with the playful babies who coquetted over the shoulders of their carriers with women on the benches behind them, the men standing more stiffly and attentively in attitudes of prayer. The school children's choir sang, a strange orchestration in which the girls' voices came in late and irregularly after the boys'. There was always a sermon, preached by one of the two community-recognized preachers. But, even if they used every piece of liturgy they knew, church took less than an hour.

Sometimes there was a wedding, more often double than single, conducted with great simplicity. The brides, who had been seated anywhere on the women's side, and their female sponsors, usually very young people like themselves, and the bridegrooms and their male sponsors got up and walked forward, standing before the preacher and repeating their vows after him. Then the preacher would preach a sermon about marriage in which the marriage of Joseph and Mary was the model, with the fostering role of a father toward children, whether his own or adopted, and the role of St. Joseph toward the infant Christ heavily underlined. When the wedding was over, a double line was formed outside the church, men and boys on one side, women and girls on the other, and the newly married pair went down the line shaking hands with members of their own sex, bridegrooms more abashed than brides.

When there was a wedding in the morning, there was usually a wedding feast in the afternoon; and the arrival of kin folk to help with the feast and of visitors drawn by the fact of the feast kept the village in a pleasant hum all day. Again, however, the weather was the deciding factor. When it rained, there might be hours of waiting before the tables could be carried out into the square. During the feast the bride and groom sat side by side on chairs, relatively easy in manner and unembarrassed, for the general belief was that the marriage was consummated physically before it was consummated religiously. So there was none of the high tension of vicarious participation in the bridal night, traditional in European weddings and formerly carried to an extreme of angry embarrassment in old Manus. The cigarettes which had been rolled before the feast, the big bunches of bright green betel nut, the bowls of food—an equal amount of each contributed by the bride's kin and the groom's kin, so that all future obligations were eliminated—were distributed. Plates were given publicly to visitors, to the preacher, to the teacher and the schoolboys, to us, and then to each clan, and were either carried away to be eaten in private, or were eaten hurriedly and without style by the children on the spot.

When small feasts were held inside the house—as in the small feast given by the bride's near kin for the groom's near kin, or in the feast given by the new father for his wife's kin in appreciation for the care of his wife and new baby—the old feast manners survived. Children sat cross-legged and relatively sedately among their elders until bowls of food were handed to them, and then they would eat quietly. These same manners obtained during the day before a big feast in the square. Children drifted in and out of the houses of their kin, eating when they were given the scrapings of a pot or a special tidbit, begging only from their own parents. But, at the end of a feast in the square, there was an unseemly scramble, accentuated by the desire of everyone to empty the feast bowls and return them at once. In Old Peri, the feast bowls were publicly distributed to people seated in canoes, who took the food home to eat and then returned the bowls the next day. Or, as happened more often, the giver would angrily come to demand back the bowls.

This old insistence upon returning the empty bowl could be interpreted as part of the uneasiness about giving and receiving, perhaps an uneasiness that there was any pause or block in the exchange. There were only two forms of giving with which the Manus were completely at ease: giving with an immediate repayment or giving with no thought of repayment. But they had little tolerance for an interrupted or incomplete exchange. Even in these new feasts in which no one's contribution is rigorously counted, for which no one can be dunned or upbraided, there is still the uneasiness, the haste in emptying the bowls, and a few moments of angry altercation over the return of the bowls, which are first piled up empty on one of the tables. Here again, the long preparations, starting during the week with visits to markets, bringing home the special feast fruits which trade friends have promised, fishing for extra shellfish, getting together the tobacco, and the cooking through the hot Sunday mid-day, with old women kneading oil into the taro, and little family groups of children happily licking the pot, were leisurely and pleasant, but the consummation was a hurried, uncomfortable scramble to finish— a little like the final moment of group fishing or childbirth where the interim attentiveness is rewarded by a catch or a new baby.

If the day is fair, the feast will be over early, and then there may be ball or dart games and much visiting and singing before the canoes of visitors leave for the neighbouring villages. If all Sundays in New Peri had a wedding, the day would be one to look forward to and would demand enough preparation and provide enough activity to make it a day of joy and rest. It is the Sundays with no wedding, which start with rain, that threaten the peace of the village, for whereas the anticipation of an event makes the Manus relaxed, the absence of any goal makes them tense and cantankerous. Characteristically, prolonged constipation is reacted to with something approaching panic, as a state in which nothing will ever happen again.

In the case of feasting, a man's house is inviolate against the entry of onlookers, unless he himself invites entrance. So people know whether they are close enough kin or close enough friends to go in where food is being cooked—which must be shared with those who come into the house—or to stop and linger where feast food is being eaten. Even where adults have full rights, as for instance when the money for the bride price—five pounds for an unmarried girl, three pounds for a widow—has been changed into shillings and the shillings are laid out in little piles to be given to each close relative who helped with the feast, very often the close relatives are not there. In these days, when the help and goods should all be free gifts, given from a desire to give and without weighing returns, the old conscientiousness and watchfulness against what others have received as return is out of place, but has not yet been fully unlearned. The old women, in particular, are likely to spring into acts of histrionic rage over a shilling or a small cluster of betel nut or even half a stick of tobacco. To such distributions it is safer to send just the children, and so today such distributions tend to be made to children who will carry home a bowl of food, three or four shillings, or a fish from the full catch. In the old days children were unreliable messengers. They could not be trusted to assert themselves angrily, either demanding or tauntingly promising full returns. Today they are the perfect messengers, shy in the presence of the giving but proud of being able to participate and walking about the village holding the fish or the feast bowl entrusted to them like a special badge of distinction.

But when trouble enters a house, all the barriers go down. Where there is illness or death, the gathering is one which draws people together in sympathy and warmth. Where there is quarrelling, which frightens and repels them, they draw near in fascination—to watch or, in the dark, to listen. If a child cries on and on and no one comforts it, anyone passing by feels free to enter and see what has happened. The sounds made by a wailing wife must be listened to even more carefully to determine if she is really hurt and afraid to run away. So people will cluster about outside a house when a first shout of anger or of hurt is heard, ready to enter at once at the suggestion that the situation is out of control. The minute that such an entrance is made, usually by a relative or a close neighbour intervening in a difficulty, others follow as audience, although restive, slightly embarrassed, and poised to leave as soon as the quarrel has calmed down. The preliminary listening takes into account whether or not the situation is out of hand: Has the child's grief or fear possessed it? Is a husband so angry that he may hurt his wife? Is a wife so grieved that she may do something foolish? Is a man so angry that he may destroy his house or attempt to hang himself? If not, then people stay unobtrusively away, except for the hordes of children who gather wherever there is any excitement and are hard to shoo away, especially in the new-style houses, which they can enter at almost any point, by climbing over the sides of the verandah, and escape from as easily.

But though in the case of temper tantrum or quarrel, one must wait and listen before entering another man's house, grief opens every door. The sound of the keening which is begun as soon as anyone loses consciousness, as women cluster close around the stricken person, is a signal to everyone in the village, and people begin to gather rapidly. In Old Peri canoes would come flying from all parts of the village. Today people seem to seep in through the open sides of the house until there is a frieze of spectators on all sides, practically no air, and an atmosphere of tenseness and panic. When a severe illness lasts a long time, there will be a whole series of occasions when one of the older women decides that death is imminent or that the patient has lost consciousness, and commences the keening. The crowd gathers, but as the crisis passes, people slip away again and the sick one is left alone with the little circle of devoted nurses, usually only three or four for a small child, sometimes as many as eleven or twelve for an adult.

When death comes, those closest to the dead entwine their arms around one another, forming a closed circle around the corpse, and begin a wail that is swollen by the arrival of each newcomer. As the circle widens, each newcomer plunging in like a swimmer into very cold water, it may become too big for the room in which the body lies. There will be a hurried reorganization; the great heaving circle will break up, the body be carried out onto the verandah or sometimes to another stronger house; the mourning group will re-form and begin wailing again. The deeply affected mourners plunge into the group at once; others come and join the spectators around the circle, stand quietly and relaxed, and then plunge in. Except for the official mourners—grandparents, parents, children, siblings, paternal aunts, and paternal great-aunts—people will expend their mourning capacity after a short bout of wailing, raise their heads from the entwined circle, extricate themselves, and rejoin the spectators.

The line between active and passive participant, although it is continually being crossed, never becomes blurred. Mourning is enacted upon a stage, and the audience is as much a part of the spectacle as are the mourners. But, unlike the theatre in which the audience must be assembled before the play is given, which is true of the great theatrical performances of the men's cult on the New Guinea mainland, where the performance ceases as soon as there are no spectator women and children, in Manus the action is the core and the spectators form only a peripheral group, recognized but unessential. So when the quarrel slackens, the spectators go; after the new baby is born there is a little flurry of visiting women during the half-hour while the baby is bathed and the new mother fed her coconut soup. At a death, the crowd increases from the moment of death until the corpse is bathed and dressed and then begins to thin out, leaving only those closely involved and the inevitable crowd of children to march in procession to the cemetery. There, at the entrance of the cemetery, with boys forming in one line and women and girls facing them, the crucifix is held by one of the religious officials, and the men carrying the corpse follow the crucifix between the lines into the graveyard where the grave has already been dug. They listen to a short sermon. Then the children each throw a leaf or a flower into the grave, and, here today, if it is a new baby its Christian name will be heard for the first time, for the farewell is always to a Christian name, "Good-bye, Joseph," "Good-bye, Maria." And the participants scatter to bathe in the sea. Significantly, no one who has not participated in the funeral procession has to bathe, except a close relative who has held the corpse but been too ill to follow to the cemetery.

All this is over very quickly. Short mourning is one of the tenets of the New Way. After an hour or so of weeping, council officials begin insisting that the coffin be made. If the close relatives are too overcome with grief, a call will be shouted through the village, which echoes into the house of mourning, that a group of men come quickly and work on the coffin. The coffin is a new box made of sawn "American" timber, even though the people realize that their supply of sawn timber is limited and that they may not be able to get any more. The corpse may be not only fully dressed but covered with extra layers of new cloth and clothing, so that death ceremonials tend to consume, like a raging fire, the new resources of the community. Periodically someone tries to lay money on the corpse to be distributed among the kin of the deceased. Although this too is forbidden by the New Way, it is reluctantly permitted to the sorrowing. The funeral takes place so quickly that someone may sicken, die, and be buried while people are away from the village at the Patusi market. The inconsolable parent or spouse may continue to wail or may resume wailing when relatives arrive from another village. But if grief lasts more than two or three days, people become very much alarmed. The swift furious display of completely abandoned sorrow is felt as safe and cathartic. But prolonged grief corrupts the mind, leads to depression and self-destruction, and those closest to the bereft reason with him and seek to draw him out of his mourning.

Other disasters besides death are treated the same way, and those who have been shipwrecked but escaped or those who have been sent to prison are wailed over—as for the dead—and then feasted when they return to the village. A few of the old songs survive—the "cries," particularly vivid laments composed for a particular death or disaster, and when they are sung men sit close together as they chant them. And in the broken rhythm, in which there is a sharp sobbing pause marking the end of a "line" but often sharply interrupting the normal syntax, one can hear again the sound of children's voices shifting from active crying, through sobbing, to monotonous chanting, on the road to sleep.

11

the New Way

The New Way* and the words which are used to describe it are inextricably woven together. The men and many of the women of New Peri speak two languages, Manus and Neo-Melanesian, while the young children still principally speak Manus. In describing the present social system it is easier for the people to speak in Neo-Melanesian than in Manus; in some cases one finds that there are no Manus words for what they want to say. The new words that have been added in the last fifty years, either for foreign things or for changes in their social system, are almost all Neo-Melanesian. As the grammar and syntax of the two languages are the same, people slip from one to the other without being conscious of doing so. In situations where people feel they are being watched, they may either parade their Neo-Melanesian or refuse to speak it at all—this is frequently so with women or young men in "court," in which all the officials will speak Neo-Melanesian and the shy defendants and witnesses will reply in Manus. Degree of mixture of Manus and Neo-Melanesian is a

---

*For a complete technical analysis of the structure and functioning of the old social organization of Manus, see Mead, Margaret, *Kinship in the Admiralty Islands*, American Museum of Natural History, *Anthropological Papers*, Vol. 34, Pt. II, 1934; and Fortune, Reo F., *Manus Religion*, American Philosophical Society, Philadelphia, 1935.

subject of jeering comments between villages, especially by men who have been away at work for a long time and are able to make fine distinctions between Neo-Melanesian as the European speaks it and Neo-Melanesian as spoken among themselves.

So, coming to work in Manus today, one would learn Neo-Melanesian first, as one would learn standard English first to start work in a remote Yorkshire village in England or in a mountain village in Kentucky, or learn Florentine Italian to begin work in an Italian village, realizing that one would also have to learn what we call "dialect" before one could listen in on the children's play, or collect cooking recipes from the old women, or comprehend a quarrel. Whether or not such dialects are felt to be part of the standard language is in large part a matter of definition. In the case of Manus and Neo-Melanesian, grammar and syntax are so similar that the shift is little more than a shift of vocabulary, and there are a great many Neo-Melanesian words and phrases in use in contemporary Manus.

To the question: "To what do you belong?" a Peri man answers, in 1953:

"I belong to Peri village, South Coast Manus, and I am inside the New Way of Thinking and I wait Council."

If the investigator then asks for a definition of Peri village:

"This is the village that once lived in the old camp over there. Many people from Patusi village also live here. We live now on a site where one of the ancestors of Peri people once owned land and planted coconut palms on the island of Shallalou. However, most of the land on Shallalou belongs to the company, Messrs. Edgell and Whiteley."

A reflective man may also add:

"They bought it a long time ago and have long since got enough return for their money. When I build a canoe, I expect to use it until I have got a return for the work and money I put into it, then I lend it or give it away freely. We [people of Peri] have enough money in the bank that belongs to us to buy Shallalou. Once there was a government officer who was going to help us do it, but Messrs. Edgell and Whiteley wanted much money and the papers never came. Meanwhile we are very crowded here, the houses are too close together, the land is swampy, there are mosquitoes and no wind, it is not a good place to live. We would like to rebuild our village, but we would like to know first whether we can buy the land. All the people of Peri are Manus-true, but there are many other people, the people of the villages of the Big Place [Manus Island] and people of the small islands [Baluan, Rambutjon, Pak, etc.] who have come inside our New Way of Thinking. There are thirty-three villages, all or part of which have come inside."

And they would name all the villages without stating, unless one asked them, which were Manus-true and which belonged to other tribes. The Neo-Melanesian phrases "people of the Big Place," and "people of the islands," are preferred to the Manus terms, Usiai [landlubbers, or "men-o'-bush"] and Matankor [literally "eye-of-the-land"], which these two groups used to be called. People take one aside and remonstrate gently if one uses either of these terms, just as active exponents of good race relations in the United States comment on such words as "nigger" or "hunky." The tone of voice of the remonstrator, who speaks more in sorrow than in anger, is so exactly similar that the effect is almost comical.

And if you ask: "Where are the other Manus-true, the sea people?"

"Well, they used to live in more villages. They all lived in houses on the sea and they had a monopoly of fishing. Today everyone shares, the land people live near the sea, the sea people live on land, the sea people teach the land people to fish and manage canoes, and the land people have both welcomed the sea people on the shore and to gardening."

"And do the erstwhile sea people garden?"

"Well, no, not much really. Paliau has said they should, but the sea people do not like walking on the earth or working in the earth. The stones bruise their feet."

"And how about the land people and the fishing?"

"Oh, they have learned to fish, and they aren't too bad with canoes—at least inside the reef!"

"And all of these thirty-three villages are 'inside'?"

"Yes, of course not in quite the same way. All of the sea people, all of the Manus, are inside—all but one or two individuals, such as men who are still away at work and won't come home, like the Manus man who is teaching in New Ireland who won't come home and teach here although we need him badly. But these other people, well, half have come inside and half have stayed outside. Some people—the ones who have come inside— have moved to the sea coast. They belong to us. Now there is only a part of them inside the council, but when the council is settled—it has been delayed and delayed, we had an election over a year ago—when it is settled, then we will all be inside."

"And the New Peri village is all Manus and the people come from Old Peri and from Old Patusi?"

"Yes."

"And how is Peri organized?"

"We have two council[lor]s, and four committee[men]s, and a clerk—he is the schoolteacher—and a teacher in the church and an extra teacher who takes the church service Sunday. Then, there is still a *luluai* [government-appointed native headman]. Yes, the *luluai* will function until the council really comes. Now is the time of wait-council. The two council[lor]s are Petrus Pomat and Karol Manoi, and they function together."

"Do all villages have two 'councils'?"

"Noooo-oo, but in Peri it is thought better to have two 'councils.' "

"And the committee[men]s are . . . ?"

"Simeon Kutan, Lukas Ketiyap, Raphael Manuwai, and Johanis Lokus."

"But Johanis Lokus lives in Bunai?"

"Yes, that is right, but that is because his wife is a Bunai woman, and their child was sick so they went to his village. He comes back if anything affects his 'men.' "

"Oh, so 'committees' have 'men'?"

"Yes, each 'committee' has his own 'line.' "

"And the 'councils'?"

"Yes, each have their own 'line'—they in the two books."

"Which two books?"

"The Pontchal book and the Peri book."

So only slowly does one find out that Peri is divided, as it has been from the beginning of European contact, into two named parts which have separate government census books, and that the reason there are two "councils" is because there are two books, proof positive to a patrol officer that there are two villages.

"But doesn't two books, which means two government 'lines,' need two *luluais*?"

"Yes, we used to have two *luluais* and two *tultuls* [appointed native interpreters], but then they got into trouble with the post-war Administration and lost their hats, and finally the government said one *luluai* is enough."

"And who is the *luluai*?"

"Mano, the Buka [Solomon Islands]. Only he is too shy to talk. He can't speak Manus and that makes him shy, so someone has to speak for him."

Questions about clans yield no answers from the young men of twenty or so. If one knows the name of a clan, then a young man may say he has heard the name, or he believes so-and-so belongs to it. But clan membership still lies just below the thinking of the older men. So I asked Lokus if he as an absentee "committee" knew what was going on in Peri. He answered, "Only if it really affects those that are my own people." "Who?" "Those who belong to Kalat" [his clan]. Pushing further, one can easily get from Lokus the list of those who actually belong to all clans which are not mentioned and about which the younger men know nothing. Or, when one gets into a discussion of how the men were chosen to be sent away to study to become medical assistants, someone says, "We wanted to send a Kalo man, as they had no one doing anything, but there was no one to send." Later, when a new *luluai* had to be nominated, they decided to pick a man from Kalo, a clan unrepresented in the new bureaucracy. When there were discussions of how the village should be rebuilt, there were those among the older men who agitated for building in clan groups again, as had been done periodically in the old village.

Listening to these discussions and then questioning the young men, one found that to the young men memberships were reckoned by the two divisions. They knew who belonged to which "book" and to which "council," Pontchal or Peri, or "he still belongs to Patusi although he lives here." And they knew, in general, which men were brothers and which people were relatives, either a type of brother vaguely of the same clan, or *kantri* [related through the mother of one, mother's brother or mother's brother's or sister's son], or "business," a word applied loosely to the more distant descendants of brothers and sisters. If the questioner knows the names of the old clans and asks about them, the young men will usually be able to associate some names with the old clan groups. But their general picture is that people become brothers by an extraordinary number of routes, and once brothers, you help one another if you want to. But all help is free, and when there are two marriage feasts on the same day, people can choose. They can help their "brothers" or their *tambu* [in-laws], as they wish, or help both a little. The world is now divided into people to whom one has some kind of kin or affinal affiliation and people to whom one can trace no ties.

"Old women know all sorts of ties that no one else does, and Pokanau knows all of them—in Peri. Kisekup knows all of them in Bunai, and if you really need to know such things—but you shouldn't, because differences between the way you treat your relatives and your in-laws are all forbidden now—ask them. Only of course once in a while there is a 'court case,' as a quarrel over sago land which, because it belonged to Peri women who were married into Patusi, was never made the communal property of the village. Well, if there was going to be quarrelling about whose land such land was, it would be a good thing for everybody concerned to go out with Pokanau and find out once and for all where the land began and stopped, because *once Pokanau is dead no one will know.*"

So one watches the position of the chronicler and expert develop at these moments of extreme transition when the young men are making every effort to disassociate themselves from the old. (The Manus were always willing to delegate any sort of task to an expert, as their sense of autonomy contained no ambition toward virtuosity beyond the demands of everyday life.) "Pokanau knows and that is why we have given him the status of the lawyer man within the village." This is said with a general pat on the back to themselves for having recognized the peculiar virtues of the brilliant, cantankerous old man whose zest, old-time manners of exhortation and bad temper, and general survival beyond the expected age were all thoroughly exasperating. So then, one asks Pokanau, who has been boasting of how much he knows about genealogies and history, "Well, who is going to know it when you are gone?" And Pokanau insists that he has taught it all to the young men of Peri, to Peranis Cholai and Nauna, and to Bopau of Pontchal. But when one asks these young men—even Peranis Cholai, who has lived very close to Pokanau—one finds they know almost nothing even of their own genealogy. Occasionally someone says firmly that he is going to take a book and ask Pokanau or one of the old women about his own family line and write it down, but he doesn't do it.

For after all, what does it matter? The old clans, the old "houses" or lineages within the clans, the old elaborate kinship ties—on which arranged marriages were based and according to which each position called for special behaviour, respect, joking, avoidance, shame—that is all gone. Of course there are a few things that remain, like *pilay*. *Pilay* is the Neo-Melanesian word for joking with a joking relative and for the new kind of feasts which are given between cross-cousins, thus stressing that the old feasts were "work," compulsory, heavy work with strong sanctions behind them bearing punishment from the ghosts and black magic from men if one failed to observe them. The new kind of feasts are made freely as play, not work. In these new feasts, a paternal aunt or her son [a cross-cousin], who is "child of the woman's side," starts the ball rolling with a request for some food, typically a request to eat or drink some European food. "I'd like to drink lemonade and cake." The nephew or cross-cousin answers, "I don't believe you are equal to it," meaning, "You haven't the money to make the return payments"; and an interchange of this sort ends in a feast or several small feasts which are climaxed by an exchange in which one side gives European food, often in large amounts, one-hundred-pound bags of rice, tins of flour and drippings, cases of meat, and ready-made clothing, etc., often combined with a formal European-style meal, and the other side gives money. These feasts validate nothing and have no sanctions behind them. They are regarded as appropriate ways of spending money made in gambling. Whether or not near relatives give help is based on an uneasy set of motivations; between the younger men, affection and interest, but among the middle-aged and the older people all the old motivations of pride, anger, and grasping particularity enter in. So at one such *pilay*, one may find the younger people good-humouredly laughing about which side "loses," while two or three old women are having temper tantrums that shake the floor boards. The middle-aged group who often take part in the quarrel will nevertheless say, "They don't know any better, but you see the young people are not quarrelling. After a while all the children will learn how not to quarrel."

There was typically more leeway given for angry old women than there was for angry old men, and Pokanau's rages were not taken so lightly. Once during my stay, when his daughters had had a quarrel, Pokanau announced that he was leaving the village altogether and actually loaded all his possessions onto his canoe while his daughters mourned loudly on the shore. Here it was his cross-cousins, Kutan and his brothers, who, falling back on the formalities of the old kinship structure, dared to approach the angry old man and finally lead him ashore and into their house, where he was offered food and betel nut and persuaded to remain in the village.

These new *pilays*, although they follow the old pattern of feasting between cross-cousins, do not require genealogists to trace the lines of relationship. Only two generations are involved and the situation has been made much simpler today by wiping out all the old asymmetries in the kinship system and assuming that everyone is called by the same term as that which he himself uses to that particular relative. So the word which used to mean paternal aunt is now used *between* paternal aunt and nephew. The term of address for father is also used by men in addressing both their son and daughter, and similarly the word for mother is used by the mother to her son and daughter. Furthermore, boys can now call their father's cross-cousin [joking relative], known as their father's "one play," "one play" also. While relatives still joke with one another, they no longer curse one another. Whether the old kinship terms have disappeared more in response to the new democracy, which regards any special privileges based on kinship as exploitative, or to the attenuation of the ceremonials with which they were connected, it is impossible to say. But gone they are, except in the memories of the old, and gone are almost all the functions, except those connected with these *pilays*, and those performed at a funeral, where the paternal aunt of the dead and the paternal aunts of the father of the dead still hold the corpse.

One important function of the father's sister, which has been diffused, is being able ceremonially to "fasten a woman"—make the woman sterile for a period after childbirth or for the rest of her life—out of anger. So the younger middle-aged woman believes that any of those old women, no matter what their relationship to her, may, if she provokes the old woman, be able to curse her and make her sterile. Here, what was once a highly specific kinship function in which the paternal aunt and her descendants had special ceremonial powers, based on a special set of ghosts, over the descendants of her brother's children—involving both blessing and cursing—has disintegrated into what might easily develop into a general fear of old women, only a step away from witchcraft, especially since the men are convinced that the old women are all skilful abortionists and that every miscarriage is due to manual manipulations.

So there are no grounds for pleading with the young to know

anything of the past. The manufacture of fishing devices, once a clan privilege, is still a matter of clan practice but no longer guarded jealously against others. Village property has succeeded clan property, and to substantiate this in an argument it is only necessary to prove past ownership by any member of the village. Occasionally some quarrel over land will come up because there has been quarrelling, and quarrelling is a civil offence under the code of the New Way. So Kilipak, busy working on a new canoe with Kutan, told his wife and Kutan's wife to go and cut some sago, and showed them a sago palm which belonged to a Usiai friend of his. Monica, wife of Kilipak's cross-cousin, Pomat, and his titular sister through their mothers, started a violent quarrel with the two sago-working wives, claiming it was her sago tree, on her father's land. As the sago was in the Bunai bush, not in the Peri bush which has been communalized for the use of the whole village, the legal case rested on the question of whether the tree *was* on Monica's paternal land or really belonged to the Usiai who gave Kilipak permission to cut it. But, although there was some attempt to determine this—long discussion in "court" in Peri with Pomat acting as council-judge presiding with an emphatic disassociation from his wife and an insistence on the values of impartiality, and suggesting as a final plan adjourning the hearing to Bunai where the land was—the "court case" in actuality was not about the land but about what justification Monica had for starting a quarrel. If there were any justice in her claim that it was her sago or that she had a right to think it was her sago, then she would not be reprimanded and fined as heavily for quarrelling as she otherwise would be. For part of the new ethic about quarrels is specifically concerned with degree of provocation, and provoking anger in another is as wrong as expressing unprovoked anger. In a case like this, the old legalities, rigid ownership of each skill, each object, each inch of reef, which still exist in the minds of the people, are recognized as past realities which may cause present trouble. However, the people articulately assume that as memory fades about old property rights so also will their capacity to get angry about them fade. So there is no room for the argument that such knowledge should be kept green—and trouble-making.[1]

Old trade friendships, inherited from generation to generation, theoretically from father to son, actually from a successful man to his successful heir, are another case where the older forms are yielding to the new. In a group of men of middle age, everyone can trace the trade relationships which each of the others has inherited. The memory of present voyages is reinforced by the memory of the voyages which their fathers or uncles took in the past, when they as small children watched the overseas canoes come and go. But present-day trade relationships based either on godparent or godchild roles or on friendships made at work are known only to the men themselves, or possibly to a co-operating brother. They may be assimilated to the old hereditary pattern, and become so stabilized that they fit into a form which everyone can remember, or they may vanish altogether if godparents are not reinstated and work-boy friendships, part of the old indenture system, also vanish.

Among many peoples of the Pacific who have hereditary trading patterns of this sort, each man's hereditary path is regarded as his own esoteric knowledge, and when his son is an adolescent, the father, if he lives long enough to do so, will "show him his road." Such patterns of behaviour—found among the Arapesh of New Guinea, among Australian aborigines, etc.—may well be legacies from periods of extreme cultural instability in the past, when some well-known pattern was fragmented and became idiosyncratic and so unlearnable that each man became the custodian of the little piece of his own unshared past. Among the Arapesh, living scattered lives in small hamlets, such a way of handling the past is part of their culture, and there is a continuous fear that some precious piece of ritual knowledge will be lost, so that the man who wants to marry a widow will seek out the necessary charms from another man who has married a widow, and trembles before the terrifying possibility that there might be no one who knows them.

But the Manus, with their preference for rapid articulate systematization of culture, feel that the past will die a natural death as those who lived that way age and die. Meanwhile it is only important to teach the children that the new ways are still very new and must be cherished or they might not become stable enough to survive.

This attitude showed up very sharply in a council discussion of a proposal to make a documentary film of the Manus, contrasting the old way of life with the new. I asked the village what they thought of it, talking it over first with the older men before presenting it to the whole village. On the whole they agreed that they would be willing to re-enact small bits of the past, even willing to build several old-style houses, if it was perfectly clear that they had thrown away all the physical trappings, the dog's teeth, the shell money, the costumes. Furthermore, they were against making enough of these to put on any large ceremony. Little pieces of the old pattern would do no harm—indeed it was rather fun to re-enact for a tape recording a séance or an old-fashioned marriage discussion between two contracting cross-cousins— little pieces could not reinstate the whole, could not tempt the imagination of the young, or awaken a nostalgia in the old. But any large-scale re-enactment might be potent and dangerous. Thus in their defences against slipping back, they themselves articulately support my judgement that they were able to move so fast because they changed the entire pattern all at once—houses, costumes, ceremonies, social organization, law—making up the new pattern out of the accumulated bits of European civilization which they had learned through the previous twenty-five years.

It is possible, however, that they have overlooked the danger of small survivals, which, as representatives of a pattern, are capable of efflorescing into full-blown traditional patterns again. The parts of their past culture which they wanted to get rid of were those forms which linked together old-style economic exchanges of property, prolonged systems of indebtedness in old-style property or money, and old-style marriage systems and marital relationships, especially the power of a woman's relatives over her and her children. They assumed that it was safe to keep parts which seemed to be free of or contrapuntal to this old system. While joking between cross-cousins did relieve the strain of their complex interrelationships, it was the avoidances between in-laws, between prospective bride and groom, which were directly related to the old economics and the old quarrelsome system. So why not eliminate all the avoidances, let men talk to and eat with their mothers-in-law, let the two families of in-laws have friendly little family meals together, let brothers-in-law use each other's personal names and joke together, and let husbands and wives treat each other with public and unabashed affection as Europeans did, as they had seen in American films.

At the same time, since it was friendliness and good humour, as opposed to avoidances and quarrelling, which were being promoted, there seemed no reason for abolishing cross-cousin joking. But cross-cousin joking—compulsory, stylized, independent of personality, involving a degree of licence which would be resented under any other circumstances and always might also be resented here—is a form of behaviour that is very different indeed from the sort of easy-going, voluntarily sought human relationships which were being advocated between relatives-in-law. Under the old system, each individual personality was permitted a series of rigidly compartmentalized forms of expression—tenderness, protectiveness toward, and dependence on father, mother, and sister, uproarious licence toward joking relatives, respect and shyness toward seniors and relatives of the opposite sex. So, between men and women of the same age a man would have tenderness and grief for a sister and avoidance and disrespect for a wife—the ultimate form of disrespect being copulation—and for a female cross-cousin, public licensed joking and even public play with her breasts. The jesting relationship, between men or between men and women (it was very rarely indulged in by two women), could be defined as "a relationship within which words and actions are permitted which if performed in any other relationships would arouse the anger either of the person with whom one jested, the parents, siblings, or spouse of that person, or the ghosts." Such jesting was accompanied by loud, specially toned laughter, and the bystanders got a genuine release from hearing men abuse each other, intrude upon each other's privacy, or lie to each other.

This release from taboos, one of the very widespread contexts in which people are able to laugh uproariously, was quite understandably confused by the Manus with happiness and friendliness. So they kept the cross-cousin relationship intact; joking and its counterpart, confession—for it was to a cross-cousin that a man could tell his sexual delinquencies—continued among cross-cousins. They also kept the possibility of economic exchange a part of the playfulness of the cross-cousin relationship, but these *pilays*, "plays," were already burgeoning into a non-functional parody of the old exchange system. The Manus had made the mistake, which is also made by so many professional analysts of culture, of disregarding context and equating for context alone. Laughter was good, so cross-cousin joking laughter was good.

The most striking thing about the contemporary kinship system is its lack of genealogical depth. People can call each other *one play, tambu, papa*, "brother," "sister," *kantri* and "business," without knowing anything of their genealogical connections or of the elaborate old system by which each individual carried on submerged relationships to his mother's kin and to his father's mother's kin, through which living members of the group were able to curse each other in the name of the relevant ghosts and by which each group was distinguished by taboos on fish or shellfish. In the old days, within each clan, separate lineages called "houses" were discriminated; today there is a tendency, where clan membership is perceived at all, to override this distinction and say, "Oh well, they are all brothers." Meanwhile, old genealogical experts will tend to emphasize these lineages more than they did in the days when the important groups clustered around entrepreneurs and cross-cut not only lineage, "house" lines, but clan lines also. As the memory of these cross-cutting entrepreneurial groups faded because the activities they represented have been specifically legislated against, the lineages for which there is a specific Manus word, *um* [literally, house], come into focus again, available for some new structural rearrangement. Meanwhile most people are content to know relationships within the present. They have exchanged for the three-four generation genealogical ties, which held them together in a firm mesh, the European calendar; they are like water weeds whose wider roots have been cut, but which are now anchored to a post that goes to the very bottom of the pond.

The role the European calendar plays as a substitute for the old genealogical system is apparent when one discusses the calendar with them. Their first comment is, "Before, we couldn't count back, we only knew our fathers and our father's fathers." The other comment is, "It was very interesting to discover how things began. Before the Mission came we thought we just appeared, like stone and trees; we had no idea how men were made. Now we know." As they had not been taught the relationship between the calendar and the birth of Christ, the effect of the calendar has been to make them feel that they have only a place in that part of the calendar which has transpired since they learned about it.

So calendrical life began in 1946. Before 1946 the Manus did not know what date it was; they lived in a different time, a time in which each generation counted forward and backward from themselves, unconcerned with whether there had been a beginning or would be an end. As events were once anchored in a mesh of genealogical relationships, they can now be anchored in time and space, if each of these events is written down on paper.

But this anchorage can only be achieved if they themselves are literate and do the writing. Too often, a people break their old ties to each other, and so to their only measure of time and the past—a past stretching back to the time when two men had the same ancestor—and put nothing in its place. Such peoples learn that to the officials of some foreign bureaucracy something called "age" is important and to co-operate will give some age, sometimes with the most improbable discrepancies, "How old are you?" "Oh, surely you are more than nineteen!" "Well, perhaps forty-nine." Such a process, in which an uncomprehended new system is substituted for a comprehended old one, detracts from human stature, so that often the illiterate peasant or illiterate urban worker seems to have less dignity than his primitive brother. To this extent the untouched savage is "nobler," because he lives within a system which he fully understands, while the various stages of illiteracy and non-literacy which accompany literacy usually deprive many people of just this dignity.

The new Manus way of reckoning relationship in just two-generation terms and of confiding records to paper carries with it a general sense of kinship co-operation. Brothers are expected to co-operate with each other, without compulsion and with none of the old problems of dominance and exploitation. To have brothers today means that all enterprises are easier to carry out, especially if one brother is particularly good at house building or deep-sea fishing. The other co-operative unit today is composed of a man and his son-in-law, a relationship which in the past was very strained and difficult. Today a young man, either from another village or one who simply gets on well with his father-in-law, may become his father-in-law's fishing companion, go with him to market, share in most of his enterprises. Whether fathers and adult sons are also going to become a co-operative unit, it is too early to say. Meanwhile with the breakdown of the old entrepreneurial system, which provided a man of initiative with young male assistants, a man who does not have either a younger brother or a young and vigorous son-in-law has either to do his own fishing and trading, lone-wolf style—as do Pomat, who relies on cylindrical fish baskets, and Alois, who relies on crab trapping—or to adopt a young relative.

Adolescent boys now do far more productive work than they did under the old system and provide a good share of the fish that is eaten in the village (as opposed to the larger scale fishing that is traded). But the work done by the adolescent boys is complicated by their position as schoolboys, which is a continuous cause of irritation between parents and school. Twenty-five years ago, children and young adolescents did practically no work; in New Peri they participate in almost every adult activity. Meanwhile, one sees many canoes put out to sea by a single mature man of standing with the rest of the crew consisting of women and children. The breakup of the old male units of work also means much more canoe work for women, as a wife—now a companion to enjoy rather than a person to avoid as much as possible—is often thought of as the most desirable person to take to market or on some other short trip. But women still do comparatively little night fishing except in the cases of young married couples who are in love with each other. The general effect of the breakdown of the entrepreneurial system of work has meant more participation for children and a greater involvement of women in tasks which were formerly regarded as appropriate only for men.

The economic system has undergone an extreme transformation without, however, altering the ordinary everyday economic life. In the past, all capital goods—that is, the tools of production, canoes, nets, and smaller fishing devices—and most consumer goods—bowls, pots, oil containers, spoons, tables, and ornaments—were obtained within the complicated system of affinal exchanges and native trade, in which the exchanges within the village were the means of distributing objects received in foreign trade and of providing the occasions for which men sailed abroad in search of specific types of goods to meet their carefully specified affinal obligations. Today, a large part of this system has completely broken down as consumer goods bought from the European trade store with European money take their place. Thatch, once bought from the Usiai, now can be gathered freely in the bush, and many of the small land products, such as small house-posts, wood for house rafters, logs for small canoes, can be taken freely from a bush which has now been made common property, shared with the land people who have come "inside."

There was only one controversy involving a muddle between old and new systems which came up while I was in Peri, and this significantly involved a man on Lou, who was "outside" the New Way. Kilipak, representing Peri village, had gone to a hereditary trade friend on Lou to ask for a great tree to make an outrigger for Big Manus, the big canoe of Peri village. This trade friend had said he would get it from a trade friend of *his*, and later the second man, the real owner of the tree, who should never have come into the picture—and would not have under the old system—began to dun the Peri people for direct payment, going so far as to take the matter to the government. Although Kilipak was quite clear that this represented a clash between the old system of long-term obligations between hereditary trade friends and a modern system of direct payment, the case was still a vexatious one, and the Peri people had compounded the uncertainty by paying the claimant whom they claimed had no claim, thus acknowledging his claims and providing the district officer with a difficult problem.

Under the present system, people are supposed to work for money and to trade for money. This is the ideal, with each individual having the right to buy or sell, the right to work for wages, and the right to the wages when he earns them. But there are still important objects of native trade—large turtles, big tree trunks, bark for making netting string, rubber gum for caulking canoes—which are hard to obtain. Native money was ineffective under the old system of trade, and European money is just as ineffective under the new. It is necessary to invoke old ties, to promise future accommodations, etc., to get the land people to provide raw rubber nut or to persuade a Manus man to look for turtle or to make a canoe. Meanwhile people are beginning to count labour as an investment, and to calculate that it may be cheaper to work for wages in a European setting—on a plantation, in Lorengau, on the Australian Naval or Air Force station—and buy cord to make a fish net than it is to buy the bark and make the cord themselves. Fishing itself has to be re-evaluated when, instead of a bundle of coconut leaves, a pressure lamp full of kerosene is used up in a night's fishing which yields no catch.

In the meantime the day-by-day economic life goes on. The night before a market day people fish, take their fish to market and trade it for taro, sago, betel nut, and various kinds of fruit. The market looks very much as it once did. People arrive and lay out their wares in matched lots—so ten taro, which equal one good-sized fish, will be laid out in a cluster. But there is an increasing instability here also, as both groups, sea and land people, can get increased prices for their wares from Europeans. There is a shortage of fresh fish on the whole island and sago is welcomed by plantation managers with a native labour line to feed. So prices rise, and the fact that they are rising for both kinds of products, land and sea, does not comfort those who have to pay in money or tobacco. Tobacco is worth just half what it was worth twenty-five years ago; all trade goods are correspondingly more expensive. At the market, if there is more fish than taro (and this is spoken of as winning over the taro sellers), then the Usiai either have to take the fish on credit, promising taro later (which is what the sea people prefer), or pay for it with money, which seems excessively dear to them while actually it is cheaper than what it would sell for to Europeans. This latter fact makes the Manus feel that they are operating at a loss and giving the fish away to their Usiai friends, cheap. Native trading operations are thus becoming more expensive, and the temptation to buy from a European trade store increases.[2]

At the same time, people do not have to work as hard as they did under the old system, especially since working for Europeans by the month, which is the principal source of money, is regarded as doing comparatively easy work. A family may feel they would like some new object, a new cooking pot or a little suit for one of the children, but this desire for an object that is urgently needed for an occasion is very different from the goading pressure of the old system of constantly having to meet obligations. Furthermore, there is very little uncertainty in the new system; if you have the money for trade cloth, the trader will sell it to you. If he hasn't got any more of the particular colour or kind you want, nothing that you can do will result in more of it appearing on his shelves. You can't promise him to catch a turtle, or make a dangerous voyage, or work for three months; all he can do is re-order a popular line, which may or may not ever arrive. Where real daily needs are taken care of by subsistence activities, and all needs above this level are either to be obtained for money or are not obtainable at all, there is a considerable lowering of anxiety and a consequent slackening of economic effort. This is the period of transition in an economy when Europeans traditionally have come to think of primitive peoples as lazy.

At their first contact with a completely primitive people Europeans may think of natives as doing very little work because the Europeans cannot understand the rhythm and style of the native activities and may not allow for all-night fishing, for work begun before dawn—which allows a long rest at noon—or for work that is done in spurts and may necessitate many sleepless nights. Also they may not consider ceremonial activities which are time- and effort-consuming as "work." But it is the second stage of contact when, by abandoning the complexities of their old system, the natives have simplified their wants and so will only do a limited amount of work for money, that the natives seem "lazy," especially to the European short of labour, anxious to recruit a boat crew, or sell more of the goods on his shelves. The same sort of period of what looks like "shiftlessness" or "laziness" occurs when working-class people are suddenly given better wages without any real change in standard of living, either because their aspirations do not change or because, as in wartime in a modern society, there is a shortage of consumer goods. Then will come the lack of incentive to work more which was so striking in Britain after World War II and which has been a continuous problem in the Soviet Union with its persistent shortage of consumer goods.

In modern economies, which depend upon high production, these breaks in the will to work are very serious, in their effects on coal mining, for example, or agriculture. In a society like Manus, in a transitional state, while it does not matter very much whether people do or do not earn more money in the southeast season in 1953, except possibly as the temporary prosperity of one local plantation or trade store is affected, the long-range effects may be very serious because the whole state of the culture is so fluid. A general sense of lack of pressure which now expresses itself in an easy-going work pace so that two men carpenter their canoes and three men sit and watch them, can easily be translated into new leisure-time habits, gambling for lack of occupation, or styles in work which mean that only the disgruntled go away from the village to work, thus making going away to work a statement of disgruntlement. A plantation owner may be driven to playing one village off against another as a way of getting any labour at all. If the diet has incorporated many items which come from the trade store, and the stocking of these items fluctuates with the supply of money which for other reasons there may be no great pressure to obtain, then dietary deficiencies may result. Twenty-five years ago the Manus were very well nourished; today coconut oil and coconuts, with copra at an all-time high, are being steadily decreased as articles of food, so that their diet is poorer in fats although better off in fruits than it was.

The New Way rules about ceremonies and payments mean that no one is hard put to it to get married or to pay for birth care. Five pounds is easy enough to collect in contributions of a few shillings each from a large number of people. In the new ceremony of the "Bride's Dishes," in which she is given pots and pans and enamels, dishes and spoons, etc., everyone is free to contribute as they please, something old or something new or nothing.

And yet some people feel that they are driven by work, especially men who have several small children and no brothers, and this is aggravated by the New Way insistence that husbands should help wives care for their children. Petrus Kenandru was an example of this sense of being driven. He lived with his old father-in-law and provided for four small children, two of his own, and two of his wife's nephews, one whose mother was dead and one whose parents were separated. Petrus did not gamble; he didn't even watch gambling; he fished continuously and expertly, and he felt put upon. When the motherless child, whom he and his wife had cared for so devotedly, died—due to the carelessness of the child's father, home on a visit after having got into some unconfessed trouble—Petrus was bitter and resentful.

Lukas Banyalo, the extra volunteer preacher, was another man who felt driven. He worked continuously, always taking one or more of his five children with him—because they were too much for his wife—and then complaining when the children got cold or sick or hurt or interrupted his activities. In fact, Manus men at present are a little like Western career women who find it difficult and unfair to try to combine work and care of children. This sense of carping bitterness of the good men, whose goodness is translated into no rewards of power and enterprise, is a negative element in the whole system, likely to lead to self-pity and cultist and anti-European attitudes. So Lukas continually talked of his poverty as being caused by the lack of true knowledge—which the Americans possessed—as to how a real modern system worked and of how to enlist God completely on one's own side. When he preached he spoke longingly of the garden of Eden, where the trees grew so low that one didn't have to reach up to pluck the immediately edible fruit. It was easy to hear in his words the echo of a Western religion that had failed to make work a bearable part of life and had instead seen it as a curse, combining with earlier Manus attitudes of being driven and coerced.

There are several other important irritants. Under the new system, both long- and short-time saving are almost impossible. Under the old system, a man might have a house filled with sago and oil, betel nut and pepper leaf, and yet be quite able to withstand requests to share or lend any part of it. All of it was pledged for particular purposes; the people who were to receive it had already been told how much they were to get. Righteous, obeying their Sir Ghosts' behests, they could refuse without undue churlishness, and they did refuse. Today, no one can plead such protection. If a man has bought a whole bag of rice for a *pilay*, he can try to keep it intact. But even so he cannot refuse to sell some to someone whose *pilay* is coming up sooner, even though this means a two-day canoe trip to get another bag. And a one-hundred-pound bag of rice once opened is used up in twenty-four hours.

This stage in which relatives prey upon anyone with a surplus of any sort is also a familiar stage of culture contact, in which

Europeans assume that the native economy is hopeless because of the way relatives claim a share in any savings. It is instructive, however, to see that this situation is in fact a product of the present transitional state of culture contact and did not exist twenty-five years ago. What has happened is that the transition to a European type of economy has been incomplete. In fact, the very same situation which occurs between natives, who can't refuse to lend or share a pound of rice, occurs among isolated Europeans in New Guinea, who feel the same kind of obligation to share anything of which they have a surplus. Storage of scarce goods in one's home, goods that must be obtained from a distance or at wide intervals, but which can be bought for money, is not a situation which can be dealt with if decent human relations are still to be preserved. Europeans can guard one lamp mantle, a supply of sheets which are just sufficient for the family, or packages which have been bought for Christmas and cannot be replaced before Christmas, but otherwise when your neighbour runs out of butter and if you have more than you need, you lend, whether you are a German-born missionary, a Manus native, or an Australian planter. Nor is it possible in Peri to carry around a full purse of change without continual inroads being made. Among the Manus, the inroads take the form of borrowing or buying by close kin and neighbour rather than begging, but where goods can only be obtained by long voyages, both loans and sales are crippling.

Take for instance the question of mantles for pressure lamps. These mantles are delicate little bits of silken net; they cost two shillings each, and once on a lamp they are likely to break at a sudden jar from the impact of an insect or a sudden shift in temperature. Any man who depends upon fishing with a pressure lamp is well-advised to have at least one spare mantle. But if he has a spare mantle, and is known to have a spare mantle, then he is subject to pressure to sell it to anyone he knows whose mantle breaks. Here it is not so much a question of other people's begging for one's property as the simple question of availability of necessary and scarce goods. So people tend not to take the trip to

Lorengau or Ndropwa to buy spare objects until *after* they need them. With uncertain weather and many days when canoe voyages are almost impossible, this means long periods without necessary objects, particularly food. Men who know that they will need coconuts to make the post-partum coconut soup meal for their daughter's delivery keep a supply of coconuts in the next village, and the large bags of rice or tins of fat which are bought for feasts are often also kept in another village.

Twenty-five years ago a study of the Manus economy threw new light on the relationships between barter and money in an economy of scarcity, where it was possible to see compulsive barter functioning as an incentive: "I will sell you X, which I have or make or can trade for, only for Y, which I want and you have or can make or can trade for." It took both this hard-fisted holding out for particular objects and the pressure of the intra-village ghost-sanctioned exchanges to keep the economy up to a high level of food and housing. The present situation throws equally vivid light on the importance of retail stores and banks in the functioning of a money economy.

At present the Manus treat a five-pound note as a bank in the sense that a piece of "blue money" cannot be broken, and a man is able to keep a five-pound note for months. Normally no one borrows such a large amount; it is too big to contribute to a bride price or other ceremony, so it can be saved for some large-sized purchase relatively safe from importunity, just as an unopened bag of rice is relatively safe. But any smaller amount of money is subject to levies all the time. So the people have created an informal bank. Selecting the most trusted man in the village, Michael Nauna, they began to entrust to him first the money which they expected to be collected as council taxes—having begun to do this almost two years before the governmental machinery for collecting the taxes went through—and later other money as well. There is as yet no idea of interest, and Michael received no remuneration of any kind for taking care of this money. There was some vague idea in everyone's mind that handling large sums of money was equivalent to being a big man in the old-time style, and people said, "Michael is the only man who lives like a big man of old; he does not go fishing or trading, he sits at home and plans."

This puzzled me for a long time because I couldn't see what he planned. His house was next to mine, one of the two "corners" of the square, the most honorary place in the new village plan. The first time I left my house for a trip to the next village, people said, "Just tell Michael you are going. He never goes away. He will look after it for you." And I became aware of a continuous quiet alertness in the house next door, where Michael, himself extraordinarily gentle and unobtrusive, and a wife as gentle as he, and their five children were quieter than any household in the village, but also more persistently wakeful. I learned that whenever there was a shout or a cry anywhere in the village, all I had to do was to call softly across to Michael, whom I had known well as an exceptionally patient, delightful little boy, to get an explanation. At one point in my stay I began to wonder whether Michael suffered from mild depressions. He had had a bad eye infection and had seemed so unusually patient and unanxious to be up and doing, whereas most Manus insist upon going about before a major infection is over. Then gradually I realized what this was all about—Michael was the bank!

He himself was only partly aware of what had happened. He knew that when he had been an appointed official people had begun to give him money to keep for council taxes. He had wearied of the official role and asked to be released from it. But they still gave him their money to keep. Some of the accumulated tax reserves had been given to contemporary officials but with disastrous results because they were not able to resist requests for loans, and so this money too was given back to Michael. He complained mildly what a dreadful lot of work it meant; sometimes people drew money out and put it back twice in one day. The illiterate among the Manus still wanted to see their money periodically. So Stefan, who was working for me again in 1953 and who was the only illiterate among my original group of adolescent boys, would ask to see his wages at the end of the month, even though he knew that I would write down the amount and keep it for him. And Michael complained that he never could go anywhere, that he always had to be at home and wakeful, guarding other people's money. While I was in Peri I shared this banking role with him for those people who did not want other people, including Michael, to know anything about how much money they had.

But the step between the old methods of handling and storing money and the new is a very difficult one to take. Banks are a serious matter in any modern country. An administrative permission to set up a series of native banks would have to be scrutinized very carefully. There is at present no legal mechanism by which people can save their tax money for the time when it is due. True there is a savings bank in Lorengau, and natives can go into Lorengau—a trip that often takes two days—and deposit their money there, then take another two-day trip if they decide they want to draw it out. Banks two days away are obviously not the solution. If Michael were given official status in the village, and some arrangement made to reimburse him, either with interest or by village contributions of labour for him—doing his house building, fishing for him, etc.—this would be more equitable. But this official status would have to be extra-legal, in the sense that it would have to be separate from any position in the council or co-operative recognized by the Administration. For here we come up against one of the dilemmas plaguing governments all over the world. It is necessary to have tighter regulations controlling the actions of government employees than those for other people. This automatically gives them less leeway. Michael is the sort of man who should be the treasurer of any new economic venture. He is the ideal treasurer, patient, intelligent, firm, proof against temptation, does not gamble, has great prestige, and belonged to a wealthy and stable family. But if he becomes a government-recognized official, then his very necessary informal banking operations would become illegal. The irony of this situation is not reduced by the manoeuvres of young Administration officials unsympathetic to local self-government, who say articulately, "I'd like to give so-and-so a government appointment, then we'd get him."

As they need a bank, so the people of Peri need a store, a small retail store where people can buy enough of some commodity for the next meal, or enough kerosene for a night's fishing or to keep a lamp going in illness, or a new lamp mantle when their mantle breaks just as they are starting out for a good night's fishing. But this would be a store in which a given individual bought articles and sold them for money at a profit, in fact a small and rival version of the trade store four miles away on Ndropwa. Planters tend to extend their trading operations by placing little branch stores about the islands wherever there is money floating about and they can find a native who can be trusted to run the store. So a native shop of any sort gets into the category of government-licensed trading operations. To have a shop you have to have a licence, which costs money and renders you liable to legal measures if you engage in irregular practices.

The alternatives are council-run or co-operative stores, which again have to conform to a most elaborate, carefully worked-out code, to keep good books, which takes good clerks, etc. In the meantime, people do not have enough to eat because of a partial shift to a money economy without the institutions which a real money economy demands. Here we see the irony of making a rapid leap into civilization. In Bali, where the people had markets which operated on a straight money economy, using strings of Chinese cash—twenty-one to an American cent—an increased dependence upon buying food or cloth or kerosene could be fitted painlessly into an economy so old and so well-established that it could absorb the change. Such a change is in fact introduced into the existing economy of the old culture. But the only money economy in New Guinea is that introduced *from* a modern culture, with all its paraphernalia of banking laws, bonding, licensing, etc., originally extended to New Guinea to regulate the way in which Europeans conducted their affairs with each other—as Europeans—with a few adaptations to protect the natives, such as making a New Guinea shilling with a hole in it so that shillings could be strung on a string and were less likely to be lost.

When the natives suddenly skip four thousand years of development and try to adjust to a full money economy, they are caught in a transition state which definitely favours the development of extra-legal institutions. If there had been old institutions of usury, pawn shops, etc., these would have been protected by the Administration as "native institutions." So also with native courts, village work, fines for failure to do village work—all the devices by which slightly more advanced peoples have maintained some sense of community over the ages. The Administration in Samoa simply imposed new tasks on the old ones. Where the local village government ordered everyone to build a guest house or ordered the young men to smooth the village square or engage in communal fishing, the American government added the duty of collecting coconut beetles on Friday. In Bali, the Netherlands government added to the local system of village-appointed watchmen another watchman duty, a man who waited at the next higher headquarters to carry such messages as were needed. This was part of his obligation to his society, already defined in a thousand intricate ways, in relation to his age, marital status, position on a native roster, etc.

But in Manus, before the New Way, there were no communities to which any individual owed anything whatsoever, so any services performed for government were not extensions of old duties but newly imposed exactions. All through the twenties and thirties struggles went on as to what rights a government official or any travelling European had to demand help from native canoes or carriers over the mountains. The missionaries who specialized in protecting native rights often found themselves confined to the coast because they were unwilling to "force"—bully by extra-legal means—the natives to carry. Since World War II the situation has become more extreme. Work on government rest houses is now paid for by the village council; government officers make patrols pay heavily for the canoes that carry them from one place to another and for carriers who carry their official and personal gear. Thus every administrative ruling which has been set up to increase native dignity, to ensure against any sort of "forced labour" or "corvée"—those legitimate bogies of European standards of freedom—has decreased the possibility of anyone ever being asked to contribute any work for the community. But the ideal of community which lies at the heart of all Anglo-Saxon institutions, that community—in which people live together in voluntary co-operation, helping each other, caring for the widows and orphans, and keeping themselves unspotted from the world—of which the Manus now dream, is a community in which people do many, many things for the community without remuneration. No American or British town could run without the hundreds of thousands of volunteer hours contributed to local projects, to the church, the hospitals, the volunteer fire service, or in large cities to committees or royal commissions examining what is wrong with the tax-supported fire department or the schools.

The community plan developed by Paliau and the Manus provided for such genuine communities, in which the names in the government book—developed originally so that government could tax and govern individuals securely pinned down in space—were to mean something real. Each of these communities was to be governed originally by a *besman*, chosen by the community and responsible to Paliau. Later, after the break with the Mission, Paliau added the plan of having each village select a lay preacher, and he called them together to learn a common ritual. At this early period, each village already had its complement of government-appointed officials, a *luluai* [headman], a *tultul* [interpreter and secondary headman], and a "doctor boy," officially called a medical assistant. During the early days of the New Way these officials used to play an important role, summoning the village to line for the patrol officer or the doctor. They naturally took an active part in the New Way. But the New Way was based upon requiring from members of the village work on community projects, and imposing fines if this work was not done, after a trial modelled on a British magistrate's court. Native officials with government appointments became liable for coercion, extortion, and usurping of governmental functions if they engaged in any way in requiring village members to get work done. Thus we have three periods: first, the organization of the villages into communities, during which the "Hats" [government-appointed officials] played important roles. Second, the complaints, usually European-instigated, against these native officials that they had committed illegal acts, with consequent government prosecution of these officials and their imprisonment at hard labour, which has remained a cause of permanent resentment and bitterness. And third, the present phase, in which the new type of locally elected officials coexists with the old type of appointed officials in the period "Wait Council," with everyone very much confused as to who has power to do what. The present régime tended to give the local population a choice between taking a complaint directly to the *luluai*, who might attempt to settle it or send the litigating parties straight into the district office, and an attempt to settle it in a village "court" presided over by the two new "councils," and the *luluai*.

In this village "court," which had no legal standing and would have been branded as illegal, there was a serious attempt to administer the local code according to the people's best understanding of the procedure of a very simple magistrate's court such as they had experienced when patrol officers heard cases. Each side presented its case in the presence of the other. The clerk kept a record and the "councils" cross-examined the witnesses. Decisions were a mixture of harangues on the ethics of the New Way and fines, which were collected but kept in escrow to wait and see what size they should really be and who would have the right to collect them. The people knew that these "courts" rested solely on the consent of the governed, and that they were without any higher sanction, except in those cases where the offence for which they were arraigned was one recognized in the Administration penal code; then "councils" and *luluai* could threaten to turn the accused over to the government. Where the crimes had merely been against the New Way code—failure to attend "the line," failure to co-operate in village work, repeating gossip—the only sanction, aside from devotion to the ethics of the New Way, was a fear that when the council was really set up, then the men who had been elected councillors would have real power. This was the sort of interim situation in which the unstable were able to contribute new cultural definitions which would have been impossible if there had not been such a long delay.

So Stefan, who was lazy, illiterate, and unreliable, said, "These days I don't go about trading or fishing. [He let his wife do it.] This means I am very poor. But I stay right here near the village officials ready to do anything they ask me to do. [This was untrue.] Later, when the council comes they will be very powerful and they will punish a man like me if I haven't done just what they said." It is easy to see how such a speech repeated to a government official would have seemed like evidence of blackmail and coercion.

Here again it is important to realize that law courts, like government, taxation, community responsibility, school, were completely new institutions and institutions which have caught the imagination of natives all over New Guinea.[3] Without effective institutions for settling disputes—except by feuds, raids, and subsequent ephemeral peace-making ceremonies often with payments in expiation—the British court with the whole sanction of armed might behind it, which could settle things impersonally and see that they stayed settled, seems a magnificent invention, as indeed it is. We often see courts as institutions which of course we have but which don't function as well as they should, and most of our attention is centred on abuses: court cases are too expensive, or there is undue influence, or the jury responds with prejudice, or judges are appointed for the wrong reasons. We see our whole precious and dearly bought legal system as "natural," something that is always with us like the weather and like the weather is a very mixed blessing. But to the New Guinea native, newly fired with a desire to keep his society "straight," the whole legal system looks fresh and beautiful. He sees it as a magnificent invention, as wonderful as the airplane, so that far into the interior of New Guinea proper the institution of illegal "courts" is spreading.

But here again there is a serious problem which did not confront more highly organized societies in Africa, where old-style native courts existed which the European administration might limit or curb. A "court" in New Guinea is inspired by and modelled after a British court, with its well-tried rules for the administration of justice. Perhaps more than any other type of official district officer acting as a magistrate, the judges have made modern New Guinea what it is today. Their impartiality and devotion to the abstract principles of Anglo-Saxon justice have slowly taught the natives that the Law is no respecter of persons, have convinced them that a European who kills a native will be dealt with even as a native who strikes a European, or more severely. After World War II, the judges travelled thousands of miles on circuit, holding courts under appallingly difficult conditions because they believed that the restoration of order and sense depended upon the natives' realization that the rule of Law had returned. In the native courts, Australian officials were appointed to defend the interests of the accused, to exercise every intellectual resource to give the accused native, who might just be meeting Europeans for the first time, his full rights. Sometimes there are fantastic difficulties involved in attempts to observe the meticulous requirements of British law and still take native psychology into account, as when the court impounded as an exhibit the half-used bottle of an anti-sorcery potion with which a sorcerer had been washing the body of a desired woman. The husband's arrival on the scene had interrupted an attempted rape. In the preliminary hearings the woman and her husband pleaded to have the magical potion back. The judge, who had a humane and extensive knowledge of native psychology, showed them the bottle in a locked case and assured them that it would be safe there during the waiting period before the next Sittings of the Supreme Court (where a case of attempted rape had to be tried). But in spite of these precautions, the woman—convinced that without a complete washing she was still vulnerable to the original spell—wasted away and died when the bottle was not returned.

There is also the problem which arises when natives who live in uncontrolled territory have to be tried for wilful murder of a white man. The officer appointed to defend the accused native may plead that the native was ignorant of the law and had only acted according to native custom. But the Commonwealth government has enacted that its law should run throughout the Territory and over everybody in the Territory, irrespective of their colour. However, the native if found "guilty of wilful murder" can then be recommended to mercy by the judge "recording" rather than "pronouncing" the sentence. Thus the law is upheld and yet due recognition is given to the natives' ignorance of Western law and custom.

Within this system, so dearly guarded, young patrol officers are educated, a kindly and responsible senior magistrate instructing their faltering tongues and unaccustomed minds. Within this system, the senior members of the judiciary have given the whole of their lives, often in loneliness, to maintaining impartiality and freedom from personal pressure, in an atmosphere where partiality and back-scratching were the natural accompaniments of a small settlement and too great familiarity among the few European settlers. If these magistrates and judges had been asked to evaluate existing native courts and find what was good in them, what was worthy of cultivation, none would have been more ardent discoverers of latent native talent. But in New Guinea they are asked instead to evaluate the possibility of native *copies* of British courts—to be set up forthwith and presided over by natives who had had no tuition whatever in what are regarded as fundamental principles of British justice—working successfully and without bringing the whole system of law into disrepute. They are asked to believe that natives, who have been reared under native political systems in which impartiality or a failure to show partisanship for one's own kin was a crime, can become impartial overnight and that men can suddenly become impervious to pressure from kin and blackmail from enemies in their own small villages where they have lived all their lives. And the judges are understandably doubtful. There is a great deal of sympathy toward the resolution of disputes—over gardens, marriage payments, etc.—within the villages, as long as these local tribunals are not confused with courts with criminal jurisdiction. But meanwhile the natives have fastened upon one essential part of the legal system—that the law has teeth in it, that behind the careful weighing of evidence, the desire for impartiality, is the power to enforce the decision. So, without any old community machinery for arbitrating disputes, there is a great temptation to construct imitations of British courts and to claim for these courts a jurisdiction which cannot be accorded to them.

So the local "courts" in New Peri in which the officials try so hard to maintain impartiality, to sift evidence, to dispense justice, suffer from a double handicap—the lack of experience of the "judges" and the knowledge of everybody concerned that the system is not only not supported by the government but may actually be punished at any moment. The shrewd can flout the local "court," and it is only the trust and confidence of the whole community in one of the two councillors which has made the system work at all, in spite of the seriousness with which people take it, as young men come in full European dress to present the evidence which clerk and councillor so meticulously take down.

"Don't call them courts and so bring the courts into disrepute." The advice of the judicial official was echoed in the words of the young schoolteacher, who said to me, "Don't call it a school, this that we have. If you do, the government may refuse to send us a real teacher. Don't say we have a school, just say that I am doing a little something to keep the minds of the children clear until the real teacher comes." For the school, like the court, was an attempt at a copy of a school but with very little to go by. Yet there was no doubt that this was a school and that the teacher, who himself had only two years of schooling, had grasped the essentials of what made a school.

After the morning line-up dispersed, the young adolescents

were all gathered together, seated in the square, drilled for a few minutes, and then marched into the church, where a blackboard (made by painting a piece of plywood black) had been set up. There were two classes plus the young grown men who sat at the back attempting to make up for lost time. The youngsters sat with clip boards on their bare knees, working on paper from old Naval Hospital charts or on USO stationery or in copy books for which they had individually paid two shillings, and learned to read and write, to add and subtract. The atmosphere was that of a school; the children's minds were being kept clear for more professional teaching, and the young adolescents were a disciplined group, owing obedience to their teacher and special services to the community. One afternoon out of the week that they attended school, for they attended school on alternate weeks, they fished for the teacher and for the church teachers or went on expeditions into the bush for building materials. The behaviour of these young people demonstrated how valuable it was to have the whole of the Western pattern so that as the power of the parent was curbed, the community could take over with a new set of expectations and disciplines.

But in the relation between the adolescents' schooling and their duties at home, a new and ironic difficulty arose. In the past, adolescents had done very little work, but with the breakdown of the old religious fears which kept them close to the village, and the breakdown of the old economic system, their services became necessary because the former supply of economically dependent young men no longer existed. Parents had just begun to rely on the work of younger adolescents. If the school had been more firmly established before this shift, if in fact the Mission had set up a fuller schooling pattern at the time of the overthrow of the old religious system, this conflict might not have arisen. As it was, parents complained that their sons worked too much for the schoolteacher, and men, unused to the larger charitableness of the New Way, complained because with student help the teacher was able to build a larger and finer house. And the councillors complained that the schoolboys weren't available to work for the larger community, echoing the familiar plaint of employers of child labour against keeping the children in school too long.

There were even conflicts between the "Department of Education" and the "Department of Public Health," the latter represented by my clinic where boys infected with yaws were given semi-weekly shots. These appointments interfered with the teacher's scheduling, "doctor" and "educator" had failed to "clear" with each other, and bitter words ensued, not made the easier to bear by my explanation that as far as was known conflicts between departments of health and departments of education were an almost inevitable part of Western democracy.

Medicine was perhaps in the most parlous state of all. One of Paliau's first moves had been to establish as a pattern in each of the villages two "hospital" buildings, one for obstetrical cases and one for non-ambulatory cases of illness, both built out over the sea to deal with the problem of sanitation. This was an exceedingly intelligent measure, but it came into conflict with the Administration, just as the "courts" had, and as the school might. Paliau was accused of setting up illegal hospitals, of keeping patients in the village who should have been sent to the government hospital. Hiding patients from visiting medical inspections had always gone on, but not in a "hospital," and again what the natives saw as a step toward complete modernization appeared to the Administration as "subversion."

When I arrived in Peri there were no longer any New-Way hospitals. There was no "doctor boy" at all, since the hat of the former old-style "doctor boy" had been taken away for failing to take a case to the hospital. And there was no supply of medicine in the village.* The Peri man who had been sent away for training was still in New Britain. I set up a clinic, well-provided with new bulk supplies by the Administration, and integrated the timing of the clinic calls into the village pattern of bells. Through the

---

*This was not typical of conditions in the Territory as a whole but a by-product of the unusual transition conditions in Manus.

months we struggled with people from other villages who brought the sick to my clinic to be cured, and encountered another glaring need in the system, which the natives had also attempted to meet and had been rebuked for—the need for the control of movement from place to place and for the reception and care of the stranger, in fact the need for public rest houses or for an inn.

In old Manus one did not go to a village where one had neither relatives nor hereditary trade friends except on formal feasting occasions. In dire emergency a nearly shipwrecked canoe might wind its shell trumpet and ask for sanctuary, which might or might not be given. People who wanted to trade, but who had only very slight kin ties, might sail into a village in the daytime, but would cook and sleep aboard their canoe, usually anchored a fair distance away. No hospitality was asked for or given. At the opposite extreme stood the behaviour of the *luluai* of Polot, who once replaced all that a shipwrecked Peri canoe had lost. In the past, travelling had been done along well-established lines; visitors arrived and departed laden with gifts of food and with trade goods; their hosts ostentatiously protected them against the suspicious unfriendliness of the rest of the village. When someone was ill and was moved from one village to another in search of a magical cure, it was again close relatives and trade friends who played host to the sick man and his accompanying kin. When people fled from the anger of ghosts or men in the village where they had lived, they again fled to kin in other villages. The sense of being a Manus free to move to another Manus village depended entirely on having actual concrete ties in another village; a strange Manus village was as strange or stranger than a nearby non-Manus village where there were kin or trade partners.

After 1946, with the great surge of a sense of common cause and common ethnic identity, people felt that it should be possible for men to move freely within the limits of the New Way and to be welcomed and entertained as brothers. But brothers of whom? The poignancy of "There was no room at the Inn," which has wrung the hearts of millions of listeners to the Christmas story, had a new version in the Manus, "There was no Inn" and no one knew how to construct one. Here their contact with Europeans gave them no proper model. Hotels in Rabaul are called "house drink"! There was no inn of any sort on Manus in 1953, no guest house other than that part of the District Commissioner's establishment where he entertained official visitors. And this house was a temporary building, post-war salvage, and falling down. The old compulsory construction of government rest houses for travelling officials and other travelling white men was also a poor model for the new order.

If all men of their own political and religious faith were to be treated as brothers, how was this to be done? They tried three solutions. A village official was appointed to blow the answering conch shell to welcome ashore passing canoes which would, as in the past, wind a conch shell requesting the right to come ashore. But this was a gesture without a sequel. It was all very well to wind the conch shell, but what came next? This new custom fell into disuse. The second model was the "pass" that every work boy has to carry in New Guinea if he goes from one district to another, a sort of temporary passport, an extension of the "pass" which a house boy's master gives him if he wishes to stay out beyond the curfew hour in European towns. This, by some fluke, was confused with "customs," another European device for regulating not the movement of men but of goods, and the requirement was set up that people moving from village to village should carry "passes" and go through "customs." Here, as in the case of Paliau's local "army" and "navy," there seemed to be a real attempt to have a local group take over functions appropriate to the central administration. So "customs" was frowned upon by the Administration.

The third attempt to deal with the question of hospitality on a community basis was for the councillor to act as host for the village, in a sense reproducing the role which the District Officer had played in Lorengau all these years—even to the circumstance that district officers in New Guinea were given no entertainment allowance for this continuous duty.* So the councillor set himself up as responsible for visitors and for strangers married into the village who got into trouble with their in-laws. This had several drawbacks: hospitality was very expensive and was likely not to be recompensed as meticulously as the old type; people were confused as to whether it was or was not one of the councillor's duties. And women were involved! Those who wished to defend the New Way and its leaders against charges of loose living maintained that the shelter given by councillors to runaway wives was given as a political duty, while the enemies of the movement described all such moves as part of the leaders' attempts to form "harems." The truth lay in between. But there was no doubt that if the people were to run a modern community in which kin ties were to be progressively loosened, then one of the devices for maintaining such a modern community was an inn, a place where the stranger has a right to ask for, and pay for, hospitality.

In a primitive community, sanctuary and hospitality are so intermixed that it is difficult to distinguish between them, and it is only later that the altar where the pursued can find refuge and the inn where he can pay for a bed are differentiated. This confusion is accentuated in Manus today by the indecision as to whether the leaders of the civil community and the leaders of the religious community are to be the same or different people. In his mixed position as head of the civil-religious community, Paliau gave sanctuary to runaway wives who were strangers in the village. In Peri, as the purely civil head of the community, Karol Manoi played the same role. But Thomas, the official preacher and a member of the immigrant community from Patusi, also provided a popular sanctuary for the runaway wives of angry husbands. People said it was because he was the preacher; it was because he was from Patusi and so Patusi women chose him as a

---

*Today, administration officials on circuit of any sort are given an allowance with which they can partially compensate their hosts. But near-compulsory hospitality extends far beyond the call of duty and includes travelling anthropologists, journalists, entomologists, etc.

refuge; or, when they had to explain why any Peri runaway with the slightest kinship claim would also run away to his house, it was because he would neither get involved in a brawl nor permit an angry man to enter his house in pursuit of his wife. Tomas, however, complained that this was a burden beyond his powers of endurance: "I like living in Peri. I saw all of these women [my 'sisters'] and I was glad to live among them, but now they all run away to me, and I am constantly embroiled in their quarrels, and it is too much to bear." And meanwhile strange canoes arrived with patients—a woman who had been bitten by a dog, a man who had nearly chopped off his hand—and as was right (on his part) and as part of his duty, one of the two councillors would take them in. As it needed a bank and a store, so the village of Peri needed an inn.

In our plans for change over the world there is a tendency to plan for only one set of institutions. The missions plan for churches, religious schools, and hospitals, the government for courts, schools, and medical services, and the traders for plantations and shops, but meanwhile a community needs the entire set of governmental, religious, and secular institutions. Without them those rights of individual freedom, of choice, of autonomy, which we offer as the reward of the democratic way of life, cannot be maintained.

12

"and unto God the things that are God's"[1]

Old Manus could not be called a theocracy because there was no community to be governed by a god or gods, but if there were a comparable term for a household ruled not by the father or the oldest male, but by the ghost of the most recently dead male, that would have described their pre-Christian state. In old Manus, the child depended upon a human father who in turn depended upon his Sir Ghost who ruled on the ghostly plane.[2] As the child's growth and activity was fostered by the human father, so also a human ward's material prosperity and health were fostered by the Sir Ghost. A child grew, learned to swim, walk and talk and count, fish and handle a canoe. A man fished and married, engaged in financial exchanges, built large houses, big canoes, and made long trading trips—and human father and supernatural guardian were gratified. The child and the ward offered efficient activity to their protectors, and the father and the Sir Ghost responded with gratification and protection. No human father was too tired to seek out a wandering child and carry him home in his arms or to hold a sick child in his arms all night, and the Sir Ghost was believed to be equally tireless on his ward's behalf.

But the Sir Ghost cult was a cult of individual households at most; a group of Sir Ghosts of a male line might be treated as a collective set of sponsors, or the ghosts of remote ancestral lines might be invoked by the women of those lines, in blessing or cursing other descendants of the same line. The morality which the ghosts supported was a wholly private morality, which functioned however to make the whole economic system work, as each man was chastised by his particular Sir Ghost if he did not work hard, fish, build, trade, invest, and pay his debts. The Sir Ghost cult stood for human economic effort and ghostly vengeance. When ghosts struck down mortals, hitting them in the neck with an axe so that blood spewed forth from the mouth, it was not correct for the father of the injured child to take an axe in turn and cut down the ward of the injuring ghost. Instead, either the Sir Ghost of the household of the injured child had to retaliate, by sending sickness on someone in the house of the attacking ghost, or expiatory payments had to be made, from attacked mortal to ward of attacking ghost. As parents took up cudgels when their children quarrelled, so ghosts took their wards' part.

Toward other households, the ghosts acted with pure human malice and capriciousness. It was only within the household, or group of co-operating households, that the Sir Ghost acted morally and responsibly, punishing where there had been sin of omission or commission, prospering the hand of him who worked hard. When the men of all the households banded together for war, then the war magic, the *ramus*, bought from other tribes and supported by other supernaturals, took over and the Sir Ghosts were, in a sense, in temporary abeyance. When all the men of the community acted together, it was possible for them to commit with impunity acts of lust and violence which, if performed as individuals, would have got them into serious trouble with their individual Sir Ghosts. Group rape was "something that belongs to us men, altogether, something that belongs to all of us." So, in the old system, the rare communal action was non-moral and magically sanctioned, and only household and lineage and occasionally clan action was moral and religiously sanctioned.[3]

Under the old cult, there was an exceedingly external practical definition of those sins which got one into difficulty with one's Sir Ghost. The Sir Ghost was interested in words, especially careless light words spoken by married women—those hapless creatures who lived precariously between the chastisement of a brother's Sir Ghost and a husband's Sir Ghost with no Sir Ghost of their own. The Sir Ghost was interested in the smallest sexual acts, a hand placed lightly on the thigh of a member of the opposite sex in a temporary melee while a snake was chased or a house collapsed. A Sir Ghost was interested in a failure to make economic plans and a failure to pay debts, sins of omission but not sins of the heart or spirit.

There was in fact no real conception of a *soul*, or a *mind*, that could sin by thought as well as by word or deed. Thought under the old conception of man took place in the neck-throat; to remember was to "neck-throat to it." The verb for *forget* also involved the same part of the body, and to change someone's mind was expressed by the phrase to "twist his neck-throat." Sorrow dwelt in the eyes, anger in the belly, and fear in the buttocks. A man had many names, as many names as there were mother's "brothers" who had participated economically in his birth ceremonies. When someone fell sick it was because a ghost or ghosts, Sir Ghost for moral chastisement, or ghost of another in revenge or malice, had struck him down, and taken some portion of his soul stuff. Sometimes portions were held by many ghosts, and it might take all night during a long séance for ghostly messengers to range far and wide collecting the purloined soul stuff, returning it to the bowl of water that stood in front of the medium, which, when it finally contained all the soul stuff, was poured over or drunk by the sick person. Someone who had a slight headache might speak of a ghost having taken a little bit of his soul stuff. There was a definite lack of any sense of a spiritual core of a human being until after death, when his soul, attached to his skull, or a substitute skull made of a coconut, did assume a definite single personality, just as his body, together with the soul stuff, had in life.

It has become customary to talk about moral standards being

internalized when a child in the course of growth takes over the moral standards which he has been taught and becomes a sort of self-executor carrying out those precepts in the absence of witnesses, while standards are said to remain external if an individual stops following a code when he is alone, away from witnesses, and in no danger of being caught.[4] The old Manus represented a curious halfway stage. Standards were certainly internalized, men drove themselves far oftener than they waited for ghostly goading; indeed in the absence of illness or bad luck, there would be no ghostly goading for months. Offences against the Sir-Ghost-administered code were the result either of daring—willingness to take the consequences however terrible—stupidity and instability, or of skepticism as to whether a particular command had really come from the ghostly plane or was the expression of a personal whim or the self-interest of a medium. Manus men away from Manus committed readily enough the sins which they were forbidden to commit against Manus women, but this is not inconsistent with internalization because Manus Sir Ghosts were not interested in the chastity of women of other peoples.

The attitudes toward property, toward accuracy, and toward excretion and exposure, toward responsibility and initiative, which were inculcated in Manus children by direct parental admonition, by example, and by punishment, survived in any environment and were more stable than the moral framework which was enforced on the spot by outraged ghosts. Sometimes individuals, especially young people who had become involved in a sex offence, were extraordinarily stubborn about confessing to a detected lapse, but they usually capitulated when one of their kin fell ill and the continued refusal to confess was blamed for the illness. So the adult system might be summarized as a set of external sanctions—Sir-Ghost-enforced and illness-validated—against words, acts, and lapses antithetical to the code. Individuals were driven by a fear of failure to do what they ought to do and by the danger that some small sin would be discovered, but the evil that they wrought could only be called sin in the sense that it was supernaturals, not men, who were deemed to be offended. The bad conscience of a Manus adult was not unlike the bad conscience of an individual in a police state where it is believed that everyone is an informer, that wires are tapped, that "crime will out." A man who was in debt, who fell ill, would go over and over his discharged and undischarged debts in a delirium, just as might the man in a modern police state who is fearful that security forces will catch up with him.

When the way in which the adult code was enforced and experienced in Manus is compared with the persistence of the type of behaviour learned in early childhood that could be said to be genuinely incorporated or internalized, it is clear that the only difference between their adult moral code and the criminal code of a modern society is in the type of sanction and type of enforcement, and not in the moral or spiritual position of the individual. The old ghosts were police, and police who acted narrowly not in the name of an over-all state but to preserve private vested interests; and the people's feeling that if they could pitch them all out at the same time it would be quite safe was like a disarmament pact among modern states. Although individuals like old Pokanau were proud of their Sir Ghosts, this was a continuation of a relationship established in life, not a relationship to a venerated spiritual power. The Manus of twenty-five years ago had strong characters, strong wills, and an external system which enforced an exacting, constructive, positive moral code, but no spirituality in the sense that spirituality is cultivated by the higher religions and no mysticism in the sense of a specially experienced relationship to a supernatural.

The steps by which this system was transformed into the present system are obscure. They learned from the Mission that man has a soul, given him by God, which goes either to Purgatory, Heaven, or Hell at death, and remains there. They learned further that a man's soul has a relationship to God which is disturbed by sin, that it is sin which separates man from God, and the remission of sin which restores his relationship to God. It seems probable also that at some point the Mission succeeded in explaining the Western point of view about civil and religious offences, using the text which is so often on the tongues of Manus leaders today, "Render therefore unto Caesar the things which are Caesar's; and unto God the things that are God's."[5] And thus the Mission sought to teach the people the difference between the offences which God punished (and His method of punishment) and those for which one would be brought before the District Officer as a misdemeanour or a crime.

Today the Manus have a view of human nature in which that which is owed to God and that for which one is accountable to man are sharply dichotomized. Man has a mind-soul—the distinction between that which thinks and that which goes up to God at death is very unclear, and for most purposes this mind-soul is thought of as one. Sin consists in thinking the wrong thoughts, while all overt offences of word and deed are considered to be civil offences punishable in terms of fines and imprisonment by the civil authorities, local village, or district administration. To lust after another man's wife is a sin but to have sex relations with her is a civil offence; to steal is a civil offence but to think about a theft before or *after* the theft is a sin. A sin is something which disturbs a man's relationship to God by "fastening" his thought, by keeping it from running clearly and freely. In this respect, a sin is strictly analogous to the Manus view about any injury to the body which "fastens" the blood—a bruise or swelling which keeps the blood in one place, prevents its free life-giving flow. In the same way, man's mind-soul should be able to move, aware of his relationship to God, relating all that he does to God, free from undue obsession of any sort. So too much desire for a woman, too great a desire to obtain some material object, to make a journey or hold a position, may "fasten" one's thought. Confession is enjoined as a way of getting past crimes off one's mind. If a man steals from another man's tree and is not detected and punished and does not confess, then "Every time he passes that tree his thought will be fast to his theft and he will not be able to think of anything else."

So, in the development of a spiritual theory of human nature, the Manus model has stressed the importance of freedom of movement, a prime essential in the relationship of children and adults to the physical world and to one another, the importance of control over the body, which must ever be free to act, the importance of freedom of thought and of control over one's thinking. As any impediment in freedom of the blood to move is felt as an injury or a sickness, so also any impediment to the smooth flow of thought is also a spiritual sickness. Physical sickness is still a punishment sent from God because one's mind-soul is in the wrong, as it was once a punishment sent by Sir Ghosts because one's body had done the wrong thing or failed to do the right thing. This new conception of man, although it is so much more spiritual and mystical, also accepts the unity of the personality and the unity of spiritual and material causes of disease far more than the old theory did. It is now accepted that illness comes from communicable disease, from contact with someone who is diseased, from infection, from mosquitoes, but it comes only to those who have something wrong with their thoughts. Illness may be combatted by Western medicine, and medicine is good, but the medicine will only work if one's thought is clear, if one is not obsessed by an unconfessed sin and if there is no anger in one's heart.

For anger is the emotion which most corrupts the mind-soul, which seizes one in its grip until one is helpless, which cuts one off from God and which dooms one to illness, misfortune, and death—for the individual so cut off from God will die, that is, his mind-soul will be taken by God. Just why death is the inevitable punishment of being cut off from God is not clear, unless it is simply an extension of the idea that illness is caused by sinful thoughts. But, whereas the old Sir Ghosts were seen as possessing extremely human emotions and weaknesses, in fact in many ways were more human than their mortal wards, God is seen as a distant administrator of Law, and indeed partakes of some of the personal majesty of the Law. When one becomes ill from sinful, obsessive thoughts, one is cut off from God, one falls a prey to illness, and, if one continues in this course, God, so the feeling goes, has no alternative except to take one, i.e., let one die. Placating God contains none of the elements which were once necessary to placate the ghosts: confession and expiatory payments. Apparently at various points there were attempts, whether made by the catechist or the converts it is impossible now to tell, to include expiatory payments to God, placed on His altar and distributed to His poor, as part of the procedure for recovering from illness; but this is now gone. The relationship between God and man is now seen as a lawful one in which it is man's duty to keep the channels of his mind-soul open to God, and in which neither prayers nor offerings can be used as substitutes. Introspective accounts of prayer are hard to get, but the general conception seems to be of standing and turning one's mind-soul to God.

Throughout this whole change—from the traditional cult of protective household Sir Ghosts, through the period of conversion to Catholicism, into the present period of the New Way—confession, the need to rid oneself of the evil by telling about it, has been a recurrent pattern. The original choice of Catholicism was partly based on the advantages of auricular confession, the desire to escape from open confession which was bruited abroad by drum beats to the whole village. One of the early patrol officer's reports on the New Way included the statement that the people felt auricular confession gave the priest more power over the community than any man should have. In the early days of the New Way, the emphasis was upon those who had sinned recounting their sin to the whole assembled community as an example, a kind of religious parable, "So we sinned, our child died, we tell you so that you may avoid the same trouble." The old function of the cross-cousin as the proper confidant of sexual sins was retained so that either a cross-cousin, a civil leader, or a religious leader might be privately confessed to, or a sinner might speak openly before any group.

Today the relationship between unconfessed sin—with the sin being angry, resentful, and obsessive thought, and, by extension, the acts which lead to such thought—and illness forms a continuous background of pressure for reconciling kin or spouses who endanger their own or their children's health by continuing to quarrel, and provides an explanation for all illness, including difficult labour and death of young children, too young to sin in their own right. Here an old New Guinea problem—how to deal with the deaths of infants when illness and death are intricately woven into the system of adult human relationships—crops up again. In old Manus, the deaths of infants were explained according to a system specific to infants. Young people and adults died because of sin, their own or those of their kinfolk, but infants died from black magic, openly practised between men who were rivals in trade. Similarly, among the mountain Arapesh, adults died because of the way angered people had delivered them into the hands of sorcerers, but infants died from simpler magical causes. With the high infant death rate, placing the burden of responsibility for infant deaths on the parents means putting them under an intolerable load of guilt. This is a point at which the present religious consistency of the New Way is overstraining individual capacity to tolerate blame.

When a delivery is difficult, both the labouring wife and the husband, who is now constrained to be present, although he usually keeps his distance at the other end of the house, are exhorted to confess anything which either one has nourished in anger against the other. As the young wife pauses between pains which are not mounting in intensity and frequency as they should, the insistent voices of the old women rise, ostensibly talking to her, actually talking also for the benefit of the husband, who may even feign sleep to avoid the pressure. They will say, "Now think. Have you not been angry? Was there not a time when something was not done? Was there not a time when you wondered why he did not come home? Was there not a time when you heard him say something which made you angry because the meaning was not clear?" On and on their voices go, pausing as the expectant mother sits up to meet a labour pain, picking up again when she relaxes. Part of this fear that anger between the couple will interfere with the birth has been conventionalized in a ceremony of shaking hands when the pains start. And if something goes wrong at the birth or the infant is born dead or deformed, the pressure increases to make the couple confess in order to assure the safety of other or future children and the safety of the parents themselves, who are obviously in danger.

Twenty-five years ago it was possible to take a list of all the infant deaths and to approach any moderately well-informed person to ascertain who had killed the child, with what magic, for what reason. The magic had been openly threatened and openly proclaimed, and the child's death remained part of the saga of the quarrel between Korotan or some other powerful man and one of his economic adversaries. Now most of the magic which had made such sorcerizing of infants possible has disappeared. There were men who still clung to their sets of charms up to 1946, but at the time of The Noise, it is said that all of the surviving magical paraphernalia was pitched into the sea. All except the magic of Poli, a magic which had no ritual objects associated with it, which was merely a form of words and a series of ritual acts. And how could one throw away something totally incorporeal? So Poli, as an unwilling magician, remains as the sole alternative to blaming the illness and death of every infant on its parents' sins.

There are at present a variety of points of view about Poli's magic, each of which contains the seeds of a whole system of magical belief and practice. There are those who claim that Poli's magic may strike any child who has not been preventively charmed by Poli's counter-magic, and the parents who believe this routinely pay Poli to charm their newborn children. This is expensive, costing from ten shillings to a pound, and it entails endless trouble, because a child who has been charmed by Poli has to keep a whole set of magical prohibitions. It cannot eat certain foods and, most tiresome of all, it cannot eat food cooked on a fire made from two common types of firewood. Most households cook with these woods, which means that there is constant danger that a charmed child will be given food which will magically harm it. So there are good, practical reasons for refusing to accept Poli's protective magic as necessary, and people argue that Poli's magic will only hurt children of those against whom Poli has been angry, or possibly even against whose grandparents Poli was once angry. Other people say, "Wait, see if the child is fretful and ailing, and if nothing else works, try Poli." Still others argue that Poli's charms are only related to Poli's anger, and if one lives in charity with Poli then neither black magic nor counter-charm will work. Thus the circle is rounded and the cause of illness is brought back to an angry heart in the parent of the child.

Meanwhile, Poli's unique possession of magic is known far beyond Peri, and occasionally parents from other villages send for him or bring their children to him, especially to deal with other small magical events which are also outside the formal religious system, as when someone is pulled toward death by having gone near the place where a close relative died a violent death, "a place of blood." This was an old belief, of little importance twenty-five years ago, and unintegrated into the Sir Ghost cult; people were simply well-advised to stay away from a place where a relative had been slain or drowned. Today, with the extreme pressure to blame all death on sin, leaving no room for caprice or malice, this belief has burgeoned, as has old Poli's small magic, into a much wider supplementary explanatory system, and many illnesses and deaths, especially of infants and children, are attributed to relatives who died suddenly or violently and who try to take their living relatives with them. But, whereas in the past the attribution of infant deaths to open black magic was a public matter, well-recognized by everyone, today these alternative explanations are frowned upon and hidden.

The orthodox New Way version of Christianity says that all children's deaths are due to their parents' sins. The new movement has made a much heavier investment in the health and life of all young infants than did the old system. In the past individual parents tried to save their children. Today it is the duty of every Manus not only to try to save their children, but to have as many children as possible so that the people may be many in the land. The new sense of ethnic identity carries with it an expansive desire for many more Manus. Women are exhorted to have children; abortion is heavily forbidden; and that woman who never menstruates because she is always either pregnant or breast-feeding a child is regarded as most patriotic and virtuous. So one part of the system, the attribution of all illness to sin, and another part of the system, the social demand for more and more children, play into each other to keep alive the belief that children die for their parents' sins, and to discourage the growth of magical practice.

At the same time Poli's magic is a threat to the whole new system. He has taught his magic to his son-in-law. He cannot get rid of it because there is no way to get rid of it. He and his son-in-law both think of the evil effects as automatic and the good curing effects as within their power to manipulate, upon request. This single surviving bit of magic might easily become the basis of a new magical theory of protection from disease in which the use of medicine would fit far less well than it does within the new orthodoxy which insists that illness—as defined by Western culture in terms of germs, infection, malnutrition, exposure, etc.— enters when one has sinned or when one's parents' sin has made one vulnerable, and that rectification of one's thought, the elimination of anger and hostility, renders one more responsive to drugs and medical practice in general. The new orthodoxy is congruent with the most modern psychosomatic thinking, but the reintroduction of old magical practices would at best provide a background for using drugs on an equal footing with charms, instead of as complementary to appropriate religious practices.

Meanwhile, the inclusion of difficulties in child-bearing and infant illnesses and deaths within the system of sin does put an almost intolerable burden on marriage. When a child is ill, if one searches one's own heart and finds no anger sufficient to have caused the illness, then it *must* be the other parent who is guilty. Then one has the choice of secretly suspecting the other, and so becoming guilty of anger, or openly accusing, and so risking a quarrel in which both parents will be angry and thus endanger the already weakened child. Despite this, during a delivery or a child's prolonged illness or the difficulties of getting an infant with a cleft palate to suck, the pressure continues. There *must* be some unconfessed, unrectified anger. He or she must still be cherishing a grudge over that old confessed and paid-for adultery. Or perhaps there is an unconfessed old affair which had better be confessed, although actually it happened before marriage and no one has a right to be angry about what has happened before marriage. So also there is the danger that if one is not angry now, something may be brought up at any moment which will arouse old angers in oneself or someone else.

The full poignancy of laying the blame for children's deaths on parents was dramatized at the birth of Ngalowen's infant. Ngalowen was the sister of Michael Nauna, the banker, the gentlest, most trusted man in Peri. She, as a small girl, had been the darling of her adopted father, her brother's chosen playmate, the pet of all her relatives. Beautiful, reserved, aristocratic, she was the model wife of Tchaumilo, the most virtuous of the four brothers of the house of Nane. Tchaumilo never gambled, Tchaumilo was a steady, hard worker, Tchaumilo cared for the young children of his less stable widowed brother, Posuman. And when Posuman's wife died, Ngalowen took the tiny new baby, Kanemon, to rear. But Ngalowen became pregnant almost immediately, thus posing a problem of where milk for the adopted baby was to come from, for the milk of a pregnant woman is believed to be poison for another child. Little Kanemon remained a starveling. At eleven months, when I arrived in the village, he was no bigger than a three-month-old baby. Tchaumilo and Ngalowen were untiring and devoted in their care of him. When I offered milk and vitamins, one of them brought him faithfully twice a day for his ration.

Ngalowen grew heavier, her body assumed a stranger and stranger shape, her abdomen became pointed, the time for the birth of her baby came and still no baby was born. People worried. Tchaumilo and Ngalowen had only two living children. Something had been wrong, something was going wrong again. True, they seemed to be free of all sin, devoted, quiet, hardworking, gentle, but still there must be something. Finally the night came when Ngalowen's labour pains began. She lay beside the fire in the little sea house which belonged to her dignified old mother and her mother's second husband. Tchaumilo stayed devotedly nearby, holding little Kanemon. There was no need to seek him out and drag him unwillingly to a ceremonial reconciliation with his wife. The house was very quiet in sharp contrast to the house just next door where I had witnessed a birth at which two vulgar, noisy old women had dominated. Here no one spoke loudly, the firelight flickered, Ngalowen's devoted brothers came and went. Nauna came to take home for the night one of his children (who usually slept with his grandmother). Ngalowen's younger sister had come home from Ndropwa, and sat with her two little children beside the fire. Ngalowen and her mother spoke together softly, and the night passed like a muffled dream, interrupted by the occasional bouts of labour, but with never a moan issuing from the expectant mother's lips.

Just at dawn there came the sudden rush and mobilization of everyone's effort, which meant that the moment of birth was at hand, and then, a sudden shocked paralyzed silence while an old woman said to me from the spot where the newborn infant lay, "She has given birth badly." Looking down, I saw an anacephalic monster of the lowest type, its truncated head open to show the cord, its skin a pallid near-white, its great protruding eyes staring up at me. Fortunately it was quite dead. The body was perfectly formed, and the awed women murmured, "A male." The dead creature was lifted up and carried across the road to the big house of Tchaumilo's older brother, Kutan, the head of the family. A little old woman, Tchaumilo's father's sister, a woman who was deaf and mildly feeble-minded, sat, with her legs stiff out in front of her, while the monster was dressed like a human being for its burial. As the sun rose, word got around the village, and people came in and stood silently, ashamed, not meeting one another's eyes. People seemed unable to communicate to one another what had happened. As Peranis said afterward, "They simply told me the baby was dead, and its body was all right, but its head wasn't." And Karol Matawai said, "No one told me anything except that it was wrong. I came in and looked from its feet. Its body was all right, and suddenly I saw its eyes. I felt this was an evil thing." No one uttered a word of mourning, and the absence of the mourning sounds, which ordinarily unite them in dramatic catharsis, seemed to shout in the silence.

And then Kutan, the eldest brother, in whose house the child lay, rose and said sternly, implacably, "We all know why children die. We all know that their deaths are the fault of their parents. This child had done no wrong. Something that its parents have done is responsible."

The monster was buried. No news of the event was spread beyond the village. People went about still not meeting one another's eyes, and Ngalowen lay still, failing to rally as she should have after the birth. Very gently her brothers watched over her, but the talk had been repeated at the grave; in the graveside sermon it had been reiterated, things like this happen only when the parents sin.

In conversation with the people, everyone admitted that it was very difficult to find any sin, any act of anger or hostility in this gentle, devoted couple. But one *must* have occurred. Ngalowen lay listless day after day, and Tchaumilo too began to sink into a depression. Meanwhile, people had told me that there had been one such birth before in the village; long ago the wife of Ngandiliu had borne a similar monstrosity. I decided that if Ngalowen was to live, and Tchaumilo was to be saved from depression and possible suicide, intervention was as definitely called for as it would have been had they both lain ill with pneumonia. The story of the other anacephalic idiot birth gave me my material. When Ngalowen and Tchaumilo were little children, the elders of Peri had given us careful genealogies to prove that one of their number had come from a line filled with mental defectives, saying, "You see his crime [that of a peeping Tom] is not his fault, all his ancestors have had trouble too." I checked over the genealogies and found that the wife of Ngandiliu had also been a member of Lo, Tchaumilo's clan. So I plotted out a chart and called in each of the responsible people—Ngalowen's brothers, Tchaumilo's brothers, Pokanau, the town clerk, the two preachers—and explained to them separately that when two such events happened in the same descent line it looked very much like heredity, and so not present sin but some long-past event—over which they had no possibility of control—should be held responsible. I added that there should be no more talk of Tchaumilo's and Ngalowen's sinful responsibility, for there was real danger that this talk was making them sicken and would result in their death. They were glad enough to accept the explanation I offered. The pronouncements ceased, Ngalowen began to eat, and Tchaumilo smiled and worked again. But the whole event dramatized the dangers in the system which were placing too great a responsibility on parents, especially in a society where so many babies died for lack of proper midwifery, proper medical care, and proper food.

The way the system worked for an adult illness showed up sharply in the illness of Teresa of Patusi. Teresa was a young married woman with a child a little under two years old. She was brought to Peri in the late stages of cerebral malaria, too delirious to do more than resist medicine and worry about whether or not her body became uncovered. A frantic group of relatives held her up in their arms, while Alois, her sister's father-in-law, and his young adolescent daughter set off all alone on the long journey to Lorengau to fetch Teresa's husband and her sister from work. Two days dragged by while Teresa's father and mother watched over her and took turns holding her in their arms. By the time her husband and sister arrived, Teresa was in a coma from which she never awakened. Meanwhile her sister sat beside her and talked of something that no one else knew. Teresa had continued to nurse a grudge against their mother for a trivial quarrel over some betel nut in which she, the confessing sister, and their mother had been involved. Teresa had sympathized with her sister and remained angry with their mother; now she lay stricken, dying because of her concealed anger. Her sister had arrived in time to tell the story, but not in time to save her. This attempt of others who know of one's sin to save one's life by confessing for one is reminiscent of the old days when eavesdropping children would "confess" a sex offence on which they had spied to save the life of the offender who lay unconscious and dying.

The attempt to eliminate all anger and all continuing lustful thoughts results in many complications. Sometimes it is felt that the best way to eliminate anger is to express it, expression being regarded as a form of elimination. There is no doubt that nursing the grudge secretly is felt to be far more dangerous to the angry person than expressing his anger openly and freely. So Raphael was genuinely surprised when I asked how he justified a furious speech made out in the open against Kilipak over a payment for part of a canoe. "But," said Raphael, "I didn't hide my anger. I went at once and told them, Kilipak and Kutan, just how angry I was, and what I was angry about. I stood in front of their houses and told them."

But another part of the complicated new ethic forbids provocation to anger. This prohibition is directed particularly against provoking sexual jealousy, and includes admonitions to a spouse who has been unfaithful in the past to abstain from arousing the retrospective jealousy of the husband or wife by any careless or flirtatious act. It is the duty of each partner to a marriage to "guard the thought" of the other, to protect the other from encountering any circumstance which arouses anger. Prohibitions of slander come under this same rubric, which contains in it the legal rule against hearsay evidence. Anyone who tattles in any way about an affair likely to arouse the anger or jealousy of another is subject to rebuke and fine from the village court. This results in a considerable degree of caution before calumnious remarks are made, and, just as twenty-five years ago people were exceedingly wary of handling the personality of another, even verbally, for fear of an angry explosion, so today it is considered better policy and better ethics to disclaim all information about the affairs of other people.

By a further extension, leaping to conclusions is disapproved of. So a woman hears her husband ask another woman whether a third person was there and assumes that the subject of the question is another woman. She flies into a rage, accuses her husband of plotting an assignation, while he claims that he was referring to the woman's husband, who, both the woman and the said husband testify, *was* there. The wife is then rebuked for leaping to unjustified conclusions.

The most troublesome aspect of the attempt to legislate anger out of the hearts of men was the way it was reflected in rules about extra-marital relationships. Traditionally, under the Sir Ghost cult, all extra-marital relationships, even to the very slightest level of a lewd word or gesture, were forbidden, for all of these women were foci of ghost-protected economic activity. During the mission period, when the publicly expressed ghostly sanctions were dropped, sexual lapses became a matter between man and God, to be concealed from men and confessed to God in order to obtain absolution. It is impossible to tell whether this actually meant any loosening of sexual morality. There were still enormous investments in marriages. But the sanctions against sexual immorality and the sanctions which protected the financial arrangements had become disassociated from each other. This may well have been one of the precursors of the great spiritualization of the idea of sin, for a sexual sin, even though it was not found out, and therefore even though it did not threaten the financial structure, nevertheless still had to be confessed to restore man's relationship to God.

Another cardinal principle of the New Way is the elimination of kinship partisanship in either a good or a bad cause. One of the very first laws promulgated said that one could not help one's brother in a quarrel no matter what the circumstances. It was wrong to "pull" the quarrel of another. Simultaneously, women were emancipated from the control of fathers and brothers, who were forbidden to embroil themselves in their sex affairs. This meant that unmarried girls of marriageable age were free to do as they liked without anyone interfering—neither father nor brother had a right to be angry, and it was anger, not sex, which interested Paliau and the other leaders. Pre-marital sexual freedom became part of the system as a by-product of giving individuals the right to make their own choices—a not infrequent event in modern history where pre-marital sex freedom has resulted from permitting women to go away to school or work or to become economically independent—and with the introduction of the idea of the "age of consent."

The Manus still exercise rigid control over those whom they regard as too young for sex, but today a young woman may use her own judgement about what favours she accords to an unmarried lover. Even if she becomes pregnant by an unmarried man, which is disapproved behaviour, the main emphasis is upon the care of the child. The girl is formally exhorted to make no attempt at an abortion and to take care of herself and her baby conscientiously. Here community and individual coincide in attitude, as interest in the baby and getting possession of the baby often over-rides interest even in a wife's infidelity. So one prominent leader insisted that his daughter's son was illegitimate and so under his care, since he had been born during a period when the daughter had left her husband, and by another man. The legal husband wanted to take his wife back as soon as the child was born, in order to get possession of her child.

Adultery is another focus of trouble. For Manus husbands and wives are jealous and possessive. Adultery, in addition to being an offence punishable with a fine or imprisonment or both under the Territory's Native Administration Regulations, is certain to lead to anger and quarrelling. If anger and quarrelling had been the only concern of the New Way, then adultery might have been forbidden to prevent anger. But it is not only anger which corrupts the soul, be it remembered, it is also any obsessive desire. Here the Manus equate desire to retaliate against an enemy and desire to copulate with a woman, retaining the old association between sex and rape. A woman who resists one's attentions deserves to be raped, just as a man who angers one deserves to be attacked in some other way. For this impasse—the need to stop adultery because it leads to quarrelling, and the need to permit it because a prohibition would lead to obsession—Paliau proposed a curious solution. "If you desire a woman," he said, "involve no one else in the matter. Go and ask her yourself, and if she agrees, copulate with her once, pay her, and have it done with." And Paliau also decreed what would be a fair payment. The woman who failed to yield under these prescribed circumstances could be seen by the man as endangering his mind-soul. His thought would remain "fast" to her, and his relationship to God be disturbed. In addition, if the woman tells of his unsuccessful suit, he may be shamed before his neighbours.

Around this rubric a whole set of extraordinary misinterpretations have grown up among the Manus themselves, and have been elaborated and embroidered upon by the enemies of the movement, especially those land people who have remained faithful to the Mission. One conclusion Manus men drew was that if their wives slept with other men and were not paid, then they had a right to be angry, but if the wife was paid the correct fee, then anger was no longer permissible. The more orthodox a man was in the New Way, the harder he tried to deal with the separate rules of this unwieldy code. If he suspected his wife, he might question her. If she told him the truth and had been properly recompensed, then he could not be angry. But if she had not been properly recompensed, or if she lied to him about any detail of the affair, claiming more or less pay than he had reason to believe she had received, then he could, and did, become furiously angry. Furthermore, if there had been such a brief affair in which the wife had been recompensed and the lover's thought set free for other and theoretically more laudable meditations, then, any signs of continuing interest between the two could, and would, set the husband's anger off into uncontrollable fits of rage. The effort of self-discipline required from an active, possessive, jealous people was so great and so conscious that a continuous search for loopholes which would permit the expression of jealousy went on. One young man had prided himself on the completeness with which he observed the New Way code, for had he not carefully hidden his own affair from his wife, to whom he was devoted, and had he not accepted a lapse of hers with good grace and remained friends with her lover? He was proud of his good behaviour and of his ability to obey the rules, proud with all the arrogance of the *unco guid*, however odd the goodness involved may seem to us. And then his wife yielded to the importunities of a lover who paid her nothing, and the young man flew into a rage and attempted suicide, his carefully controlled world falling in ruins around him.

Others interpreted the rules as an admonition to commit adultery. If you were told how to do it, was not this the same thing as being told to do it? All this was complicated by Paliau's coming from Baluan, a culture which the puritanical Manus had always regarded with deep suspicion. Matters were not improved by the sexual adventures of the leaders, by "court cases," and by the use of the law against adultery for political purposes by government officers opposed to the New Way, and in cases of political rivalry within the New Way. The rule against go-between by which Paliau had attempted to isolate adulterous incidents, although administered vigorously, did not really work. The go-between had after all committed only a "civil" offence, whereas all the regulations about anger, anger arousal, and lust and lust seduction were designed to deal not with man's body but with man's soul and thus belonged not to Caesar but to God. "We are trying hard to learn the ways of Western man," said Kilipak, "but," he added with a sigh, "one thing we have not learned yet, and that is how to commit adultery properly!"

Caught within a series of extrapolations which followed from the freedom of women, the need to keep one's thoughts in a clear, unencumbered relationship to God and to God's work, the desirability of recompensing a woman for yielding sexually to a man—which the Manus still see as primarily painful and distasteful—the Manus confusion about the handling of sex mores leads to quarrelling and anxiety. When a Usiai landsman is involved, it is difficult to distinguish between the husband who demands that his pride be maintained and his anger quenched by his wife being properly recompensed, and the husband who is seen as a procurer, benefitting from his wife's favours. All passion, requited and unrequited, past, present, and to come, is regarded as obsessive and dangerously evocative of all the states of mind which interfere between man and God. So the old association between sex and sin is retained even through the most drastic reorganization of the social system. To the Mission, the Manus natives today appear as a thoroughly debauched people, wallowing in sin from which they have no desire whatsoever to be rescued. To themselves, the Manus are a people struggling to master an unfamiliar Western code and failing to do so properly.

Meanwhile, other theories of disease contend with the official one. There are diseases, called mysteriously "diseases of the ground," i.e., the local community, which the European medicine cannot cure. By an extension of the present blend between keeping one's thoughts straight and the use of medicine, this diagnosis is interpreted to mean that a disease which is incurable medically was not caused by one's own sin but is to be attributed to sorcery, sorcery which has its origin outside Manus, among natives from other parts of New Guinea. Men who have wasting complaints which have proved impossible to diagnose or unresponsive to treatment, and who can think of any situation where they have worked with non-Manus natives, will themselves diagnose their illness as sorcery, and then the only recourse is to counter-magical measures. There are a few lively practitioners who collect huge fees for using counter-magic, and the tendency to blame disease, whenever it is intractable, on a source wholly external to the society, and to reserve the tractable and curable diseases for confession, reform, and medicine, might well grow. These attributions of sorcery to work boys from other areas are justified on the grounds that other peoples may not have "thrown everything away yet," and by the statements of the inability to help made by European physicians and employers who send hopeless cases home to their villages to die.

Another conflict rages: whether auricular confession is necessary or desirable. One of the emphases of the New Way is upon the ability of a man to set himself straight with God. The Mission claims that there was a period when the Manus still wished to be admitted to the sacrament of Communion but refused to go to confession, claiming that they could order their own relationships to God. No Manus with whom I talked admitted to this heresy in just this form, but the leaders did admit that it was a moot question whether confession was necessary or not. Together with a social choice as to what kind of socio-economic life they wished to live, and the great sense of autonomy which each Manus derives from this choice, went also the sense that each man was the captain of his own soul and could, if he wished, straighten out his own mind-soul. Confession might be a wise precaution, it might be the only way to stop thinking about something, but it was no longer necessary. Men who got into bad tangles with the community would withdraw for a few days from community affairs and "work to straighten out their thoughts." So a young Manus, facing me across a table after his second suicide attempt, said over and over, "But there is no use my lying. My thoughts are not yet straightened out. In this state of mind I can promise nothing."*

The trend in Peri over the last few years has been definitely away from confession and toward the privatization of man's ethical life, with the mental agonies which he goes through exposed only when the village rules against open quarrelling are broken and the peace of the village disturbed. Even when death comes, there may be no confession to account for it.

Meanwhile, a new belief in reincarnation has slowly been asserting itself. Infants are named after recently dead siblings and are believed to be direct incarnations. People ask God to send back to them their dead children. From a civic point of view this is seen as increasing the number of Manus by not permitting Manus souls to remain in Heaven where they contribute nothing to the New Way and its need for a larger population. So far this belief is applied only to children, but there seems no reason why it might not be extended to adults so that death might become a temporary sojourn in the other world—a punishment somewhat

---

*In other villages, the objections to auricular confession and the insistence on public confession were stronger.

more drastic than illness—from which the erring individual might ultimately return.

Shorn of any orderly contact with the religion from which their present religious ideas are derived, without religious books of any sort except their copybook versions of remembered bits of liturgy which they had learned by heart and which have been further systematized by Paliau, the whole system of religious beliefs is extraordinarily unstable and responsive to interpretations given it by any individual leader. It may well be argued that these breaks with the missionizing religion, which are such a frequent accompaniment of a "nativistic cult" and result in encysting bits of a great religion within the memories and control of men who have a most imperfect knowledge of either theology or practice, are one of the important factors in shutting a native people off from further participation in Western culture, even where the desire to participate in Western culture may have precipitated the wave of mysticism and the original break.

"The Mission told us the truth, but they did not show us the way," the people say.* With every day that passes, their idiosyncratic interpretations—the increased number of remarried divorced people who now have children, such beliefs as reincarnation, the development of new mortuary practices contrary to Christian practice, such as the elaborate involvement of property in the burial—make their version of Christianity more difficult to reintegrate with Catholicism again. Yet they think of themselves as Catholics and are not receptive to Protestant missions. The identification of their local version of Christianity with their sense of themselves as a people grows, and just as the development of a costume which sets a people off decreases their chance of assimilation into the larger society, so the development of their version of Christianity tends to confine them within a circle which closes them off from the intellectual and spiritual values of

_____

*In the earlier days of the cult the missionaries were accused of deliberate lying.

the Christian Church and limits them to local mystical prophecy or local political manipulations as their only sources of inspiration and rationalization of innovation.

The old local religion, the Sir Ghost cult, self-contained and consistent with its level of social organization, was extraordinarily adequate in the old culture, relying as it did on administration by competent members of the village who understood all the issues involved. The present religion is a bad fit in which most of the tenets of Christianity—with the single exception of the idea of the brotherhood of men and the fatherhood of God—have been distorted and impaired in a fashion similar to the way peasant peoples all over the world have tended, when living in partial isolation, to reduce their own stature as they took over part of a world religion and combined it with previous local practices, perpetuated in a situation of complete or partial illiteracy.

13

rage, rhythm, and autonomy

We may now turn to the way in which children born to Manus parents develop a character within which anger is both so ready and so disapproved that exhortations against anger can become the focus of a whole local religious system.

In describing the way Manus character is formed, it is not possible to go back to a comparison of the way the newborn was treated in 1928.[1] For, in 1928, no woman who had not herself borne a child was permitted to witness a birth, and I believed then, as I do now, that in field work it is essential not to deceive those from whom one wishes to learn the truth. I had never had a child and so I saw no childbirth. None of the accounts which I was given told me anything of importance about the way in which a child's introduction to the world foreshadowed the future course of its development.

When I returned to Peri in 1953, having fulfilled the necessary requirements for 1928, I found them no longer necessary. Husbands and even small children were now freely admitted to a birth.

For the birth of a Peri child, it is necessary to have three women present besides the expectant mother, one of whom is a representative of the husband's family, two from her own family.

Of these three, two should be experienced older women. Inexperience may mean an incorrectly cut cord or a long delay in expelling the afterbirth. No preparations are made for a birth except to accumulate coconuts which will be made into a special dish for the new mother after the baby is born. These coconuts are provided by the mother's brother, and cooked by her brother's wife, one of the three women who ideally participate in the birth.

Today, children are born in the little old-style houses which are built out over the sea for the old women, who are given a chance to live as they have always lived, close to the shifting tides, and who also provide hospital care both for obstetric cases and for any illness which makes it difficult for the patient to walk about. These little houses do not differ from the houses of long ago, except that they are very tiny and flimsy and the fear of the house collapsing if a crowd gathers is more intense. The fireplace is still a framed square of ashes on the floor, and the expectant mother lies on a mat beside the fire; a second mat is hung up to shield her from the chance gaze of the men in the house. The walls of these houses are made of leaf thatch attached to very slender supports. For something against which the expectant woman can push her feet, an extra post or board will be fastened firmly in the wall.

Then the three helping women seat themselves, one on each side and one behind the woman in labour. Each old woman has a slightly different style, but the pattern is the same. The expectant mother hooks one leg over the leg of the midwife—this gives her human support—and presses with the other against the wall. The women behind and beside her support her and, as each pain passes, she again relaxes and lies quietly on her mat, while the three women also relax, smoke, chew betel, or possibly one of them nurses her own baby, laying it down or passing it to someone else when the next pain comes. From time to time, the midwife suggests that the expectant mother lie on her other side, and she and her vis-à-vis change sides. The expectant mother then hooks her other leg over the midwife's other leg in the reciprocal of her previous position. This may be done twenty or thirty times during the course of labour, each position being the replica of the previous one until one has almost the sense of an impersonal machine-like exactitude, a mobile unit in which each piece plays a set role.

When the experienced old midwife realizes that the moment of birth is near, all three women go into action, backing up the mother, literally and emotionally, in the act of getting the baby born, efficiently, quickly. At the moment of birth they form a hollow square, each part flexible, determined, strong, within which the baby arrives and is picked up by the woman on the other side. Cord uncut, the baby is held facing the mother and completing the square. Then, two things happen together: the midwife and the woman behind the mother redouble their efforts to get her to expel the afterbirth, and the nurse, holding the baby, begins to sing a lullaby in time with the baby's birth wail. As the baby wails and the midwife and her assistant exhort the mother to greater and greater effort, the nurse's voice rises into a crescendo and the new baby becomes part of the rhythm of the world about it. Only if the afterbirth is delayed for ten or fifteen minutes or more are more drastic methods necessary, and the cord will be cut before the expulsion of the afterbirth. Once the afterbirth is out, the cord is pinched clear of matter and tied so that "no blood will run back into the belly," then cut, and the helping women—by this time there may be several more—set to work on the post-delivery tasks, bathing the baby in sea water, preparing a hot coconut soup for the mother.

The baby characteristically shrieks at being bathed, and is then settled into a nest made of an old grass skirt, a nest that is slightly prickly but soft and yielding and from which it can gaze about in the flickering lights of the room, lit from the doorway and by light that comes through the slats in the floor. Whenever it cries, there is the lullaby echoing its cry in exact rising cadence, interspersed with the words, "Some day you'll be a fisherman, you'll sail the seas, you'll catch fish," if the baby is a boy, and with, "You'll grow big, you'll bear children," if a girl. The new baby is not fed until it "cries for food," which may be several hours after birth, when it is given milk by the woman who has given birth most recently. Thus the Manus baby's introduction to the world, its first contact, is not an oral one, but is muscular and auditory, as its nurse moves and (I hesitate to use the word *sing*, for the lullaby is more like a noisy roar modelled on an infant's wail) lulls it, taking her time cues from its voice. Instead of initially communicating with the world through a nipple from which comes food and comfort, it moves with a world that moves with it. And the cue the child gives is a real cue, not an imputed one, a real cry which the adult imitates. It has begun a life of autonomy in which others will respond to its strength, its initiative, its will to move, to make sounds, to grasp food.

So from the moment that a Manus baby is born, it is caught into a system which emphasizes the active rhythmic reciprocity with the world about it, de-emphasizes differences in size and strength and sex, and stresses its existence as an independent organism. Before the cord is cut, the old woman who takes care of the newborn starts to lull it in time to its own crying, so that the sound it makes and the sounds it hears are as nearly one as it is possible to make them.

Nursing is not done in a way which makes the infant into either a passive receptor or into a demanding little monarch to whom the mother passively offers herself through the breast. Instead, the complementary relationship, in which the strong, large mother with milk gives her breast to the small, passive, hungry baby, is converted by the Manus into a kind of reciprocal exchange in which the breast is treated more like a connecting piece of tubing between mother and child than as a part of the mother. From the earliest suckling the mother handles her breast in this detached way, and the child soon learns to treat it in the same way, pulling, dragging, stretching the breast about to suit its needs.

During the first month or so of a child's life, the mother stays close to the baby. It is held on her outstretched legs, or laid near by to sleep, always on its side, propped up by pillows, or picked up, fed, kept warm and clean, and given every attention. The hands that hold it—twenty-five years ago they were women's hands only, today fathers also take part in the care of very small babies—are alert for every sign of strength, of potential autonomy, for its ability to rear its head, reach out its hand, sit with support, go through the gestures of walking with support. Every slight advance is taken advantage of.

This treatment results in very marked differences between babies who continue to gain rapidly after the first couple of months and turn into very active assertive happy babies, and the babies who grow less well and are not as easily stimulated into activity, who tend to develop into more passive, whining babies, less willing to go to others, less attractive to a crowd of young aunts, uncles, and cousins. These weaker babies are more of a drain on the mother, who is crosser and more irritable, thus in turn reinforcing in her child its pettishness, fretfulness, and intolerance of others.

The active, growing baby forges ahead, refuses to be hand-fed, but instead insists on taking its lumps of taro in its own hands, is active every single moment that it is awake, and is terrifically tiring to care for. Where the fretful, passive baby demands to be carried most of the time, the active happy baby will also prefer human arms and attention to sitting on the floor or playing by itself. Twenty-five years ago, babies were left more to themselves, laid on mats on the floor, while their mothers did elaborate hand work of beads or string. But today, the mothers do no hand work, most of the year they do not even have to go for water, and their housework takes at most a couple of hours a day. For the rest of the time they have no excuse not to hold a child except that they are already holding another one. It is a common sight to see a woman with a small child in her arms and a knee baby* holding on to her back. Before they can walk, children become accustomed to adults turning themselves into human pedestals and transportation systems. If the baby wants to see, it is lifted up; if

_____

*Second child, displaced child, or "child whose nose is out of joint."

it leans toward something, the adult leans with it; if it wants to go toward something, the adult carries it. At every turn the adult makes up to the child for its lesser strength and size, holding it high, carrying it because its legs are too weak or too slow.

This desire to move is underwritten by the care with which the adults seize on each forward movement, and yet never force a child, never make it sit without support before its back is strong enough. However much it is put through walking paces, its arms are held firmly and no weight inhibits the playful walking movements until the legs are ready to carry it. Every bit of motor freedom is anticipated, rehearsed with enthusiasm. But there is no forcing and no pushing, no attribution to the child of a strength which it does not have. This is in strong contrast with the behaviour of the American mother who says she can do "nothing" with her two-year-old, confusing her desire for him to be wilful and independent with a totally spurious physical strength. When it is necessary, Manus adults simply pick up children of any age and lift them to where they want them to go, in spite of their articulate howls and kicks. There is no real attempt to keep them from howling and kicking, although the parent will repeat the reason why the force is being exerted. But any discipline is a genuine trial of strength, not a spurious one, so that the child is not confused about its own strength, which has been tested again and again, realistically, against the hands and arms and wills of adults who know their own capacities. As a result, a very angry three-year-old girl can often be handled only by an adult, and a twelve-year-old girl or boy who attempts to lift and carry a screaming three-year-old child may be kicked and unable to keep the child from squirming away. The adults teach the child autonomy, conveying by their every act the admonitions: Be as strong as you can. We will back you up, give you every help, admire every step, turn away our eyes from your mistakes. So a small child's learning to walk has also the qualities of a coming-out party; there are always willing hands to help it practise and a group of spectators to exclaim with delight.

This combination of a premium on autonomy, on realistic

precocity, and a willingness to stand the strain and work involved in letting a child be as active as it wishes to be, produces a child who is highly sensitive to activity in others and extremely ready to imitate the behaviour of others who have, in many cases, picked up its behaviour to imitate initially. All action—walking, swimming, talking, dancing—is a reciprocal interweaving between adult and child in which every discrepancy in size, strength, and knowledge that can be minimized is minimized, as adult and child exchange the same word, fifty or sixty times, and the child sees itself mirrored, not in a looking glass, but in the strong, sure accents of adulthood. This mirroring has as an accompaniment a tendency for one to respond immediately to the acts of others. Any Manus group is an organized tense pattern in which people on one side of a room respond, either by direct imitation or contrapuntally, to people on the other, or in which groups are formed around some activity, like sailing a canoe, delivering a woman, caulking a hull, etc. Response to the activity of others is built into the activity system. The minimizing of differences between adult and child minimizes any tendency toward a refusal to imitate. Manus children can be stubborn and mutinous to verbal commands and respond with negative activity, but this does not seem to be accompanied by emotional withdrawal. As they experience imitation as a free act, so they also feel free to go their own way. Later, in adult life, the slightest diminution of ability to move, a strained muscle, a sprained ankle, a swollen shoulder, immediately arouses anxiety, brooding, and depression. The sense of the self as a zestful, active person is dependent upon movement, upon the ability to move freely and to get out of any uncomfortable situation by running away. The adults in turn work to circumscribe the area of safe running away; all political differences used to be solved by part of the group running away. But in contrast to running away, people saw themselves as freely choosing everything that they were free to choose. The old Sir Ghost system, which simply channelled and drove this activity, was felt to be externally coercive.

In developmental terms,[2] the Manus emphasize as the tasks

of the toddler control of the body and sphincter control. The dominant emotional mood is anger. Even fear is expressed as anger, so that if a parent takes a child into a frightening situation, the child expresses its fear by beating the parent with its fists, demanding to be lifted off the ground or taken out of the frightening situation. Demanding assertiveness up to the limits of one's strength pitted against adult strength evokes in the adult a conflict between the desire to indulge the child and fatigue from the child's demands, and leads to frequent outbursts of anger. A mother sitting with a fifteen-month-old child in her arms and two older children playing about is engaged in a continuous active struggle as the children climb, push, and pull at each other with her body as the stage. Not infrequently, the observer is unable to distinguish between love and anger, play and punishment, as a mother breast-feeds a child, paddles it on its little behind, or shouts a lullaby which imitates the child's recent screams. The amount of screaming that goes on is roughly in inverse relation to the number of hands free to hold and amuse the child.

In Raphael's household there were five or six people always ready to engage in roughhouse with the vigorous, demanding baby girl Nyawaseu, and in Michael's house there was another galaxy of delighted old and young nurses for Pwochelau. These babies, boasted their parents, never cried; they were healthy, rapidly growing, active babies, and they had an older sibling or an adult at their disposal every minute. To rear a child who realized the full potentialities of this Manus system of giving the child every inch of autonomy possible, took most of the time of at least two adults for every child under three. This meant that in practice there would be a great difference in personality of the children of isolated, poverty-stricken families, where the parents were over-worked and the children had to be penned in and left to scream. The children in such families were not popular candidates for adoption. In the more established families from which adoption was frequent, many children had, as a result, four parents—their own and their future parents—to carry them about and play with them (Plate X). Such children were sunny and delightful, and usually accumulated many more admirers and temporary steeds among other adults also.

Compare, for instance, the households of Michael Nauna (Plate XIII), seaward of my house, and of Petrus Pomat, on the inland side. Michael and Raphael were of *lapan* rank and were actively co-operating "brothers." Raphael's father had been adopted by Michael's father. In the preceding generation, Raphael's father and mother had adopted Ngalowen, the little sister who came between Michael Nauna and his younger brother Ponkob (Plate VII). This had meant that Ngalowen had the adoring care of her adopted father and mother and of a much older adopted brother, Raphael. Before Michael married, his mother, a remarried widow, had adopted the eldest daughter of Raphael, Anna, who grew up as an eldest daughter in Michael's household. In 1953, Michael had six children, the eldest a girl of twelve. His wife, Rosina (Plate XIII), was regarded as the best mother in the village, endlessly patient, good-humoured, tireless in helping her baby learn to walk. All of the children were handsome, gay, bright, well-mothered. And also there were many hands. Anna was there to help and so was Otto, an orphan from Patusi about two years older than Anna, who had been adopted by Michael. Poten, the second little girl, and Pomat, the knee baby, both lived with their grandmother, which meant there were three people to minister to Pomat's whims, lug him to the clinic to have a sore dressed, carry him about whenever he chose to shift from being an independent canoe man to being a demanding little lord of the universe. And the third child, Lapun, lived with his father's younger brother, Ponkob, who had no children of his own, and there received undivided attention. Then Anna married. She and her young husband became part of the household of her real father, Raphael, and he gave her as a future adopted child his six-month-old baby, Nyawaseu, a gay happy little girl who had been cared for from birth by a large household (Plate X). So Anna's experience as the indulged child who then had taken part in rearing a group of happy babies was now further reinforced in the capacity of a potential good mother, by being given a baby who expected to be well cared for. The hereditary position of *lapan*, combined with habits of leadership and wealth, reinforced the necessary conditions for the optimum development of personality in little children, and this was picked up later in four- to six-year-olds in their identification with parental style of behaviour.

The household of Petrus Pomat represented a strong contrast to that of Michael Nauna. Pomat had been the youngest of the six children of Kemwai, a substantial, reliable, but not very bright commoner of the Lo clan, who married the most brilliant woman in Peri, a member of a *lapan* family of the leading clan. During his lifetime Kemwai played a considerable economic role, cooperating with his parallel cousin, Nane, who was also married to a *lapan* woman of Shallalou, the *lapan* section of the clan of Lo. Pomat was the child of his father's effective middle age, and, in addition to having a group of four much older sisters, had also been adopted in his early childhood by some of his father's higher-ranking relatives, where he had been adored and indulged. Equitable in temper, with a pattern of devotion to children and a habit of sure command—a compound of his father's moral authority and his mother's high intelligence—Pomat had married a Bunai woman with very few living connections. All of Pomat's sisters married away from Peri. The sons of Nane all lived in Peri and formed a solid group of four brothers, cooperating happily together. Furthermore, the eldest, Kutan, married the sister of Michael Nauna's wife, and he and Michael built their houses side by side. Today, with no more entrepreneurial affairs or clan-minded Sir Ghosts in the picture, there was hardly any relationship between the cousins Kutan and Pomat, who would have co-operated actively under the old system. Functionally, Pomat was virtually without relatives.

Pomat and Monica had four children, all of them thin, underdeveloped, fretful. No one wanted to care for them, and Monica and Pomat had their hands too full. Pomat supported the family by deep-sea basket fishing, a lonely and patient occupation, and Monica scolded shrewishly because of the demands made upon her. Meanwhile the two older children who could have been a help with Lomot, a solid, demanding two-and-a-half-year-old, kept as far away from home as possible. In the evenings Pomat used to wander about, staying as far away from the house as possible, while Monica egged on the little four-year-old knee baby to scream for his father. Monica, pregnant and further hampered by a serious cut on her foot which crippled her for weeks, would leave Lomot at home. Lomot spent an average of two hours a day screaming because she was left alone instead of being carried somewhere, while Pomat continued to carry the slender little four-year-old about.

Here we see the cumulative effects of the opposite sort from those which made the household of Michael Nauna such a happy one, with a cluster of beautiful, well-nourished children. It was quite impossible to put a finger on any one cause; it was even impossible to say that Pomat's marriage was not a successful one. On the TAT card* which was so often interpreted as a man whose wife is dying or dead, Pomat commented touchingly on how helpless a man is who loses his wife. He was believed never to have been unfaithful to Monica, a rare reputation in contemporary Peri. He was intelligent, responsible, patient, trusted by everyone, and Monica, for all her shrewish loquacity and tendency to interfere in "court cases" which her husband was trying, was a brave and intelligent woman. But the children were puny, retarded, fretful. The situation of the home meant that they had been provided with an insufficient amount of the kind of care and support which Manus children need to develop their potentialities.

There could also be other sorts of complications. When Peranis Kiapin married Lucia, her former husband had claimed the child which Lucia had given birth to, not while with him but while with a married lover, during a temporary period of estrange-

---

*Thematic Apperception Test in which the subject is asked to tell a story about a picture which is designed ambiguously.

ment from her husband. Lucia was now pregnant by her new young husband and she and Peranis went away to work on Ndropwa, leaving the wife of Lukas Banyalo—Peranis' putative adopted "older brother"—pregnant by a fifth child. Before this child was born, and when she herself was pregnant, Lucia and her husband proposed adopting the new baby of Lukas Banyalo when it would be born. This baby was then born some five months before Lucia's baby, and during this period Lucia and Peranis took a great deal of care of young Ponowan, who as a result of this care flourished into a great, bouncing, fat baby, terribly active and heavy. Lukas' wife was badly over-worked, with a family of five and no relatives—they were immigrants from Patusi—and Lucia and Peranis' help was a godsend. Then Lucia's baby was born and proved to be very frail and sickly, with something wrong with its eyes. Peranis Kiapin was a model of the new Manus husband, actively and intrusively interested in the care which his wife was to give her child. Even during her pregnancy he had fussed about her going into a mourning group, or into any situation where she might run a risk of infection or strain or physical injury. Under the pressure which he exerted, she became a slave to her ailing baby, and neither she nor Peranis had time any longer to take their adopted child Ponowan off the hands of his over-worked mother who was still breast-feeding him. Consequently Ponowan sickened, developed boils, and changed from the gay, vigorous baby he had been, reflecting the change from a baby always held and cared for to a baby frequently left alone. As a younger baby, in spite of his great obesity, he had been strong and active; now he became flaccid. The greater strain in the household of Lukas Banyalo was further reflected in less attention being given to the next two children, and the two months that followed were months of continuous illness and accidents among them.

Thus this present-day system of underwriting the child's autonomy entails many hazards when it is combined with the present emphasis on the importance of a high birthrate, the breakdown of wider kinship ties, and the disapproval of any measures for limiting the population. The child becomes dependent on the constant stimulation of willing, involved, adult hands and on adult accession to its demands. In many ways it is comparable to modern American family life, where there are also too few hands available.

Within this setting, anger becomes a recognized and trusted means of expression. Whereas children in most societies are soothed by a lullaby which is definitely contrapuntal to their rage and despair, in Manus the lullabies merely echo the child's rage more loudly, in rhythmic agreement and rising crescendo. Babies are expected to cry and to cry hard whenever they want anything which they do not have. And these wants are expected to be real wants, not artificial ones.

The importance of this point is one which I would not have seen if, between 1928 and 1953, I had not studied another society in which rage in children was valued and provoked.[3] Among the Iatmul of the Sepik River, there is a deep, pervasive fear of passivity, including an articulate fear of passive homosexuality in men, and a continuous attempt on the part of the mothers, older boys, and men to make children be active, cry for what they want and fight back when attacked. Mothers put their babies on the other side of the house and refuse to feed them until they have cried hard, or, after cooking pancakes, hang the pancakes up out of the children's reach and go fishing, leaving the children to cry. Little boys who playfully imitate the disallowed sodomy, to which they hear frequent verbal reference, are armed with long poles by older boys and forced to fight. As adults, temper tantrums are applauded as the way in which action can be initiated and most action has to be instigated by calculated insults. The Iatmul so value rage that when a man is destroying a year's work, chopping up a house-post or a mosquito basket, there is a large, happy, grinning audience. Rage means the ability to display the strength which will make others yield to one's demands. Among the Iatmul, rage is good. This is similar to many modern attitudes toward child rearing in the United States, where mothers worry because their little boys aren't aggressive enough, don't stand up for themselves, and theatrically enact situations in which the child has the upper hand: "It was so cold out by the sand pile, but Bill (age two) just wouldn't come in." This parallels the behaviour of the Iatmul mother chasing a fleeing midget with a ten-foot pole and uttering loud threats, but being careful not to catch him, giving him a false and contradictory picture of his own strength.

Against this background of Iatmul and, to a certain extent, American childhood,[4] the Manus treatment of children stands out in sharp relief. The Manus are not afraid that their children will not show enough aggression or self-assertion; they simply cherish autonomy in their children as something to be valued and cultivated. Even toilet training, which so often is an area in which the autonomy of a child is invaded by the demands of society as mediated by the parents, is carried out with respect for the child's control of his own behaviour. Patiently, untiringly, the parents push the child toward the limits of its ability to take charge of itself. Children are expected almost as soon as they can walk alone to get up and walk away from their sleeping mats and to lift a slat in the floor if they have to urinate or defecate at night. They are expected to walk a little distance if there is a light in the house, but only to get off the edge of the mat or blanket if there is no light. They are expected to point out next morning just where they have defecated so that the parent can carry away the feces— an onerous chore now that the people live on land instead of over the conveniently shifting tides of the salt lagoon.

In every way—the way it is held, fed, lulled, put to sleep, and awakened—the child learns that the adult values its autonomy. Each autonomous move is supported, dependency moves are, at best, reluctantly acceded to. Anger which is part of autonomy is valued also, so that the child who cries vigorously and definitely because it wants something does not disturb a Manus community. If a child decides to stand and scream in the middle of the square, a parent may come along and pick it up and carry it off, but almost without emotion and without rancour. At the same time, every child's cry is initially attended to and there is instant response if the cry reveals anything except anger—fear or pain cries bring immediate rescue. And, in the long temper tantrums that may last an hour or so, people listen, alert for the moment when the child is no longer in possession of itself, no longer in control of its rage (Plate XII). The instant that the lusty autonomous cry changes to the rhythmic, uncontrolled sobbing of the child whom rage has possessed, someone steps in, soothes, appeases, wipes the sand off the body, the tears off the face. This same sort of tidying-up behaviour also follows a fall; a first rescuing step is always to brush off and clean up the tear-stained, sand-encrusted child (Plate XII). Sometimes, as in the case of individual older children who are notorious for temper tantrums, one can see an older brother or sister standing quietly waiting for the moment to intervene when the child's rage slackens or takes possession of it, both cues that intervention should occur. So the child learns that rage which it can control is good, will be respected, and often will get results, but that loss of control is something which others hasten to rescue it from.

This learning shows up all through life and forms the background of the present fear of any outburst of temper. In the group, as in the individual, anger is good when it is under control, when one person can exhort or declaim or harangue the rest without his anger so taking possession of him that it becomes contagious and results in the whole group getting caught up in an irresistible group tantrum. But such contagion is very likely to occur; any real quarrel between two people tends to assume a mounting tempo, drawing others in, generating side battles, renewing old grievances. All the tendencies to active reciprocation of movement and mood come into play whenever someone becomes really enraged. If the group succeeds in resisting the contagion and the flurry of involvement dies down, leaving one infuriated person raving on and on, then, as in the case of the child who cries in protest, the group assumes a spectator role, waiting quietly until the aggrieved person has cried or muttered himself out. In this case, the anger dies off into a kind of rhythmic crying, not the sobbing in which the body is shaken against the will but a kind of crying which comes increasingly under control and which small children often change into singing.

The old-fashioned chant, *enrilang*, is translated as "cries" and there is a close relationship between the type of sharp aspiration which occurs at marked points in the chant, not coincident with the place where a break would come in terms of meaning, and the sound of a sob. Children often finally sing themselves to sleep in this half-chanting, half-sobbing voice, and as often wake from sleep already crying lustily.

In the crying over the dead, control is again emphasized. A contrast is felt between the grief which the individual wills to express and grief that continues to make him cry in spite of himself, which is felt to be bad, dangerous, and likely to lead to madness and suicide.

In mourning and in shouting matches in which no one is beside himself with rage, there is a similar phenomenon of demonstrating control by very abrupt starts and stops. Someone will be sitting dry-eyed and quiet on the outskirts of the mourning circle one minute, and the next will plunge into the group wailing at the top of his voice. This wailing may stop as suddenly as it has begun. This is equally true of quarrels. Each participant breaks off from a burst of imprecation in a tone which gives no clue as to whether any more will ever be said. So there are several different mechanisms of control present; the training in keeping possession of one's own rage, rhythmic devices which may supersede, help to slacken, or stylize the anger, and finally breaks which prevent any sense of mounting climax.

So far I have discussed Manus today, the way the child is reared, and the way anger is treated. There are several marked contrasts to the old form of child rearing and the old treatment of anger. Under the old system of child care children moved from the house floor, where they learned to crawl and walk, either into the shallow portions of the lagoon, where they learned to swim under the supervision of an adult, or to a canoe. As soon as they were able to swim, they were trusted to play in the water. There was every incentive to both parent and child for swimming to be early and efficiently learned. Once in the water the child had great freedom of movement. Adults had to lift very little children up and down ladders; for the rest these very small children played in the house, sat in canoes, or played in the water. Today children do not learn to swim until they are around three; the houses are smaller, and instead of going about the village in canoes, one has to go on foot. Both parents and children are impatient with the pace at which children walk, and this leads to much carrying.

Twenty-five years ago, while parents took much the same general attitude toward small children, they were much busier, much more rigorous in their demands that children respect property. There was more disciplinary conflict between parents and children and less of a tendency for children to make continuous effective exactions from parents for care and transport. In the society at large highly controlled histrionic displays of rage were expected from the leaders, but there was the same dislike of anger which might possess one, the same preoccupation with devices which might reduce the kind of quarrelling which involved whole chains of people. People were faced, then as now, with a kind of character in which one kind of anger is approved and another disapproved. And today there is little outlet for rage. Quarrelling is forbidden, black magic is gone, wife beating is interdicted, brawling is censurable for disturbing the peace. There are some indications that this is leading to the turning inward of anger, to self-destructive fits of rage, and to depression. The young men who go to work for the European are even touchier than they used to be, and are liable to leave in an instant at some fancied slight or insult.

So Josef Bopau was transformed within a couple of months from a creature who bounded over railings, leaping high in the air on his way to the beach, an incarnation of gaiety and perfectly spontaneous movement, into a bitter, depressed, sullen boy because the European for whom he had gone to work had accused him of stealing a pair of contemptible old shorts and fed him worse than the chickens. In childhood the lesson of alert defence of the self, of one's position, one's autonomy, one's spontaneity, is learned to the accompaniment of noisy anger. This alert touchiness remains a double hazard, to the self as the trigger for a depression, to others as the spark that may set off a devastating quarrel.

The quick capacity to mobilize energy, to support will with anger, is also expressed in the sex attitudes of Manus men— among whom potency and anger are closely associated. Women are characteristically regarded as objecting to sex, and whether a man is subduing an unwilling mistress, avenging himself on a woman who has scorned him, disciplining an angry wife after a quarrel, or angrily copulating with his wife so that she won't suspect he's just copulated with someone else, anger runs like a constant theme through the potency which seems to be reliable and untroubled. This, in turn, provides another route through which unexpressed anger and dammed-up movement of any kind may lead to depression, and it fits into the emphasis of the New Way ethic on the dangers of suppressed rage, angry thoughts, and sexual desire that is not acted upon. Far from being the sensual people deeply interested in sex satisfaction which their critics represent them to be, the Manus seem to use sex as one expression of the will. Sex is not something to be delighted in for its own sake, certainly seldom a part of any tender relationship, but is treated like a whip held in a hand tensed for movement, likely to stiffen if not used. In the very young marriages there is an appearance of greater tenderness, but whenever a disagreement occurs the old attitudes which associate sex and anger seem to reassert themselves. Illicit love affairs, affairs of choice, are, significantly enough, described as situations in which people need not have sex relations if they do not wish to, but can simply sit and talk and laugh together.

14

new working of old themes

Manus adults are living today in a world which in its ideas and values represents thousands of years of development from the ideas and values of the world in which they were reared. Today they are friendly where formerly they would have been harshly competitive; they are actively concerned with the prevention of types of behaviour which they would formerly have regarded as natural and desirable; they are relaxed and unworried where they would formerly have been tense; they are rearing their children with a kind of indulgence which would have been unheard of twenty-five years ago. Yet it has been a thesis of modern anthropology that there is a systematic correspondence between the institutions of a society and the character structure of the individuals who embody those institutions.[1] When every institution has changed and we find new institutions—the small two-generation family, freedom of women, political democracy, sense of community, ideals of human dignity and stature, a psychosomatic theory of disease, a capacity to deal with boundaries in space and time— what can we say about the personalities of these adults? When their behaviour has changed so radically, how have they themselves changed? If we try to derive a character structure from their institutional practices and from their child-rearing patterns today, what will we find?

Let us take a second look at the children of twenty-five years ago. We have seen that then as now the Manus child was an active, competent, high-tempered, demanding, imperious youngster, sure-footed, responsible in all physical situations, and intolerant of delay or frustration which involved human situations. It was subjected to very early toilet training and an intense respect for property and name taboos, which were handled as the property rights of others. These children indulged freely in temper tantrums which the adults encouraged by trying to fulfil the demands on the basis of which the temper tantrum had been thrown; so husbands beat their wives for not feeding a crying child and fathers attempted to gratify their small children's slightest whims. The children maintained a contemptuous disrespect for the exactions of the adult world. As six- and seven-year-olds, they blithely ignored parental threats of ghostly dangers, and often had to be dragged home at night time protesting all the way. They learned to imitate every physical activity of their elders with great facility, to aim a spear, to paddle and punt and sail a canoe, to fasten and bail and right a canoe which had over-turned.

There was also, especially during the period of four to six years of age, a high degree of identification in stance, tone, posture, degree of initiative, between children, especially little boys, and their fathers. It was possible to trace a striking parallel between the style of personality expressed by a boy or young man and the degree of economic initiative which his father or foster father had or was exercising. Children also showed an almost compulsive tendency toward bodily imitation of movements, even attempting to imitate the intent posture of a white man sitting on a chair writing. They watched every step in an operation such as lighting a lamp with breathless attention, absolutely still, except for their eyes, and later were able to reproduce each operation perfectly. When they asked questions, the questions were based on previous observation of "first he did that, then he did that," etc.

But there was one noticeable peculiarity about the children's play twenty-five years ago, which I recorded but did not understand. Imitative as they were, their play was empty of any content which imitated adult social relations. Whereas the adults exchanged property all day long, the children did not exchange property. The impressive ceremonials of adult life, birth exchanges, betrothals, marriages, mourning, war dances, were missing from children's play. While children in some societies imitated adult ceremonial by the hour, Manus children simply ignored it. They paid no attention to adult ceremonies, they were ignorant of the details, and they made no play with them. In some cultures such lack of imitative play is found to be related to the extent of the children's role in adult activities. So Samoan children did not play at cooking, because they actually took part in making the family oven, nor did they have dolls for there were more than enough real babies to be carried about and cared for. But they did play at the intricate social ceremonial of adult life. Arapesh children spent almost all their time in small family groups participating in their elders' lives. The small girl wore a carrying bag just like her mother's; the little boy carried a miniature bow and arrow when he followed father or uncle on the hunt. These children who actually participated in the workaday world of the adults played no games which imitated this work, in contrast to children in other societies like the Plains Indians,[2] or the Iatmul of the Sepik River, where, excluded from adult activities, they faithfully reproduced, in mimetic microcosm, the details of the pattern of adult life, of marrying, having children, hunting, going into trance, etc.

The Manus children's failure to imitate the lives of the adults could not be attributed to their participation in the workaday world, for they did not work, often even refusing such small tasks as diving for an object which had fallen through the house floor into the lagoon. True, they learned to paddle large canoes by being allowed to transport their fathers about the village, but this was a form of exhibitionistic display of their achievements and not work in the sense of making any useful contribution. Later, after the little girls were betrothed, they were absorbed into the lives of the married women. They learned to do beadwork, not cunning little play beadwork, not even a cross between work and play like the samplers made by little girls in colonial America, but real, hard work like that over which the women tired their eyes and backs.

This lack of either imitative or real participation in the whole superstructure of adult life seemed to me in 1929 to be one of the reasons why Manus adolescents hated to play a role in adult life, and assumed their responsibilities so ungraciously, one of the reasons why such strong sanctions were required to bring them into line. I did not, however, realize the opposite and reciprocal point: that this very hatred which adolescents and young adults expressed toward the social superstructure might be one of the conditions which led to the children's refusal both to participate in and to re-enact the adult ceremonies. When the extent of the adult rejection of their adult roles is taken into account, and the extent of the children's identification with their fathers, then the relationship between the two superficially contrasting positions becomes clear. The children imitated their parents' every physical move, they imitated their degrees of initiative and autonomy—children whose fathers were still young and shy behaved quietly, children whose fathers were arrogant and certain of themselves were much more self-assured. They also imitated their fathers' sense that the whole of the adult life was a burden, heavy and exacting, enforced by a coercive set of Sir Ghosts, from which any sensible person would escape as long as possible. These children also saw their older brothers and cousins run away from the village rather than become prematurely involved in the adult life, and this further reinforced their sense of the lack of rewards in the adult forms.

This interpretation can be partially tested out by the behaviour of children today. They are still learning physical skills early and capably, although they now learn to swim later than they once did. They still respond with great speed to cues given them by the activity of others. In any group rhythmic waves of motor imitation, sometimes a replica, sometimes a mirror image, are constantly passing through the group. They still alternate between a state of attention in which every muscle is quiet and only their eyes watch and a state of active imitation. But today they both participate in adult activities and engage in mimetic play. Children play house, they build very tiny houses and also houses big enough to get inside, and play at housekeeping. Children imitate the games of older children and adults, holding up small squares of cloth against the light of pressure lamps, as their elders have learned to do to make shadow-plays. As significantly, they now participate in adult work. When there is a party scraping paraminium nut to caulk a canoe, there is a whole cluster of children about scraping their small shares. Small boys stand in line with the older men to move big canoes, sit beside the basket-makers making some smaller object. More significantly still, they attend with enthusiasm the adult ceremonial life which is superficially so much duller than the old ceremonial life which they used to ignore. They are the first to arrive at council meetings, sitting in solemn rows like swallows on a clothes line, quiet, intent, good. Swarms of them appear wherever anything is going on, and it is almost impossible to shoo them away.

In church they follow the pattern established by the Mission of children sitting in the front rows and being the first to march out at the end of the service. The smallest ones who are not yet supposed to go to school get into the school and try to learn to read and write. As they once participated in the adult rejection of social life and imitated only those sections which the adults enjoyed—physical movement, fishing, sailing, and behaviour involving self-assertion—so today their behaviour follows the lines of adult feeling, sharing the adult enthusiasm for the new way of life, the feeling that work is no longer too heavy or too long, that there is time to rest, that it is fun to make a new canoe, and that council meetings are tremendously important affairs to which it is a privilege to go. Where participation occurs, they do not engage in mimetic play.

Watching today's children, then, throws new light on the childhood of today's adults, who as children could look forward to no life which kindled their imaginations. The children of twenty-five years ago engaged as little in flights of fancy as they did in exact imitation of adult life. They made little models of canoes and of European pinnaces, and sailed them about; they sailed in their own canoes, they learned to handle bigger canoes, they fished with miniature spears and bows and arrows and threw miniature darts. Nowhere was there any elaboration. When they drew, they repeated these same active scenes—men paddling canoes, spearing turtle and dugong, men fighting. But, intelligent and lively and curious as they were, they were definitely not imaginative. And this same quality, this inability to go beyond the actual, is seen today in the responses of these same children as adults to TAT cards: "I see half a man and half a woman. I can see both hands of the man and only one hand of the woman. This drawing is very indistinct. I do not know any more about it. If I knew something about the man who made it, I might be able to tell you more."* There speaks today the man who as a boy thought it was more fun to use colours to reproduce accurately the red and blue ornaments on the canoes than to use them in any free way, and who was delighted by a chance discovery that it was possible to take small objects like spoons and to draw around them and get a most accurate reproduction. These children, too, were perfectly willing to draw the same man or men, the same fish, the same canoe, over and over again, day after day, delighted by activity. The materials—paper and pencil and coloured crayons—tempted them into no land of fantasy. Reproduction of reality over and over again, and the reproduction of the loved activity-type of reality—paddling, fishing, struggling—this was what they drew.[3]

Today, their play is more imaginative in several different ways. Instead of small boats which are simply accurate miniature canoes or pinnaces, which they can sail about the shallows day after day, they make fleets of more fragile, sketchy little models, with sails of green leaves, and launch them off into the deep, letting the first wind carry them out to the open sea. They make

---

*On a Rorschach Test they will enumerate the ways in which the ink blot does *not* look like a man.

endless play with airplanes, airplanes whose propellers are modelled on a pinwheel, airplanes with small disk-shaped wheels so set
in the carriage that when the wind whirls the propeller the little
airplane, which cannot fly, will run briskly along the sand. This
airplane play takes many forms. They make propellers without
planes and carry them in their hands; they make tiny planes and
set them up in lonely places on high posts, the propellers turning
in the wind, unwatched by human eye, until another group of
children find them and go on with the game. Given art materials,
they respond with enthusiasm, the smaller ones using colours
decoratively and freely, while the older ones are most interested in
being able to write the names of objects as well as draw them.
There is less representation of activity itself in their drawing and
more enumeration of the objects of the New Way—so the gate
which is the symbol of the village, the railing around the village,
the gong which calls people to work, the new-style house, the
church, are drawn over and over again. They are more static,
more diagrammatic statements of the new order, catching perhaps some of the adult cultist attachment to this set of symbols.

In retrospect, it is possible to suggest that twenty-five years
ago the sense of an undesired, unlovable, and unfree future lay
upon their youthful spirits much as the hopelessness of their
future social position has been found to lie upon and depress children of partly acculturated native peoples or minority groups,
who, after a promising childhood, lose all incentive to use their
minds or respond freely or imaginatively in school. In a sense
also, in the period of first culture contact twenty-five years ago,
Manus youth were delinquent, vacillating between running away
altogether, against their parents' will, to the adventure of working
for the white man and a series of sexual escapades, which the
community only partially succeeded in regularizing as permitted
youthful licence. So they beat against the adult world, not only
by presenting a wall of active indifference, but by noisy, insubordinate, destructive behaviour, no longer channelled into socially
approved long voyages and warfare.

When the young middle-aged men in their early thirties succeeded in building their New Way, they were doing two things which were wholly congruent with their childhood rearing. In the first place, they were copying reality as they had experienced it, rearranging a series of elements—the British law court and government, the Church and its emphasis on the soul, the American Army and its system of interpersonal relations, the Western world's type of clothing, housing, marriage, etc. They had watched how people moved in these situations, listened to the judge as he made a hearing into a learning experience both for some junior patrol officer and for the natives who were first experiencing British justice, listened to the priests and the catechists as they preached from the chancel step, listened to the tone of voice of the Americans who abandoned huge amounts of property so easily and yet went to such pains for one another. They caught the spirit, the style, the movement involved in these institutions—what a school was, what taxes were meant for, what participation in a meeting really was—by active identification with those whom they saw going through the forms and rituals of civilization. In the second place, they showed a continued rejection of the traditional adult way of life which they had learned from their elders to hate and chafe under. Neither activity was new in form; only the content was new. They had experienced a highly systematized culture which they had learned to dislike. They built, along imitative lines, a new highly systematized culture which contained substitutions and corrections for the rejected elements—arranged marriage, exploitation of young men by their elder male relatives, taboo restrictions between relatives-in-law, slavery to acquiring forms of native money which were only useful for increasing the turnover within the system, clothes and houses which differentiated them from the world of the white man. They constructed a new set of social forms congenial to them, building a world which was like the world they had learned to escape to, and unimaginatively reproducing the form of something which they perceived as a pattern. They had had an opportunity to see *how* the Americans lived within the Western system, and they set to work to reproduce the system.

In this reproduction there was all the surge of excitement, of discovery, of newness. These were men who had from childhood avoided their future mothers-in-law, who had lain embarrassed and inert under mats on the platforms of canoes because of the presence nearby of their affianced wives, who in the early days of their marriages had had to treat their wives, whom they had not chosen, with avoidance and embarrassment. So they were experiencing the full force of the change in expectation, the relaxation of being able to live in a world free from already experienced taboos. This sense of release, of newness, of weights lifted off the shoulders and doors opened where only barriers had been anticipated, provides the conditions of zest and ardour with which human beings adapt to a new situation, as immigrants to a country where their footsteps are no longer dogged by political police, as women who walk the streets with bared faces in a land where only yesterday all women were veiled, even, in situational terms, as discharged servicemen who, after a long period of rigid discipline, are free again as civilians to treat every man whom they meet on his merits. It is this zest which patterns and allocates the energy needed to build a new society, to build a whole village at once or sit attentively through meetings which drone on through many hours of unaccustomed clumsy deliberation. It is because the people who do these things are the *same* people that they receive such a sense of freedom from the change.

The great problem then becomes how this sense of a dearly won and so valuable freedom is to be incorporated in the next generation, in the children who now play on safe beaches, less rigorously trained to care for property now that objects are no longer so breakable and difficult to replace and parents are not so economically driven. The results will also be different because it is not merely a matter of parents transmitting to their children their own sense of release and freedom but because the generation who were late adolescents at the time of the change have to be taken into account.

These young men in their early twenties represent a particularly difficult problem because the war cut them off from both the continuing teaching they would have received from the Mission and from the ordinary sort of long-term work for the European in which their elders had been schooled.[4] They were just reaching adolescence when the Japanese occupation started, and very few of them were old enough to do much work for the Americans. Their knowledge of Neo-Melanesian is inferior to that of the older men, and they do not have the same sense of free communication with Europeans which their elders learned as work boys. Their encounter with the Mission after the war was brief and stormy, and in their first experiences with the return of Australian civil government there were many negative aspects— unsympathetic patrol officers who made fun of the New Way or ordered the beautiful gateway made with such effort to be destroyed. The only Australian communities they have known at all have been post-war communities, put together with a patchwork of wartime salvage materials, and a harassed set of terribly over-worked post-war officials faced with a thousand new problems and trying to cope with a group of people who were defined as subversive.

These young men can hardly remember the pre-war world, either at home or abroad. They had taken part in the exciting episode of pitching into the sea the catechist who had wanted to carry away the altar of the Peri church, and they had spent weeks of ceremonial weeping and feasting to celebrate the return from prison of their leaders, who had been imprisoned for abuse of authority in giving beatings to a quarrelsome young married couple. They were the generation who with no preparation at all for any relationship to women except the old highly patterned ones—freedom to tease and jest with a cross-cousin, respectful dependence on a mother and sister, avoidance of all affinal females—were now suddenly told to form the new kind of marriages, walk about with their wives in public, "enjoy" their wives' company.

It was in this group that one found men with the most rigid ideas, who wore their clothes, not with the practised nonchalance of an American on leave, but stiffly, as a kind of cult uniform.

These were the men who recoiled from pictures showing people in the old costumes, pictures over which the older men and women (who had worn them) laughed with amusement. "These appear as unpleasing in my eyes," said Peranis Cholai, stiffly. "I see nothing good here," said Peranis Kiapin, coldly. This group also had the poorest memory of their childhood and much worse memories for their childhood playmates and elders than had people ten and fifteen years older. They were definitely children of the new order, clinging to it rigidly, almost angrily, and without the flexibility and resiliency of their elders. They had had, in fact, the same early childhood as the older men, but had lacked the kind of late childhood and adolescence in which the older men's habits of companionship and friendliness, and so their capacity to feel free, had been born. They were stiff and difficult fathers, with less tenderness and indulgence toward their children than the older or much younger men. In them could be seen clearly the first effects of a far-reaching change: the new life, which the older men embraced with a depth of thankfulness that was never completely gone from their voices, has become to them something barely attained, thin, brittle, likely to be destroyed by inimical forces, something to defend in anger, and to live by rote. They were the ones who got into complications with their parents-in-law, demanding as a right that their in-laws "talk and eat and laugh with them," in sharp contrast to the tender smile on the middle-aged Raphael's lips as he had looked at his diminutive mother-in-law and said, "And once I would not have been allowed even to say her name, and now she can be here with us and help us eat the delicious scrapings of the feast cooking pot."

In the brittle, dogmatic pride and orthodoxy of these young men, there is limited refusal to accept anything new. This is characteristic of the prophets of "cargo cults" or other reforms which claim to have got the whole answer. One of the features of these "movements of enthusiasm" throughout Christendom is that they are blueprints of a heaven on earth, to be attained overnight by following the right set of rules. So the Manus preachers almost invariably begin a sermon: "For a long time now you and I have known the whole truth, have known *everything* that we must do." There is little flexibility in these apocalyptic cults, and there is a corresponding rigidity and lack of any receptivity to ideas of change to be found in the attitudes of some of these young men.

Such a type of rigidity is to be found very often in the second generation of those who emigrate from their own country to another country, and it is echoed in the American dependence on law and legislation to make the world good. So in the United States, the optimism of those who designed their own new constitution, not knowing how new it was, and the rigidity and doctrinaire orthodoxy of immigrant children who learn in school in 1955 that the world was made new on July 4, 1776, are found combined in our continuing belief that we can make the world into our image, with an over-insistence on doing it by blueprints of the good life.

It is also notable that apocalyptic movements—Quakers, Shakers, Hutterites, etc.—have flourished among people just glimpsing freedom, and especially on the frontiers of the world, with California the most recent frontier. Any discussion, either of those narrow cult movements which seek to turn twenty farm houses into the New Jerusalem by cutting off such iniquities as printed matter or television, or of the narrow dogmatism of successful revolution, as in the Soviet Union or Communist China, must take this generation into account. This is the group who never experienced, as adults, the release and invigoration from the changes that the new faith or the new political economic life promised, but instead experienced only the moment of struggle before the hardly won success—they never knew the long trek through the wilderness, the terrors of civil war.

When this special type of experience is complicated, as it was in Manus, by the quality of late childhood and early adolescence, which had actually supplied part of the incentive for the new order of brotherhood and sharing, then it is easy to see why the young men who entered the movement in adolescence are the least prepared to carry on its spirit, bringing with them as they do only the experiences of early childhood, without the early adolescent counterpoint which inspired and mellowed their leaders.

There was one man in Peri, Karol Manoi, who contained within himself the whole series of these attitudes which were otherwise seen embodied in different generations (Plate XIV). He would shift from a mood of relaxed gay companionship in which every muscle softened to the ukulele that he played with delight, every posture expressing a sense of freedom and ease, to a mood of rigid, watchful, unadventurous determination "to get it right," in which his clothes (which a few minutes before fitted him like his own skin) seemed to crawl up his neck till his collar was too high, his sleeves too short—the whole an unbecoming and unaccustomed uniform—his brow furrowed with anxiety and his mind become hampered by his certainty that there was a right answer and he didn't know what it was. An hour later he might have shifted to an active, competent man dealing with some problem, turning his house into a reception centre for the sick, or discussing with glowing eyes how Paliau had insisted so skilfully in court that men who had been told to obey the Japanese could not be later tried for having done so. But, while the memory of such a flexible, intelligent conversation was still fresh in my ears, there would be another shift. Karol, looking as savage and tense as a leopard, every muscle subtle with a rage which was designed to destroy, or if not destroy, to punish terribly, would be walking across the square, a stick in hand, and I knew that he had caught someone doing something for which he felt he had a right to be angry. The uninhibited rage of his father, Pwendrili, the old war leader of Patusi, which he had experienced in his muscles as a small boy, was now combined with the moral fury of an official of the New Way, ready to pounce on the two luckless women who had "stolen" two of Messrs. Edgell and Whiteley's coconuts. And it might be only two days later that he was off gambling, hoping to retrieve the money which he had "borrowed" without proper authority.

It was both painful and exciting to watch his very good mind struggle with these shifts, trying to work out what had got into him that made him see a thing one way on one occasion and quite differently on another. Hauled into a village "court" for beating his wife because she had refused to make their child limeade, he repeated over and over: "Here in court when she repeats it as evidence, it sounds like something of no moment. Here she says, 'I only asked the child, "Is it your lime?" ' in just an ordinary voice. The wind of the saying is different. But before, when I heard it, when she said it to our son, it sounded very different. I know it was different. It sounded as if he was not her child." And the local "judges," with minds less complicated than his, wrinkled their brows in perplexity. What did all this talk about the wind being different—he didn't even have a word to express what he meant—have to do with the quarrel?

Characteristically Karol did not know to which age group he belonged. One minute he was playing a ukulele with the schoolboys, the next flying into a rage and impounding their ukuleles because no one had volunteered to run a message for him as head of the village. Nor did he know on whose side he was. He and John Kilipak had been enemies and rivals ever since he had been instrumental in getting Matawai, John Kilipak's adored younger half-brother, put into prison—a case where Karol Manoi had been pursuing private and personal family goals of his own. John Kilipak and he were usually on terms of formality, each chipping away at the other whenever possible. Yet, when Karol acting as "judge" saw John Kilipak sorely pressed and endangered by the possibility that his violent mistress, angry at the way the evidence was going, would say something irrevocable in "court," it was Karol who slipped in a device for giving the angry woman a chance to score off everyone and go away somewhat appeased. A temporary complete identification with John Kilipak possessed him and he responded intelligently, expertly, to save the man who was actually so very much like himself, but less violent and less complex in his moods and identifications.

Karol Manoi bridged the gaps between his dead father (whose mood types were directly present today only in the sud-

den rages of old Pokanau, his near contemporary), his own age
mates—the generation who had made the New Way—and his
juniors, like the young teacher, Peranis Cholai, who adhered to
the New Way with such stiff, anxious rigidity, and the schoolboys, who were the beneficiaries of the greater joy in work and
song which was the heritage of the New Way. Also he had something of the intelligent flexible mind of Paliau, the rigid, apocalyptic religious ideas of Lukas Banyalo, the self-elected preacher,
and the reactive criminal potentialities for adultery, gambling and
graft, which the New Way also provided. His personality summed
up the conflicts between old and new, and the people of Peri
themselves vacillated about him, one minute following his lead,
the next, angry and sullen, or baffled and perplexed.

But for the most part the different attitudes were more
sharply associated with generation. Old Pokanau (Plate VIII) had
something of the dogmatism of the young men, perhaps because
he had to superimpose on a long life of traditional behaviour the
new admonitions about gentleness and friendliness. In the old
system he had been an idealist, a stiff believer in the smallest
orthodoxy of the Sir Ghost system, eternally opposed to Joseph
Lalinge, who after meeting the old system with skepticism, had
remained a Catholic opposed to the New Way up to the time of
his death. Pokanau had also opposed the first moves for change
led by Mateus Banyalo, not, he now claimed, because he was necessarily against them, but because the young men were doing
something themselves instead of letting the older men take the
initiative as was proper. He had become an enthusiastic orthodox
supporter of the New Way not for the reasons of the younger
men who had never known his seniority and authoritativeness,
but because he thought the New Way could be more efficient
than the old way in making people behave—behave the way they
should behave, systematically and consistently. The New Way
seemed to him to have better sanctions behind it. Unwavering in
his support of the new system, he was also extremely dogmatic,
and his dogmatism had reinforced that of young Peranis Cholai,
whom he had partially reared.

Thus, above and below the middle group who carried the new movement, there was rigidity, and for comparable but different reasons. Pokanau—and to a lesser extent the few other less important old men who rigidly obeyed the rules of the community, attended the council meetings, stood up in "the line," obeyed without question group orders for work—was accepting a system of which he heartily approved, but to the dynamics of which he essentially lacked the key. These men had been too old to get the sense of release. According to their status in the old system, they took a new but less well-understood position in the new system—Pokanau as a leader, mainly by force of his zest and intelligence, Poli, the magician, and Polin as a man skilled in the old arts of building, etc., and the rest as dependents, men who had been dependent and bullied all their adult life and who now accepted the new order in much the same unenterprising spirit. Pokanau, with the habits of leadership, the years of uninhibited expression of anger behind him, was likely to be seized with uncontrollable rages, and these rages had formed a bridge for Peranis Cholai from his terrible old grandfather, Korotan, to his late-adolescent rigidities in dealing with the new system. Peranis identified with his whole proud earlier tradition, mediated by Pokanau, and he treated his old grandfather Korotan both as a symbol of the deeply hated past and as somehow embodied in himself. In the photograph recognition test,* he alone of all those of his age and older refused to recognize Korotan.

Peranis Cholai (Plate XVI) in his way was almost as complicated as Karol Manoi and, like Karol Manoi, he had the childhood memory of a violent war-leader ancestor and the intermediate tutoring of Pokanau. But there were several differences besides their age difference of some twelve years, for Karol Manoi was the same age as those in the main leadership group in Peri in 1938–1940, while Peranis at twenty-seven was the youngest leader in the village. Karol Manoi had been rejected by Pokanau after he himself had, as Pokanau's foster son, succoured Pokanau's

---

*A test which I made up from photographs taken in 1928–29.

son Matawai when he came to Rabaul as a work boy. But Peranis had remained the apple of Pokanau's eye. On Peranis all of Pokanau's ambition and pride of Old Peri times were centred. Matawai was ailing and weak. Bopau, his brother's son, had forsaken him. With Karol Manoi he had quarrelled. Peranis only was left. When I arrived in Peri, Peranis and his young wife and child were living in Pokanau's house while Peranis was building another with the help of the schoolboys. He was an active functioning member of the community, the schoolteacher, the leading preacher, the organizer and owner of the school canoe—the largest functioning canoe in the village—the builder of a new-style larger house at a little distance from the village. He was the best-versed man in a knowledge of the old tradition—although this meant far less than Pokanau claimed—and he also was the most literate member of the village. A month after we arrived he was elected as the clerk of the whole South Coast "wait-council" organization. From every point of view he seemed to be the most promising man in the village, despite his being a little too rigid, with a precarious insistence on the small details of the New Way, and treating the rules of minority expression or minority compliance with a kind of fanatical respect. Occasionally a completely fanatical note would come into his eye or into his voice. And then came an explosion—a quarrel with his wife—followed by threats of suicide, of leaving the village, of destroying all that he had so laboriously built up. His kin responded to the first of these threats with deep anxiety and affection, and old Pokanau lovingly prepared a meal and set it before him. When they were repeated, the response became one of impatience and rejection, "Why should we call him to come out of the bush? He doesn't heed our voices." His rage—with no open road of expression in actual combat or open black magic—was inappropriate in a community devoted to the proposition that anger was the greatest sin, and it cut him off from his fellows. He had no recourse except to turn it against himself.

This, like the birth of Ngalowen's anacephalic child, was another situation into which I felt justified in intervening—keeping careful account of my intervention. Peranis stood between the heartbreak of old Pokanau, for whom he represented the future but who had committed himself to the belief that anger was wrong, and the expectations of his immediate senior generation, who expected him, as the precocious favoured child of the New Way, to behave like a new Manus man. He had never been a work boy; he had gone to school; he had been favoured in office and power. What had a quarrel over his wife's still casting eyes at a former lover, over his wife's failure to guard his serenity and prevent him from the sin of anger, what had such a quarrel to do with upsetting the whole equilibrium of the village? He also had had an affair, long over and paid for, with a bright and attractive woman, and it was his wife's jealousy of this other woman that had tipped off the last series of rages. In his suicide threats even his former mistress turned against him. She was anxious and worried when it was feared that he might be dead, but she had no sympathy for the excesses of his self-destructive anger. Even Benedikta, Pokanau's daughter and so Peranis' foster sister, Benedikta, who was herself the prey of moods of complete lack of control, was impatient and unsympathetic. So Peranis Cholai swayed back and forth, one day a model of the new leadership, the next ready for self-destruction; one day explaining with patient clarity the beauties of parliamentary procedure, the next falling a prey to some resurgence of the new Ghost Cult which was spreading from Johnston Island.

Peranis Cholai was about twenty-seven. In the next age group were the men who were just marrying, who had been too young to experience the early struggles of the New Way, and now formed the group of just-married men. Such a young man was Josef, the husband of Anna, daughter of Raphael, whose marriage we witnessed right after the return from the evacuation. He belonged completely within the new régime, while Peranis Cholai's marriage had been originally arranged under the old. Josef and Anna had been school children when they decided to marry, and had been bidden to leave school, there being a general feeling that school should end when sex began. Josef had only been out of the village for very brief work experiences at the Australian Naval Base at Lombrum, which was not comparable to the work experience of an indentured labourer which his elders had had. He had never been taught by a European teacher, nor lived in a European house. He had not heard a European preach since he was a child. He had been taught that he was free to choose a wife, free to sleep with her before marriage, and that he should be on friendly relations with his parents-in-law. And so he was. He and Raphael fished together, and Raphael, luxuriating in his freedom from old taboos, even shouted lightly jokes in questionable taste from his kitchen to the adjacent house where Josef and Anna were sleeping.

Josef's parents were sober, solid citizens. His father Stephan Kaloi would have been independent under the old system, and was now a rather enterprising, individualistic supporter of the new form of behaviour. Kaloi gave up his house to his newly married son, and moved first into the kitchen. Later he built himself a stilt house over the sea in which to relax.

There were occasional ripples in the tranquil life of the younger couple, otherwise punctuated only by fishing and teaching their adopted baby* to walk. Josef's father-in-law, Raphael, had a young wife whose previous marital history was a compound of the old and new systems, who had at one point run away from Peri with another woman and fled to Baluan, whence after a variety of adventures, she returned and married Raphael simply by entering his house and with no ceremony. As a divorced woman, once sacramentally married in the Church, she could not remarry in church. Raphael, uxorious, slightly pompous, emphasized that his conduct was well within the law; each wife had died before he married another. He was a little over-defiant about his present position, and Elisabeta responded with temper and irritability toward her step-daughter, Anna, who was so demure and so com-

---

*See pages 338–39 for earlier history of Anna, who died in childbirth in 1954.

plete a member of the new society, and toward Ngaoli, her co-mother-in-law, gaunt, old, married under the old system, a devoted church member, righteous mother of five children, who fortunately had no complications in her life.

So there were occasional sputterings, bickerings, and even a "court case" over what Ngaoli had said to Elisabeta about what Elisabeta had said to Anna about a canoe which Raphael had given to Elisabeta but then loaned to Anna while he was away on a trip. Formally, this young marriage was idyllic; every pattern of the New Way was embodied in it—a feast given by both sides to the whole village, no quarrelling over the redistribution of the five pounds which had been given by the bridegroom's family to the bride's. All the decorum of the *lapans* of Peri had gone into the proceedings; the little new ceremony of the bride's dishes, complete with the new pots and pans that Raphael had gone all the way to Lorengau to purchase—with the proceeds of a turtle he had caught and sold—had only been enhanced by the gentleness with which Peter Kenandru had apologized because he had nothing to contribute.

Yet this young marriage, conforming perfectly as it did to the new way of life, would still be beset by the instabilities of those who failed to follow the new patterns, by slightly older young men who would beat their wives as their fathers had done, by slightly older young women who had been mixed up in the promiscuity which followed the misunderstanding of Paliau's prophylactic rules about adultery, by conflicts between the old midwives' notions about the care of infants and the new uses of cloth and possibly even a feeding bottle. Above all, this marriage was in jeopardy because of extraordinary ignorance, because of the thinness of the new culture.

Even such people such as young Peranis Cholai and Anna's young stepmother Elisabeta had grown up and been betrothed under the old, highly complicated kinship exchange system, a system which however onerous gave a sense of form and style and momentousness to life. And they had as children been tutored by

European priests and trained catechists. Raphael and Stephan Kaloi, the two fathers-in-law, had both worked for many years for Europeans; they were experienced in the ways of the white man's world, knowledgeable about ships and machines, trade stores, recruiting, courts, the intricacies of large plantations and government stations. Josef knew none of these things. His Neo-Melanesian was inferior to that of his father and his father-in-law. His acceptance of the New Way was not underwritten by any of the conflict and violent adherence of Peranis Cholai or Peranis Kiapin. He had none of the rejections nor the enthusiastic acceptances of those older than he but only the simple precepts of marital choice, freedom to work as one wished, admonitions to help in the village and in the wider group, to participate in meetings, to enjoy his wife, to help advance the new society.

Gentle and pleasant as this was, it was a thin brew, and it was hard to see what situations would develop Josef into the leader his father-in-law was, or even into a responsible second-string man, as his father was. He represented the successful embodiment of the New Way, with its less assertive character structure, its devotion to home and family and village and lack of desire to leave the ordered friendly life of the village, combined with a shallow unelaborated culture which at present received practically no infusions from the life of the Territory. When a proper school with trained teachers is established, he will be too old to go back to school. He knew how to read and write, a very, very little, adequate only for the limited needs of the village.

In Josef and other young men like him, there is another slender stratum, embodying another phase in the rapidly shifting local scene which will contribute to the final form of the new culture. And beneath his quiet, orderly adjustment, there is a babyhood in which whim and tantrum were indulged and in which there were in his father's and his mother's voices, when angered, echoes of the rage which they had experienced all through their childhood among the elders all around them. These could be picked up again; Josef's ignorance and quiet unfanatical devotion to his new community could easily be exploited by a political or cult leader. The more quickly Peri is linked with the larger world, with active council and co-operative, with active diffusions of New Guinea–wide patterns of modernization, the safer the Josefs will be, the more chance they will have to learn some content that will give weight to their pleasant and precarious formal character structure.

Just a little younger than Josef and Anna are the school children, who have been welded into an age group with duties and privileges. School is kept on alternate weeks, and when "the line" is made, schoolboys stand together as a group, schoolgirls as another. They form a tight corporation, with two divisions, and the smaller boys and girls are desperately anxious to share the activities of the older ones. On school days, they form into a line and drill before entering the church. A request made for their help, if granted, will mean that all will have to be assigned to a task, and that in fulfilling it each will round up the others. There is genuine co-education, boys and girls studying together—however, with only one girl, the adopted daughter of Mateus Banyalo, in the top class—and boys and girls take part together in games and stylized non-touching dances.

The village is proud of them, proud of having a school, and older men say proudly, "When they grow up they will understand everything." Meanwhile, they are far shyer than were their predecessors twenty-five years ago, enfolded as they are in general village approval instead of roistering on the edge of sin and delinquency. They fish; in fact, the schoolboys do much of the fishing for each household, swimming underwater with goggles and a modern type of fish spear. Given an English primer they will sit up until late hours poring over it, chanting the English words in unison. Casual, friendly, deeply attached to each other, treating the girls with a measure of contempt, but without antagonism, more socially disciplined and less anxious than their predecessors, the boys are still ready to learn many things that Josef, removed from school, probably will not learn. If the new council and the co-operative work well, most of them will remain sober, pleasantly gay, and a little bored, within the confines of their villages, running away to work on some nearby plantation only in case of a marital quarrel, divorce, or a death in the family, or perhaps taking their wives with them for brief working periods to accumulate a little cash. With more education and closer ties with the outside world, these boys, so intelligent, so perfectly co-ordinated physically, so adept at handling materials, might easily be transformed into valuable civil servants or mechanics to play a role on a New Guinea–wide stage. Without such education and such contact, but with their well-absorbed image of themselves as members of a new society deserving of respect, they may be hard put to it to maintain the slender framework, the slight content of that new society.

So, we can consider the varying character structure of members of the different generations. Even today's adolescents were reared initially in the sea, trained to an unceasing alertness—so that they could spot immediately a shift in a camera lens direction focused from one hundred yards away—taught to treat material things with respect, allowed and even encouraged to angry displays of temper. The present personalities embody, in ways that vary with their age, experience, and temperament, the rapid phases of the developing New Way. This again is a transformation rather than a change unrelated to the past. In the old character structure—as variations on a common theme of infant training, early childhood, and adolescent experience within a shared social system—it was possible to relate the phases of the parent's economic position and the phases through which a child was passing, and trace in later behaviour the particular exposures of childhood. The tendency to a continuing, active identification with parent, with older child, with the whole group, with the situation, remains as a constant factor, contributing to the way in which each generation embodies the recent rapid and revolutionary changes which have taken place. But we still have to ask what of those who find no place in the new order, who belong in spirit to the past that has been destroyed forever.

15

the Sunday that was straight

Old Manus had been a culture without a ceremonial calendar.[1] All events of importance were triggered by individual happenings in the lives of human beings, a birth, menarche, ear piercing, marriage, death; even large inter-village feasts, which occurred very rarely, were built around some event like cutting the hair of a leading man's son. Ceremonial was anchored to the rhythm of human lives, and set against the ceremonial were the recurrent illnesses and misfortunes, in the name of which the ghosts demanded and redemanded the ceremonial. Ceremonial was work, dancing was work, even feasting, set as it was against a background of carefully weighed exchange and of aggressively matched feast bowls, was onerous and lay heavy on people's spirits. The day of a feast there was simply more quarrelling than on other days. So in old Manus, people lived between states of more or less tension. At times of illness and death, when the whole village was over-shadowed by fear of the next ghostly reprisal, tension was at its height. The gentle expectancies of a calendrically arranged life, in which people may look forward to the harvest, to Easter, to the New Year, all were absent. One could merely hope for less trouble, or less tension, for fewer coincidences between sin and ceremony, for more times when the whole village was not so filled with supernatural danger that it was unsafe to let children play in the moonlight.

One of the distinctive innovations from the Western world was the Christian calendar, with its recurrent day of rest and its high points at Christmas and Easter and Pentecost, occasions which were presided over by God and uncontrolled by the whims of ordinary men. Today, set directly antithetical to the pattern of event sequences is the expected peace and harmony of the Sabbath and the great religious feasts which to the people stand for the New Way and are calibrated to periodic harmony rather than to the restless triggering of events by personal ambition, illness, and death.

But this new reliance on the calendar for providing recurrent moments of harmony and rededication has an earlier model which lies, like all old Manus models, in genealogical time. The old model of harmony was based on reciprocity through time of balanced social roles, in which *lapan* and *lau* fulfilled their parts, in which sister wept for brother, and leader provided for follower, a model composed partly of roles and partly of balance among real personalities, in which the more aggressive assertive women played appropriate parts, and quiet men could use status instead of loud-voiced anger in their dealings with their fellows. There was, in Old Peri, a pattern within which there was a very real sense of the fit between a man's or a woman's personality and the role they played in the village, a sense of fitness in the way in which a dead man's debts and commitments went not to a specified descendant but to the young man whose personality was such that he could assume them.

In the Peri of 1953, struggling with the extreme problems of transition, the two themes of how to take full advantage of the new expectancies of the calendar and of how to fulfil the demand for the old type of fitness between man and role were sharply revealed. For there was one very serious lack of the old type of harmony in Peri in the person of Kilipak. Chosen and trained from infancy for the role of Talikai, who had been the inheritor of old Korotan's role as war leader, Kilipak was at odds with the community, in it but not of it, stalking aloofly about, the proudest, the most brilliant, the natural and correct leader, but refusing to participate.

We have become so accustomed to seeing the way a community functions in terms of who leads that it requires a certain wrench to consider instead how significant it may be when someone who could lead, who should lead, does not—how absence can be as definitive as presence. When a leader dies, we do, for a while, acknowledge the shape given to later events by his absence—"If Lincoln had lived . . ." is a way of lamenting the Reconstruction. But the havoc that is wrought when the more gifted stand aside and remain terrifyingly visible to discourage the ambitious, is indeed like a living death.

In 1928, Kilipak, at fourteen, was the most gifted of his age group and the most loved, quick as an arrow, vibrant with active, intelligent zest for life. Where he led, the others followed; his were the hardest tasks—the tasks of organization, of planning, of conceptualizing. The fluid system of successorship meant that this adopted son of Talikai, this favourite nephew of Talikai's sister Isali, the leading medium and planner in Peri, was acknowledged by everyone to be the perfect heir for the most important position in Peri—the son of the most aristocratic clan. In the rebukes which he received when he transgressed, in the sound of his name when his age mates called to him, the sure expectation of his future role sounded and resounded.

In the first moments of searching for old friends in Peri, I had called for Kilipak and he had come, taking command with all his expected sureness, bringing his pressure lamp to light the house. Direct, gay, intellectually as alert as ever, he seemed—although his face was extraordinarily lined, almost as if it had been carved—very much as I expected him to be. In the first excitement of settling in, and before I had had time to map out present roles and functions, Kilipak was about a great deal, and the constant shouts for "Johanis," as the middle-aged called him now, or "John," as the young called him, seemed an expected echo from the remembered shouts of old and young twenty-five years

before. There was pride, sureness, and command in every word
he uttered, and others followed where he led—or so it seemed.
He organized the big canoe that evacuated me from the village,
and my possessions, and the newborn baby, and the specially ill.
And, up on the mountaintop, he played the same role, acting and
speaking authoritatively. He took the posts of action and of
urgency, setting himself the task of keeping communication open
between our mountaintop and Ted and Lenore Schwartz on
another mountaintop with the evacuated village of Bunai.

And then, when we returned to Peri and normal village life
was resumed, I began to realize that Kilipak had no role in the village at all. He had been assigned a house in the front row facing
the square—as one of the most important men—but it was
falling into disrepair, lived in by casual emigrants from other villages or some wandering young couple. Soon, he said, he was
going to break it altogether. He had started to build himself a
new house, spectacularly larger and better than any of the others,
far outside the close-packed, tightly designed village.* Meanwhile, he and his family were living in the house of an old widow,
close to the site of his new house. He held no position of any sort
in the new bureaucracy. Neither he nor others, when they started
to tell me of all the exciting events which had led from The Noise
to the setting up of the new social structure, described Kilipak as
having had any role. It was only after many weeks that I learned
he had been first a religious leader and then a civil one. After the
evacuation he built a big canoe for us—he was the most skilled
canoe builder in his generation—and we learned that he had been
the captain of the village canoe, but had resigned his post in a
huff because Karol Manoi, the councillor-elect, had tried to give
him orders about when to take the canoe out. "The Captain of
the Montura makes his own decision," said Kilipak, quietly
equating the Peri canoe and the largest ship of his boyhood.

One day there was a marriage feast, and suddenly it was Kili-

---

*See diagram on page 236.

pak who had arranged it all, on a grand scale; but when the feast speeches were given, he was away in Bunai, gambling, uninvolved. Occasional "court cases" brought him into contact with the village life; he enjoyed every detail of a trial, was proud of the skill with which he dressed for different legal roles. When a case was to be tried by a white government official, Kilipak wore only a simple, not too new, *laplap*, preferably black, neatly fastened by a belt, and spoke very quietly without brilliance or fanfare, shorn of any suggestion that he might be a new Manus man. But, when it was a village "court" and he wanted to carry things along rapidly, he dressed with equal care in whites, worn impeccably. Characteristically enough, he was almost unknown to the various officials who had come into close touch with the village, and yet he was the only one to go in to Lorengau and buy—through all the intricacies of bureaucratic bidding—a large supply of sheet iron for his new house.

But he did not go to church; he did not attend council meetings; he did not "line"—and no one tried to make him—although he often stood somewhere near the end of "the line," listening to what was going on. He gambled continuously and shrewdly, making a virtue of control, setting aside part of each big win, often stopping his play altogether and merely watching, conspicuous in his ability to take it or leave it. He was not one of those who were so afraid of gambling that they had to stay away from the sight of the cards for fear of plunging into an avalanche of self-destruction. The peace of the village was periodically disturbed by the repercussions of his long-standing love affair with the most brilliant woman in the village, Benedikta, the daughter of Pokanau, and the wife of an exceedingly stupid, irascible man. Episode after episode highlighted the continuance of a tie that had once been publicly exposed and paid for, in a village "court case" in which Benedikta had insisted on paying half the fine, saying proudly that she was glad to pay for what she had chosen and enjoyed, thus flouting the Peri convention that even chosen extramarital sex is painful to a woman. One day it would be a quarrel between Benedikta and her husband. She would run away, he would take an axe and start to chop down his house. Long complicated reconciliations would follow. Now it would be a public quarrel between Benedikta and Kilipak's wife Joanna, which ended in violent, screaming invective. Now it would be the conspicuous accident of Benedikta and the mistress of Kilipak's best friend appearing simultaneously in new, very modish skirts of the same bright blue. Flagrantly, an affair, which must have counted very few real rendezvous, was being flaunted by Benedikta, while Kilipak vacillated between delight in her brilliance and in the intensity of her passion and fear over the social and moral disintegration into which the whole affair promised to turn.

Through the months I watched these episodes, watched the restless way in which he went back and forth to Bunai, worked by fits and starts at his house, spent days with us as our best informant, an informant with whom one could discuss any question, from his personal perplexities to the present social functioning of the New Way. Periodically some event would occur in which he would intervene, authoritatively, suddenly, and always the same almost miraculous falling into place of disparate elements would occur. The father who had been screaming at his wife as he held his unconscious child in his arms would cease as Kilipak remarked, "How do you expect the Piyap's medicine to work while you hold that child in anger?" Quarrelsome old women, bickering over how much money each had contributed to a feast, would quiet as Kilipak sat down, took all the money away from everyone and firmly recounted it.

As far as his relations with his fellows went, he might have been two people—an arrogant, reckless gambler, flaunting his passions in the face of the community, flouting its dictates, going his own way, persisting in his liaison, neglecting his family, a man too dangerous to cross and impossible to restrain and so let alone, a man contemptuous of the repetitive preaching of good behaviour in the absence of any economic plan, a man primarily interested in making money, who had gone away to fish in Lorengau at a time when he had been holding an office in the new system—

this, on the one hand, recognized even in the comments of Petrus Pomat who loved Kilipak and deplored these aspects of his behaviour. But, on the other hand, he had, in flashes, the gentle sure touch of the man who saw most clearly and cared most for the conduct of life in Peri; he was a man who abhorred violence and loved virtue, who passionately wanted the New Way, was proud of his homely little wife who was always the first to arrive on any scene of illness or disaster, proud of his three beautiful children—he had spaced them properly so that his wife would not be over-worked, he said.

And through both roles there shone a kind of youthful zest and gaiety that made him in many ways a fitter companion for adolescents than for his own age mates, who were fleeing from temptation into pontifical and denunciatory behaviour or becoming trapped by situations of gambling and debt and by quarrelsome, meaningless bits of adultery. Kilipak's curiosity was endless, and he was willing to spend many hours with us in active interchange, alternating between doing projective tests, commenting on some aspect of the language, exploring his own visual imagery, and asking us about California—was it really as superior to the rest of the United States as Californians claimed?

Yet Kilipak was no double personality. The man who sat gambling, tensely watching every card, could stop to rescue a crying child, look up with the same sunniness as could the Kilipak who was for a few brief moments playing the role of trusted leader in the village. He was simply a man who got caught short by the very rapid changes. Where the New Way had offered to Raphael Manuwai and Michael Nauna the same type of roles they would have played in the old society, had given Banyalo and Karol Manoi new roles which would not have been available to them—as cultural expatriate and immigrant with shallow ties—Kilipak was caught between two worlds. His strengths lay in a sure, arrogant sense of aristocracy, perfect control, and fierce gentleness; his weaknesses, in his brilliant, restless, impatient mind, his superficial recklessness which combined with a deep fear fed by his active imagination of the uncontrollable impulses of other people, especially of his wife and his mistress. His sure hand on the steering paddle of a sea-going canoe in a storm—every muscle alert to the shifts in the paddle beneath his hand—was counterbalanced by the passionate despair with which he could throw his head down between his arms, seeking some sort of refuge against the devastation wrought by the passions which he had aroused.

He was the one who had said wistfully: "One thing we have not yet learned, and that is how to commit adultery properly." He had been betrothed as a small boy and married under the old system, married after his first long bout of work away from the village. His wife was both homely and rather stupid, but she was good and devoted, always ready to hold the sick and comfort the dying. His tall son, of whom people smilingly said, "He has his mother's features but his father's body," was the most gifted of his age group. He described how he and Joanna had been the first adults in the village to meet the challenge of the catechism, which adults had to undergo to become church members. "The others waited and watched," said Kilipak, "because if I had failed there would have been no hope for them." He had welcomed the discipline of Catholicism, enjoyed the structure of dogma and ritual, cared for and maintained his family as a traditional Manus father was expected to do and without any closeness to or enjoyment of his wife. He had enormously enjoyed the Americans, the gay give-and-take of fishing expeditions, the long talks about the details of American life. And he had welcomed the New Way, been one of the first teachers and later one of the first officials of the new bureaucracy. All his preferences were for active virtue, a life in which adventure lay in long and arduous voyages, and in close romantic friendships, both distant and near.

He was thus trapped by a social system which was too static at the moment to answer his craving for excitement—as the older system of long voyages and astute trading would have done—and which promised a new freedom to the young who, not having been married as Catholics, had a different sort of marriage pattern—marriage by choice, and divorce by choice. Essentially, he still accepted marriage as final, and when Karol Matawai divorced the young wife for whom Kilipak had paid—just before the coming of the New Way—Kilipak never really recognized the divorce but continued to treat her as a member of his family.*

Kilipak definitely did not want to marry his mistress—whatever claims she or others might make about the paternity of her youngest child—and the possibility of blowing the fires of dissension and jealousy into some impossible conflagration never left him. He kept money hidden against the possible need for flight from the village that he nevertheless loved so much that he had come home before any of his age mates. At the time of the second eruption of the volcano, a group of us were staying in Bunai. Sailing home in the early dawn, Kilipak strained his eyes toward Peri, and sighed with enormous relief when the rooftops came into view. During the night it might have suffered some hurt—who knew—with the volcano steaming up again, with the recent suicide attempts of one of his gifted protégés, with the great sore on his son's leg having only just healed. His body relaxed for a moment, the tiller at rest in his hands, as he breathed, "It exists, safe."

The whole village suffered from Kilipak's behaviour. He was so much the best orator that no one could speak best among those who deigned to speak. His distinction made Raphael seem vulgar and greedy and made people remember that Raphael wasn't really a member of the Peri clan, although actually it was Raphael's grandfather who had been adopted, while Kilipak himself was an adopted member. His control made Karol Manoi's fits of intemperate gambling and cruelty all the more conspicuous, his gaiety and style made Michael Nauna, for all his sculptured beauty, seem dull and unenterprising; the clarity of his perception that the whole community was in a state of doldrums made the pious repetitiousness of the preachers of the New Way seem the uninspired stuff that most of it was. His position pointed up very sharply the vulnerability of a closed society from which people do not go away and in which those who do not play the part

---

*For the story of Karol Matawai's marriage, see Chapter IX, pp. 215.

expected of them—by virtue of inheritance, personality, skill, or gift—in a way prevent the position they might have taken from being filled at all.

The one perfect Sunday in Peri, in 1953, when complete harmony reigned—the harmony that people vainly exhorted each other to seek and maintain—was the Sunday, at the beginning of the Advent, when Kilipak came to church after having taken up again the night before his natural role as leader. That Sunday there were no quarrels, no gambling games, no raised voices—the ideal of the New Way was experienced, fleetingly, by the whole village.

The night before there had been a meeting at which attendance had been large but attention negligible. Speaker after speaker spoke to no purpose. There were little meaningless asides, indeterminate exchanges, between the Solomon Islander who had worked for the company on the island for long before the people of Peri moved ashore, and the husband of Benedikta who complained that the land all belonged to him—and here Benedikta, addicted to any show of rage, supported her husband. People argued purposelessly. Then Kilipak spoke—for the first time since the evacuation—summed up the argument, proposed a plan for final settlement.

And as if this participation had given him some kind of absolution, the next morning he walked into church, perfectly dressed, early, gravely attentive. The Christian calendrical expectation which has been kept by the New Way was that with the beginning of Advent men faced a new dedication. Tomas, the young preacher, chosen for his gentleness, lacking the dogmatism of Peranis and the ranting fanaticism of Lukas, spoke, and his radiance reflected the light on the people's faces, the easiness in their posture, on this Sunday when the town meeting had reached agreement. Blessedly there was no wrong and everyone was there.

"Now, all of you, hear my talk. This is Sunday. On the Seventh Day God rested. He had finished all his work. He had made heaven and earth and also all mankind. He finished work on the

Seventh Day, and he rested. You and I must also rest on the Seventh Day. He spoke about Sunday. These words you and I have heard completely and often up to today. This Church to which we belong did not simply arise now at the present time. It arose before. A long while ago the Church came among us, the Name of God, and God the Son, and God the Holy Ghost, all three, They had come among us. For a long time the Church has been here, and you and I have been accustomed to follow it. However, you and I did not understand well about the path [of men] on earth, about our well being and about our ill health—how were we to know about these? You and I have been members of the Church for many years. At the time the Mission was still here we were Christians, through the years, up to today. In 1946, this New Way of Thought arose among us. There was no man who persuaded us, who lied to us or coerced us; it is something which is our very own. You and I knew that it was true; you and I held strongly to it; you and I together came inside the new movement. You and I have made it unto today. You and I know that here on our ground, in our village, evil does not simply come, for you and I, human beings, know how to set things straight. If you and I, human beings, sin, then the sin attaches to our place, our place will draw wrong from our sins. So you and I must think hard of good works and good thoughts which will make our place straight, so that this our place can have well being. You and I have heard completely, evil does not arise from nothing. It arises within us men. If you and I do not hear the word, if you and I are cheeky, then this sin will enter into our community.

"It is this truth that straightens out us men, that makes us as one; you and I sit together; you and I eat together. Without this truth holding us together, one would follow his own ideas, another would make his own work according to his own desires. It would be as it once was [among us]; one man would not listen to another, one man could not teach another, and quarrels and fights arose. This truth is the chief truth, the most important, among us. If you and I do not hear it, then all the ways of the past will arise again.

"We, human beings, cannot win [alone]. The angels did not win; Adam and Eva, they two did not win; all the men of the past failed to win; even unto the present so it has been. We also cannot win. God is angry with all the men who came before us, and this [anger] is with us who follow after them. Of this, you and I must think. All of this we have understood.

"Now for the words of Christmas. Today is the twenty-second of November. The days of this month will finish; the last week will finish on Monday, the first day of December will fall on a Tuesday. Only four weeks remain. Now, you and I must think lest Christmas come and find something wrong within us. You and I must search for this sinfulness. You and I must work to straighten this out so that the straightness is within us. Each of us knows that our thoughts are not straight yet. Inside our souls we must straighten our thoughts by the time that Christmas comes. Why? Because we do not know what the New Year will bring. In this year which is finishing now, we have heard [God's truth]. Now we must care for this truth at Christmas; we must worship truly. We must not follow the [erring] ways of this past year. As the season passes from Christmas to Easter to Pentecost, the service of God must not be ended. Some men worship truly, some simply follow after the words of others, some simply sit, and some blaspheme Jesus. This has been the way of church-going in the past. These ways we must not follow. We must follow a good fashion, as of today, so that we come together at Christmas, in peace with each other, with feasting and joy on the Christmas that is coming. You and I, we must not become angry, we must not quarrel, and sin will not come up among us. This, then, is that of which we must think. It is finished."

And the people streamed out into the sunlit morning, their hearts set on a Christmas that, like this Sunday, would also be straight.

For that one day, even without sufficient formal occupation to engage the attention of the people, the spell of harmony lasted. Small family groups set off in canoes to visit in other villages; clerk and councillor caught up with their laboriously penned notes; there was a ball game played on the new ball field, and people walked about talking of Christmas. The remembered balance—which had been one of the great values of the old days—of the occasions on which the fitness of leadership by a man born to be a leader had gone unchallenged, and the sought-for balance of the new, when an orderly and calendrical life would impose its own order on the community, coincided—for this one day—as Kilipak played, for twenty-four hours, an old part in the new scene.

In tracing the sequence of events which has separated Kilipak from his place in his group, a group still woven together by the complex relationships and expectancies of the past, we may find another dimension of the process that goes on in small communities which must change their way of life—together, rather than by scattering to the four winds, to master the new conditions one by one. For the strength of the closed group who have lived together for generations is also a vulnerability if any one of their number fails to play the expected role, if the part remains unfilled, the lines unspoken, the play incomplete.

16

women, sex, and sin

The Manus didn't know what to do with women twenty-five years ago, and they know almost as little today. The whole ethos is an essentially masculine one, in which the protective capacities of the male rather than the specifically maternal capacities of the female are the ones woven into the idea of parenthood. The ideal of personality is active, assertive, demanding, with great emphasis upon freedom of movement. There is likewise a very low interest in biological parenthood, in the breast-feeding tie between mother and child, or in any softness of feminine sex responsiveness which would yield too easily to evoke a measure of masculine anger.

Twenty-five years ago, the most valued women in the village were dominating women, even those who dominated their husbands, women who had strong clear minds, and who, as mediums, controlled a good part of the public affairs of the village. The woman who was regarded as the most dangerous woman in the village was a good-natured, easily responsive, slightly stupid widow, who was said to have been responsible for the deaths of six good men. Young women who were recalcitrant at marriage could be disciplined into shape, if necessary, as had been done in the case of one Peri wife who was finally shaped into compliancy on one of the smaller islands by a week end of rape in which her husband and a group of his age mates participated. The pliant, the warm, the responsive were simply so many danger spots—girls who might be persuaded into running away or simply yielding to seduction. As daughters, as sisters, as wives, and as widows they were regarded as both dangerous and essentially unattractive.

In the long years between betrothal as a little girl of eight or ten and marriage, the girl of Old Peri was not being "good" in the sense that she was expected to be pure in heart and mind and never let her thoughts wander into areas of lust or even of desire, like the traditional expectation for unmarried Catholic girls in southern Europe. She was, it is true, expected to be circumspect, to obey the rules and avoidances, expected not to say her future husband's name, not to let herself get into any situation where property that had already been expended in her name would be jeopardized. Her virginity and reputation were rather like a sack of money which she was left to guard alone in the house, and out of loyalty to her relatives, fear of their anger and of the penalties which their Sir Ghosts would exact, she guarded them. A theft, or even a slight defection which turned her head away from the main task for a moment would bring ruin on many people—perhaps death to one of her closest kin. Nor did her kin trust to her conscience; she was watched and chaperoned very severely; the slightest indiscretion brought down torrents of abuse and recrimination.

The young men were in a slightly different position. If one of them had an affair, it would bring about an awful row between the Sir Ghost of their own household and that of the girl's kin; there would be expiations and payments, and perhaps someone would die in the end, but the attitude toward the young men was more that of indulgence toward a successful bandit. Failure to guard in the case of the girl was far more serious than success in breaking in on the part of the boy. Virginity was merely important as it affected the marriage arrangements. If the girl's lapse or the boy's lapse could be glossed over, expiatory payments made, ghosts and Sir Ghosts appeased, the mere technical matter of physical virginity did not matter very much.

To the young girl growing up in the village, the one person on whom her mind could not dwell, the one person about whom she could not daydream, give a sly, quick look or a provocative nudge in a crowd, was her fiancé. Toward him her relatives focused all the feeling they had shown earlier toward any failure of the girl to control her sphincters; his name, his appearance, everything about him was considered shame-evoking. The young girl's mother and all her older female relatives shared in this attitude toward him. Where she had been freer with her father than her mother could be, once betrothed she was again bound in with her mother because her father could no longer take her with him, and because of the taboos which she and her mother shared. Her materials for fantasy were the shame-arousing, unmentionable future marriage relationship, possibilities of seduction and rape which would only bring disaster in their train, and a conscious focus on the outward and visible forms of her present and future position—how many dog's teeth, how much shell money had been and would be given away in her name, how many strings of ornaments, how many money aprons would she wear as a bride, with how many canoe loads of sago would she be fetched back home after the birth of her child.

Thus, all through girlhood the way was paved for married women to shrink from their husbands' advances and still conform to the moral code, avoiding out of sheer fear and not out of any compliancy the anger of their husbands' ghosts as they had avoided the anger of their own. They ran then, as they do today, the risk of being violently attracted by an extra-marital adventure, which presented the contrast between the appeal of danger and the inhibitions of shame. Just as in the men's lives there was an overlay of careful, continuous industriousness supported by ghostly sanctions, while underneath there was a far easier, more reckless self-confidence which was given very little scope, so also among the women a heavily sanctioned demand for circumspection and diligence screened a vigorous, reckless wilfulness, which only very, very heavy sanctions could prevent from coming into play. Meanwhile, at no point was there a chance to develop any gentleness associated with sex behaviour in marriage itself.

Into this background of active, demanding babyhood and

early childhood, inhibited and chaperoned girlhood set against the ever-present possibility of seduction and rape, and finally marriage—which was only made tolerable by emphasis upon the role of the economically successful woman who kept her husband's house and provided beadwork for her brother—came the first teaching of the Mission. Here one of the special aspects of Catholic as compared with Protestant missions came into play. When the Manus saw Catholic women, they saw nuns, not wives. For the little girls who went away to mission schools (there were two such women in Peri), sisters did not present an ideal to which they could ever aspire, but were rather earthly representatives of heavenly powers, intent, like the ghosts of old, upon making the girls quiet, obedient, and well-disciplined. They learned standards of personal neatness, learned to read and write, learned to sing, but they had no models of Christian marriage which seriously challenged the models which they had learned in their youth.

Then, in 1946, came the emancipation of women by the New Way, the removal of all taboos, the disappearance of the old name avoidances, the prohibition of child betrothals, the permission for women to consent to their own seductions, the prohibition against fathers or brothers becoming angered by the behaviour of daughters and sisters. There was the exhortation to young couples to behave in a way which was a mixture of work boys' memories of the marriages of Australian officials—in which husbands were protective of their wives, helped them on and off with their evening wraps, and hired servants to work for them, talked to them at meals, and kissed them on arrival and departure—and a model derived from American films—in which free choice of a mate and conspicuous, demonstrative public affection was felt to be the key to American marriage.

It was Manus men, and not Manus women, who had been work boys in Australian households, who had seen American films. The emancipation of the women was presented to them by fiat; no more taboos—if you have a husband, speak his name. Explore freely if you are unmarried, and, once married, by choice, of course, publicly demonstrate your affection for your husband,

and have as many children as possible so as to make the Manus, now so few, into many. As monarchs in Europe once ordered their people to follow them into baptism and membership in the Christian Church, so Manus men laid down the rules by which women in the New Way were to become emancipated and affectionate. There were to be no more taboos, girls and boys were to go to school together, young people to experiment in the choice of a mate. Having once been ordered to be compliant, to hide behind their avoidance mats and cloaks, to sit quietly and do beadwork, women were now ordered to be spontaneous, responsible, actively loving. And the men, modelling themselves on Australians, who had not expected their wives to care for children and do all the housework, having no servants, took over part of the care of even the very young babies.

The present results of this emancipation of women are both astonishing and depressing. Twenty-five years ago, the most conspicuous thing about Manus women was that they were deprived in those areas of affectionate domesticity which most societies permit to women and driven into continuous public economic participation. Today, the most conspicuous point is the extent to which they have been driven into a public display of a new form of personal relations, with little or no understanding or preparation for the new role. Manus women, twenty-five years ago, were singularly unattractive, angular, assertive, walking without any sense of the appeal of their own femininity, muting and constricting their femininity, emphasizing, with strident voice and sharp, unappealing gestures, that it might be possible to rape them, it might even be possible to seduce them—if enough risk attached—but what love and tenderness they had was already bespoken in formal terms by brothers. Manus women today are almost equally unattractive, but they look and act very differently. Where their contours were once sharp and angular, they are now softer, a little blurred. Where before, if one laid one's hands on a girl's shoulder, the muscles quivered like a taut bowstring, unused to gestures which were not menacing, stylized, and brittle, today they are heavier and slower, their bodies give a little beneath one's hand. The tense restiveness is gone, but no responsiveness has come to take its place. It is easy to see how husbands who once would have beaten them—as opponents in an unresolved contest—now beat them to get any response out of them at all. Whereas twenty-five years ago a husband's main complaints were about acts—a wife gossiped about him with her relatives or his brother's wife, a wife got up at night without her grass skirt when there were strangers sleeping in the house or was careless in feeding the baby—today the overwhelmingly most frequent complaint is that she "fastens her mouth." Some phrase, some slight act, will set her to brooding, and brooding she grows silent until her husband in a rage beats her, a beating which typically ends either in her running away or in a sexual reunion which has the elements of successful rape.

In the past, sex was something to be avoided by women in marriage, and in general; for men it was a reckless, brief adventure, usually accompanied by some kind of trouble. Women had grudged their husbands the brief encounters with captured prostituted stranger women, and did their best to spoil their husbands' pleasure by screeching taunts from a distance as they swung on bamboo swings far out over the lagoon, their laughter designed to echo into the men's house and make the men impotent. Today there are constant complaints both of sexual rejection and of wives who insist upon their husbands sleeping with them just to prove they haven't been with other women. "The one time you must have intercourse with your wife," say Peri men, "is if you have already been with another woman. Otherwise she is sure to find out and be angry." Counterpointed to this is the ideal extra-marital affair which emphasizes choice—"She paid her half of the fine; she said she had chosen me," "This is really from the desire of both." But, even more important, the wonderful thing about lovers is that you don't have to sleep with them. If either man or woman feels tired and disinclined toward love-making, the couple can simply sit and talk, and they need not have sex relations.

So, in spite of the apparent great change from a system in

which women were the helpless pawns of complicated marriage exchanges, completely controlled by fathers, brothers, and husbands, to a system of marriage by choice and freedom of consent, the crucial position of sex has not changed very much. Sex is still associated with anger, with rights, with expression of or response to various sorts of resistance, and love is defined as a relationship in which sex can be ignored in favour of affection. As women once screamed their anger and jealousy because a man took a fish from his catch to his sister's house and sat quietly beside her fire, so they now rage over comparable incidents such as a husband bringing home a piece of cloth for his mistress, or his mother cooking him a meal. Quarrels in the village hinge not on the number of actual adulteries, but on the glances, tokens, and hints of adulteries long past, or perhaps never to come. The coincidence of two people who have had an affair turning up in a distant village the same day, even though they hardly exchange a word there, gives the pair enormous pleasure and is guaranteed, if it is discovered, to throw their offended spouses into a rage.

In fact, there is a correspondence between the present chief requirement of a wife, that she should protect her husband's mind-soul from the sin of anger, and the chief enjoyment of illicit love, which is to tease and tantalize one's rivals. This teasing may go so far as, for example, Benedikta taking delight in getting her husband to buy her, with her own money, some conspicuous object, like a knife with a red handle, which she exhibited conspicuously, walking about, certain that her lover's wife would fall into the trap of thinking her lover had given it to her. Or two women whose lovers were friends would put on skirts of the same material, thus emphasizing the relationships and setting echoes going in the heads of the two wives who were their rivals. It is a game played by those who do not in any case expect satisfaction from sex, in whatever form it comes, and who get what satisfaction they can out of playing with dissatisfaction.

Appropriately enough, illicit love affairs and gambling were associated together. Men enjoyed giving women, their own or their friends' mistresses, money to gamble with; women enjoyed borrowing money from their lovers and lovers' friends, and gambling games were watched closely by hawk-eyed spouses alert for trouble. Among the occasional couples where both gambled, the style of the game was upset and inexplicable rows developed, as when Maria had a temper tantrum in "court" because her husband paid back his share of a debt to her with money she had lent a friend of his not knowing that part of it was for her own husband.

So women and sex remain associated with sin; where once they were associated with the punitive anger of ghosts, now they are associated with the jealous anger of men and women. Where once women were unwillingly circumspect and men grudgingly prudent, in order to prevent the ghost from visiting illness and death on them or their relatives because they had violated property rights or upset important economic affairs, today virtue consists in men and women leading quiet lives. Husbands and wives should reject overtures and opportunities so that there will be no anger between them, which might endanger the lives of their children, or anger concealed in the heart of the spouse, or in their own hearts because of the anger of the spouse. The responsiveness of men to resistance, either active or passive, the insistent demand of the jealous woman for sex expression, not for its own sake but as a symbol of her possession of her husband, keep this edge between sex and anger keen and sharp. From Christian teaching the Manus learned that sex was evil—a matter on which they were well convinced already—at least as far as women were concerned, but that sexual sins could be confessed and forgiven. From observation of Australian life, lightly reinforced by American films, they came to the conclusion that somehow Western white men seemed to manage their sex lives better, for there was so much less quarrelling, and that this better management came from giving women more consideration, not beating them, helping them with their work, and giving them freedom of choice, and being mildly demonstrative in public. But the type of deeply responsible, tender marriage, which stands as the ideal of Catholic teaching, they never have had a chance to see.

For the woman who is intelligent, ambitious, and active, the New Way, in spite of its nominal emancipation of women, offers no roles comparable to the part that women could play as mediums and entrepreneurs in the old system. The most that the wife of a member of the new bureaucracy is expected to do is to be a model for the rest of the community, keep her house and children in a modern way, and never embarrass her husband by being old-fashioned. This was the role played by the wife of Samol in Bunai; her baby received the most perfect infant care, her clothes were the most carefully chosen. Not until the Manus are introduced to the sort of women's clubs which have grown up around Port Moresby and in modern Samoa will the women have any glimpse of any sort of responsible public role again.

Meanwhile, there remains with them the remembrance of gentleness received from old women, not mothers but grandmothers or aunts, who, freed from the tempestuousness of active sex lives, freed from quarrelling with their husbands, were gentle and indulgent to small children. In the memories of the middle-aged men who are still dissatisfied with their marriages, who quarrel with their wives and beat them, the nostalgia for these kind old women can be heard, with its echo that some day their marriages, if they live long enough, will have in them women who are as kind and gentle as Pomat's Tchalolo grandmother or Kilipak's aunt, Isali. Something of the tenderness of these contacts used to survive in the relationship between brother and sister, and appeared again in the marriages of many years' standing, when the shame of speaking together and eating together had worn off. When husband and wife had co-operated in many enterprises, after she had borne him many children, as they came to the point of being grandparents, a gentleness could settle between them.

At present when people are still young, public opinion will side strongly with the wife who is asked to do more than her share in supporting the household, but as they grow older, the case of the sickening husband with a wife who neglects him focuses the rage of the community, as it did when Christof,* his arm shrunken and helpless, his legs mere sticks, was cut by his wife. When I arrived on this scene a few people were gathered around Christof. His wife, who had done the slashing, was sitting at a distance unconcernedly working on thatch. Suddenly their grown son, a great, husky creature, hurled himself through the room and began kicking his mother violently. People rushed to restrain him, and the councillor pontificated, "Don't add one trouble to another." His mother put on a dramatic, hysterical act to get attention from the bystanders, and even with this she received no sympathy. One of the gentlest of bystanders commented, "I am like that also, Piyap. If my mother didn't give food to my father, I'd fight her." And when I asked, "If your father would have fought your mother, would you have helped her?" I received an astonished, "Indeed, no!" One of the very young men summed it up: "I heard what it was about. They weren't giving Christof anything to eat. Yesterday he was angry, and he did not eat. Today they all went to the bush, and then they cooked food and gave him none. And he, is he a strong man? His arm and his leg are useless. He's just like a child [a great exaggeration]. Why don't they care for him? Now, Sepa [the daughter] wanted to give food to her father, and her mother was angry and cut her husband with a knife. Tomas heard this and he quarrelled with his mother. His mother said, 'All right, beat me if you wish. This food, was it something you produced, so you have a right to talk?' Then Tomas said, 'Every day I bring sago, fish, and other things to you all, and I think you don't give any to my father.' Then Tomas was angry, and he attacked his mother."

The councillor's comment was that if the family decided to bring the matter to "court," Tomas would be in particular trouble because he had broken an important law of the New Way, the law that forbade one to get involved in one's relatives' quarrels.

---

*The same Christof who had gone insane during The Noise (see Chapter IX).

When the case came to "court," the weight of disapproval was directed against the wife. This was a familiar pattern. If the blame could be firmly affixed to a woman, then any failure of the New Way among men need not be faced so directly. Women after all were still uneducated, illiterate, and undependable. For their part, the women felt many of the new procedures as traps. When they gave evidence it was written down and if, on being asked to repeat it, they gave a slightly different version, the whole weight of the "court" would come down upon them. The disapproval of indirect evidence also weighed heavily upon them, for when they would protest that they "knew" something was going on, the men would doggedly confront them with a "Did you see it with your eyes? Did you touch it with your hands?" So, in spite of their nominal emancipation, they still live in a world which in repudiating sex also repudiates women, and which in exalting fatherhood leaves less room for motherhood, except as a sort of delegated fatherhood.

17

reprieve—in twentieth-century terms

As the months wore on in Peri, the dilemma of the village and the dilemma of the village leadership became clearer, and its relationship to dilemmas everywhere in the world today became clearer also. The whole group of villages which had joined together under Paliau's leadership had a population of about five thousand*— few enough to give them a sense of being a people with a name and a style of life on the basis of which they could meet other men, white, brown, and black men, from other countries, proudly and with dignity. The shift from being Manus tribesmen, identified with clan and village, and superficially with all Manus-speaking people who felt themselves superior in most things, certainly equal in all, to the other peoples of the Admiralties, to being members of the New Way, a modern community organized in accordance with the principles of the Western world, meant a shift in the bases of pride and identity for each member of Peri village.

Instead of reciting the names of coral atolls, which had once belonged to Manus clans now settled all along the South Coast, Peri men recounted proudly the list of the thirty-three villages, all

---

*Four thousand four hundred and eighty-seven according to the UN Report for 1953–54.

or part of which "had come inside this New Way of Thinking." For this group of people, Lipan, on the island of Baluan, was the capital, and Paliau the acknowledged leader. Upon the ties among them, upon the success of regional "council" meetings, depended the continuance of the new alignment on which the members of each group that had "come inside" had staked their whole relationship to progress and human dignity. Unless this group held together, demonstrated they could work together, proved to both friendly and hostile Europeans and to the dissident minorities in the non-Manus groups—like the Seventh Day Adventists on Baluan, the sections of the Usiai villages which had refused to come down and live on the sea coast—that they were, in fact, a new political unity and a new culture, each individual Manus would lose his whole sense of the meaning of life.

The old way had been destroyed completely. However much new institutions like the *pilay* might seem, to the anthropologist, to embody much of the structure of the old cross-cousin exchanges, to the people of Peri these were brand-new institutions. If they made their peace with the Mission, which many people thought they might some day do, it would be in terms of being a new kind of people, a treaty rather than acts of individual submission from former converts. They must either go forward as a group or fall to bits. The very absence of a dissident group among them, the absence of factionalism among the Manus, made their whole position the more precarious. They had burned their religious and political boats completely, perhaps the more completely because they had conserved their old economics. There would be no hunger, no poverty due to too rapid economic change, with which to rationalize their failure or provide an ethically weighted road back.

Writing of the early days of the movement in 1948, an observant young patrol officer[1] had said, "The cargo must come," thus summing up his sense of the urgency of the people's need for modern tools and modern materials. Four years later, it was possible to say that the urgency lay not in modern tools or some way of making more money locally, although these were important, but in some form of on-going activity which bound the pace of life within the little communities to the pace of life in the outer world which they wished to join. The New Way had been born in activity and enterprise. People had worked together with unprecedented and rapid co-operativeness to build their new villages, their docks and churches, to set up their model communities, getting ready—but ready for what?

First ritually, in clothing, marching, and formal organization, they imitated, without completely understanding, the world they wished to enter. Then, after Paliau had been taken to Port Moresby and they had been promised a council, they prepared actually to have a council, to become politically integrated within the Territory. Then came the Machiavellian suggestion from an old-style government official—why not split the group in two, split the solid Manus, described in government reports as "the proud, restless, discontented, land-less people of the South Coast," and put the groups centred on the smaller islands in a council with Baluan as the capital, nominally giving Paliau recognition, and leave out the other half, the whole South Coast group along the shore of Manus Island to which Peri belonged. This split would weaken the Manus bloc inside the council, making it necessary for them to get on with the Matankor peoples of the other islands, and it would probably weaken the council sufficiently so it would fail. Meanwhile, the rest of the South Coast villages also composed of a large Manus bloc—now cut off from the other Manus near the island capital—would naturally fall to pieces if left without a council, and, even if given one later, the diversity of groups within (former enemies and many who spoke different languages) would create so much difficulty for the Manus bloc, now only half its strength, that the South Coast movement would crumble also. This was the destructive hope.

So a council was set up on Baluan, a council designed by the imaginative planners at the top to utilize and conserve the energy of "native leaders," and to turn these energies to constructive ends, but a council seen locally by most of the officials charged with setting it up as a *pis aller*. To the Senior Native Authority Officer who had initially to approve the council, it was a threat because the Manus were not technically ready for a council; they had no one educated enough to keep the necessary records and accounts; a clerk would have to be imported from another island. Futhermore, a council that was set up in the train of events following a long period of political upheaval and dissidence was heavily compromised at the start, and the Native Authority Officer was deeply committed to the success of the councils in the Territory. His reports when he visited Baluan echoed his distrust of people who had kept the accounts of the money which had been entrusted to Paliau in their own fantastic fashion, counting five-pound notes and ten-shilling notes as units, and so writing seven pounds, twelve shillings, and six pence as 1–2–1–2–6, instead of writing it properly and logically as £7–12–6! By government officials, by the Mission, by plantation owners and managers, the council was viewed as a way of clipping Paliau's wings, but as likely to be a nuisance if it succeeded—providing economic competition to the traders, becoming a political tumour to the Administration, and acting as a source for the strengthening of heresy in the eyes of the Mission.

With all of these handicaps, the choice of James Landman as the local Native Authority Officer, who, together with his wife, Marjorie Landman, was genuinely sympathetic to the people, was almost the only happy portent. The truncated council, with only a part of the Manus within it, with Baluan designated as the capital (because Paliau was after all a Baluan, and the circumstance that he had built his own house among the Manus of Mouk, whom he had caused to move ashore on Baluan, and not among the Baluan people was ignored), hinged on local Baluan politics, rather than on the Manus group as a whole. The Mouk Manus, who had been the backbone of Paliau's movement, were pushed off into a more peripheral position, except for their representatives in the council who had a chance to work with Jim Landman. So, in spite of the imagination and sympathy of the

Landmans, a positive circumstance which might not be dupli-
cated again in comparable situations anywhere in New Guinea,
the policy of splitting the South Coast into two groups, separat-
ing the island Manus from the coastal Manus, and of setting
Paliau the task of maintaining his position with a shrunken
group, confined to the local politics of his own tribe, from which
he was basically alienated, and to local politics of the Mouk
Manus, who were placed in an ambivalent relationship to him as
well as to Baluan, was doing just the harm that those who had
recommended it had hoped it would.

Baluan had been turned into a beautiful little capital. The
new buildings, school, store, and council house were designed
with taste and skill and executed in local materials. The semi-
circular table at the end of the council house was a physical model
of democracy, and the group who sat around it benefitted, week
by week, by the patient lessons in democratic procedure which
Jim Landman gave them. But still Baluan was a capital without a
constituency, reduced to the petty politics of a party caucus,
partly cut off from its sources of political strength.

Meanwhile, the question of the council for the rest of the
South Coast hung fire. The houses, so bright and new when they
had been built, began to fall to pieces; the beds and chairs
obtained from the U.S. salvage began to tear, to rust, to break. In
1952, an election was held for South Coast officers. This was
legal; it was held under the supervision of administration officials,
as a preparation for the council being proclaimed. In the meantime, the role and authority of the older form of official, the
appointed local *luluai*, declined. The councillors-elect, combining the prestige of leaders of the New Way, with which the
council-to-be had become completely identified, and the power
which the council-to-be would, they believed, give them, were in
a frustrated anomalous position. If, as elected officials of a not yet
existent council, they acted with authority, they might be disciplined by the Administration for abuse of power; if, as heirs of the
leadership of the New Way and as Paliau's political deputies
(which they also were), they failed to lead, then life would go out

of the relationship between the South Coast villages and their capital, to which they so deeply wanted to be joined.

In the confusion, the apathy, and the sense of arrested movement in which these villages lived, there was rich soil for every sort of discontent. A new leader might have sprung up to dispute Paliau's leadership, and, if this had happened, there might have been an end to the possibilities of progress. As it was, Samol of Bunai—the most effective leader of the South Coast, favourite adopted son of old Kisekup, who had been the former paramount *luluai*, first under the Germans, then under the Australians— identified himself with Paliau, worked with him at every step, dressed his wife and child as Paliau dressed his, and provided a bulwark against dissolution of the whole political fabric. Samol was practically a unique case among the Manus, a man who had succeeded a powerful father figure while the father still remained alive—a succession that old Kisekup, still vigorous but with one badly withered arm, took gracefully. Kisekup had not died when most Manus men die, in their early middle age; he had attained power very young, held it for a long period, made the big feast just before the Mission came in, retired without loss of self-respect in favour of an heir as gifted and intelligent as himself. So Samol, experienced in a totally aberrant method of direct succession, worked well with Paliau and helped hold the South Coast together.

But months went by. When the natives inquired about the hope of a council, there were no answers for them, except answers in terms of administrative shifts, such as someone was away on leave from Port Moresby—accurate enough statements of one reason why things were held up, but lending to their fate just the aura of individual accident and caprice which it was most important to remove.

The hope of an orderly, responsible New Guinea, as its Chief Justice so thoroughly recognized in his indefatigable circuits of that territory just after the war, lies in the people of New Guinea learning to respect the orderly and impersonal procedures of law and political democracy. The long bureaucratic line which linked the little aspirant community to the higher political organization of the Commonwealth and the United Nations, was—and will be in every case—a difficult medium along which impersonal and orderly decisions can flow. "This year, next year, sometime, never," the divinatory absolutes of the plucked daisy petals became more and more the mood of the people. What was the use of "lining" every day, of working on village projects, of trying to keep the school going, of obeying the high demands of the New Way, of saving money for taxes which no one ever came to collect, of giving deference to local officials-elect—only too human after all—whom no power confirmed in their roles?

There were regional meetings in Bunai, and long discussions of how meetings of regional officials were to be financed. Should each village send a consignment of sago to feed them during meetings? Should each village help build a meeting house and storehouse for the council? House building plans were encouraged. Patusi tore down every house in the village; people lived in their little kitchens while new houses were built. Bunai, with its six hamlets—two Manus and four Usiai—went into a regular convulsion of rebuilding every house in the village. Peri would rebuild after Christmas. There were discussions of what setting up a co-operative would mean, and if there should be separate stores on the South Coast, the argument getting badly confused in people's fear that this was one other way of cutting them off from Baluan. The leaders—Paliau from Baluan, Samol in Bunai, Pomat, Karol, and Peranis of Peri, Gabriel of Patusi, Kampo, the most gifted Usiai leader—did their best to think up issues about which meetings could be held, about which reports of what had been done could be read or recited in the home village. But political life was empty and essentially lifeless.

The forms the people had set up were not cathartic rituals, self-limited and satisfying, as when people week after week engage in the same dance, the same religious ritual, or the same delectable feast. The political practices of the West carry very lit-

tle intrinsic sense of reward—a meeting at which there is no battle, a measure passed without objections, an election without
deep and apparently almost irreconcilable differences, is also
likely to be a bore. People yawn and wonder why they bothered
to go, stay away from the polls, become apathetic and cynical.[2]
The whole tone of the village reminded me of certain towns in
Great Britain during World War II which had been spared by the
Nazi bombers and in which one heard over and over again:
"What this town needs is a few good bombs." Civilian defence
had been stepped up to a pitch which could only be maintained
by some real emergency to match it. Instead, night after night fire
watchers sat up to drink the pallid, not too well-sugared tea, waiting for air-raid attacks which never came. Volunteers who would
have worked with devotion side by side had there been a bombing, got on each other's nerves; petty bickering and feuds developed; people heard with envy of the heroic achievements in other
towns. Not only the bombers but the whole moral impetus of a
world keyed to maximum effort seemed to have passed them by.
In moods of this sort, a community, an age grade, an organization, is vulnerable to any demagogue with new plans for the use
of the new channels which have been prepared with an energy
which then is not evoked. It is vulnerable also from the inside, to
occasions for factionalism and disruption of the new untried
social structure which has been formed about the new purposes.

In Peri, Petrus Pomat recognized, and rightly, that without
frequent meetings the village morale went to pieces. He called
meeting after meeting to which fewer and fewer people came. A
handful of people would gather in the moonlit square, speeches
would be made about the way in which people didn't come to
meetings, and finally, with the statement that there weren't
enough people, those few who had come would drift away to
their homes.

Meanwhile, there was a complementary tendency in the village to seize on small issues and make a tremendous fuss about
them. "Court cases" about minor incidents and tremendous
arguments about some small point which could enliven a meeting

kept developing. The alternative was gambling—intervillage gambling games in which huge sums changed hands each week end, described afterward in phrases identical with those that had been used to describe intervillage warfare, "They took one hundred, and then we went and took one hundred," etc. The position of gambling in the whole social scene was pointed up by a growing tendency to stress rules in gambling, the effort on the part of the responsible older men (who also occupied positions of authority in the local bureaucracy) to establish a gambling ethic—playing with real money, no I O U's, rules about ganging up, rules about playing with borrowed money, and moralizing about the desirability of dividing one's money into two parts, one for gambling and one for real life. This move to include gambling within the social order, however, was politically doomed, as gambling was forbidden by government, an offence punishable under the Territory's Native Administration Regulations by imprisonment. Continuous enforcement of the code out in the villages was a practical impossibility, but the periodic rows were likely to bring it to administrative attention. It was possible for a single individual to blackmail the village by threatening—if he were taken to court, or if his losses weren't recouped—to take all the twenty or thirty people who had been present at the game in question into court, and have them all imprisoned. Every local official jeopardized his position by having anything to do with gambling; the involvement of reliable officials meant the weakening of the tie between the elected officials and the Administration, the danger of the development of secretive, anti-Administration cabals, the fostering of dissidence.

It was in this atmosphere of helpless apathy, with the periodic over-stress of some local issue or local bit of factionalism, that the Peri schoolboys succeeded in producing an explosion. November 3, 1953, had been a rainy day, with a small desultory "line." Lenore Schwartz arrived in the middle of the morning to do projective tests, while Ted Schwartz was away in Lorengau getting our first three months of film off on a ship. In the afternoon the low tone of village life was disturbed by a quarrel between two men over a gambling episode in which the wife of one had been involved. Then life settled down again, and after supper I left Lenore testing the schoolboys and went out torch fishing with Raphael to make a detailed study of the type of relaxation and alertness which the night fisherman maintained. We fished about a mile or so from the village, cut off from the world in the little circle of light made by the lamp held by the fisherman standing in the bow of the canoe. There was no sound except the splashing of fish, the occasional flashing attack of the fish spear, and Raphael's voice talking quietly of how when he was a child he had thought the sun and the moon were one until one day he had seen them both together in the sky and known—all the Manus enthusiasm for incontrovertible fact rang in his voice—known that they were not one, but two. We talked of what it was like to be out fishing until dawn, about the dash for the market with a large catch of fish, the disappointment if there were no land people at the market and the dash home again, over the open miles, tacking in long diagonals with an unfavourable wind, paddling in desperate haste to get the fish home in time to save them on smoking trays—and still without any food since the night before. But one thing was better now; there was tea, thick and sweet, which could be warmed up at sea and would take the night chill off one's body. Occasionally another canoe passed us, giving us a wide berth and going out to find a place where the fish were undisturbed. "It is no good bringing children out," said Raphael, "they only go to sleep. It is hard for the one who paddles not to go to sleep if there is no one to talk with him."

When we returned, the village was very quiet. I found Lenore Schwartz with a group of schoolgirls waiting to be tested and I asked casually how the group had shifted from boys to girls. She said the schoolboys had come, checked over her lists, made sure that the last boy was finished, and had then produced some girls for her to test. The girls sat, clinging like lumps of cold tallow to the verandah bench; it seemed to me they looked as if they never meant to go home. And then—Raphael appeared in the doorway, looking like an angel of doom.

"All the schoolboys have run away!"

He might as well have said, "The is the end of the world."

The whole hope of the New Way lay in the schoolboys, the schoolboys who were learning to read and write, the schoolboys for whom the government would provide a real school with a real teacher, the schoolboys who sat up late at night, around a flickering bit of rag dipped in oil, chanting syllables in unison from English primers. When people did not understand, when the old women shrieked at each other in the style of long ago, when middle-aged men beat their wives, when there was poor attendance at church, when discussion in meetings was rambling and petty, the enthusiasts for the New Way would comfort themselves and each other, saying, "When the schoolboys grow up it will be different. These others, they grew up in the old bad ways. It is not their fault that they fly into rages, that they do not know how to speak in a meeting or how to treat their wives and children. But when the children grow up it will be different." And my mind went back to the impassioned plea of the young teacher for help, for books, to which he had added, "But do not say that it is a school, or we may never get a real school. Say only that I work to keep the children's minds clear, so that they will still be able to learn when the real school comes."

And he had done this. The difference between the disciplined group of Peri boys who worked together, clip boards on bare knees, or spear guns in hands in a fishing party, and the boys of the same age in Bunai who had had no schooling was striking. Peranis, the teacher, although he had had so little training, knew what a school should be, knew what it meant to sit still and work at learning, and this he had taught them. Diffident toward strangers, shyer than their obstreperous forebears, gay as they gathered to sing on the beach in the moonlight, the schoolboys were the hope of the village, of the Manus, of the New Way, of a place for the Manus in the modern world.

"The schoolboys have run away!" The words echoed and re-echoed as Raphael stood there, his great lust for life gone dim and heavy, as suddenly he became a man without hope. By one of those curious chances which link past and future, distant village and metropolitan world, together, I had read only a few days before a nightmare short story[3] which pictured all the children in New York—and all over the world—going away to a children's peace conference, and told of the agony of the parents who realized, slowly, that those hundreds of thousands of children wearing banners on their chests, pouring into subways as if on an excursion, would never return. The nightmare of the story began to blend with the images that rushed into my mind as I followed the boys away from the village in my mind's eye—Otto so responsible, who had done all of Michael Nauna's fishing; Keraman, with his loose-jointed humour and wry brilliance; Mathias, so sturdy and almost grown up; Alois and his brother, who had kept their widowed mother's large household going; and most of all, Petrus, Kilipak's son, and Johanis, the son of Kutan, who in their ardour and their discipline had been the promise of a different day.

Petrus' father, Johanis' father, Karol Manoi, the councillor, whose whole hope was tied up with the village, Tomas, the devoted young minister, were all away. What would it be like when they returned to find the schoolboys gone? Not like the old days when boys ran away to work for the white man, ran away to an adventurous three years, to return laden with cloth and knives and boxes. Then, too, fathers had been angry and affronted, and, if ill fortune befell a boy while he was away and he was heard to have been put in jail, laments might be composed and sung in the village. If one died, then he had been mourned as if he had been lost at sea. These were individual events in the lives of individual families, and their kin and village mates united with them in sympathy, in ceremony, sleeping in the houses of the bereaved. Nothing could have symbolized more completely the change from the old order to the new than the reaction to the schoolboys running away, for now it was not the individual sons of individual men, not a son of the house of Korotan, or of the clan of Matchupal, who had run away, but "*the* schoolboys," the next generation, the "hope of the world."

A second echo went through my mind—World War I, in a little town in Pennsylvania, the Italian-born principal who taught us civics had such fierce enthusiasm for his new country that we found it less boring than adolescents usually do; his voice rang out, against the background of the Second Battle of the Marne, reading from an essay by some other enthusiast: "Boys and girls of America, you are the hope of the world." And, beneath this were memories of the Pied Piper of Hamelin, that mythical figure of our early childhood, wreaking vengeance on the city fathers.

The text of the modern short story was spread before my eyes, the story of the posses of men searching, as they came to the one place that had not been searched, "with their eyes coated with a dust that is death."

> They went into the woods as men everywhere were going into the woods and the mountains and into all the last impossible places, places which were like shallow cups of remaining hope whose pitiful last draining would mean dry despair. They moved quietly into the woods and searched and said nothing. The stars were so bright that in some places they reflected on the stones. There were clearings where one or two of the chest banners were found. They were torn and covered with dust and it appeared that they might have been trampled. But for all the rest there was only to be found a great still and encompassing quiet, an empty immensity. Everywhere over the earth there spread a void of terrible beauty and peace—just as it had said on the ribbon.[4]

At the centre of the modern world, where *The New York Times* put Einstein's newest formula on the front page, and in the rim of the South Coast lagoon in the middle of the Pacific, among a people newly come into the modern world, there was the same fear—that democracy was not going to work after all, that the children would go away forever. In the modern story, an eavesdropper reported that the children had talked about "schools and examinations and having to be born and bossed around. They said the world was a trap we made for them." What had our little group of Peri schoolboys said, I wondered. All this streamed through my brain while Raphael stood there, aging before my eyes. "They have all run away completely. Peranis scolded them and took all their books away, and they have run away to find a school in some other place."

Slowly the story unfolded, as people gathered in the square and talked in muffled voices. After supper, Peranis, the teacher, had sent for them all—as he often did of an evening—and had taken all their books, their precious English readers, away from them. No one knew why. The boys had come back into the centre of the village, checked with Lenore Schwartz to be sure the last one of their number had completed his tests—responsible for this last schoolboy task. Then they had rounded up the girls to take their places. Otto had told Michael Nauna, whose foster son he was, that he was leaving; Johanis had first asked his mother for his newly bought *laplap*, and then changed his mind and left it behind. They were going, they said, to find a better school, a school where they would be allowed to learn without interference.

The women stood, whispering in excitement not untinged by a malicious savouring of how their husbands would feel. Two of the boys were sons of men whose love affairs were notorious, and the mothers somehow conveyed that there was a relationship between individual moral turpitude, the failure of the New Way, the hopelessness of men's pretensions, and the schoolboys' decision.

I sent a messenger to Peranis' house, who returned to report that Peranis was asleep—I could guess in what a mood, after having destroyed all that he had built so imaginatively through the last four years. His rage, always so close to the surface, always ready to destroy, had won. There was no one of authority in the village except Raphael and Michael Nauna. The fact that Raphael's own odd little boy, who was a little younger and who refused to go to school, had not joined the others was no comfort to him.

But Michael's adopted son, Otto, had asked if he could go, and, out of his despair, his growing belief that nothing would ever happen, that all the money they had saved would stay forever impounded by the government, that no ship would ever be bought, Michael had said, "Go!"—the last vestige of life in the old way turned against hope in the new. Karol Manoi and Petrus Pomat were out fishing, Kilipak was in Lorengau, Kutan was on Ndropwa, Tomas was away in Patusi.

So people stood speculating: Would someone stop them? Where? With what signal? A group of inexperienced boys—no older it is true than their fathers had been when they had run away to join the indentured labour force, but then there had been a recruiter ready to snap them up—had gone away in search of a dream, a school where they could learn without interruption, a school where the teacher would not lose his temper and upbraid them. I remembered the explanation that Lukas Banyalo, our illiterate mystic preacher, had given of why he had left the mission school as a child—because he couldn't bear to see the children chastised. The boys had left on foot, taken the land road, which to every Manus is a road of stones that bruise and hurt the flesh, a road for pigs, not for men who sail the open seas so proudly. What would happen? Would they, hungry and footsore, disappear aimlessly into casual labour in Lorengau or Lombrum, their sense of unity and purpose dissipated forever? Would they ever come back?

At last, Alois Poniu, Oko, and Peranis Katiwai, three responsible young married men, decided to follow, to search for them, if possible to find them and bring them back. The evening wore on with the women still standing about in little knots. The square was empty of small children. The schoolgirls who had come to be tested sat on, stubborn and heavy, long after we had said there would be no more tests that night.

Then the three men who had gone to look for them returned to say that they had been intercepted in Bunai, that Samol, the leader of Bunai, had "fastened them," had assigned them a house, given them food, and had placed them under his personal official protection with the statement that this wasn't a matter which could be dealt with informally, "Not just anybody from Peri could fetch them back." Tomorrow Samol would deal with the matter through "official" channels. Alois and Oko stayed awake until Petrus Pomat came back from fishing to report to him what had happened. I went to sleep, still uncertain as to what had happened, or was likely to happen.

Up at 5:45 A.M., I found Tomas, son of Raphael, and Patri, another small boy, walking about the square, every muscle adjusted to show their role in the drama—they were the younger boys who had not run away—and they walked with the air of importance which distinguishes a Manus who is acting a part in a ceremonial occasion from a Manus who is not. Later, three slightly older boys, Rigat, Uper, and Bomboi, all in the lower of the two school classes, appeared. They had been asleep, they said, and so had been left behind. The widow Pinkes appeared to say that three of the runaway boys had gone on to the Naval Station, thus ceasing to be schoolboys and so not subject to Samol's direction. It was market day. At eight o'clock Peranis, the teacher, walked across the square, looked up, caught my eye—and smiled.

I knew then it was all right, that not only had the boys been found, but Peranis was somehow pleased with the whole situation. Slowly, during the day, the plot unfolded. Three of the boys returned to get forgotten spear guns, stalking across the village square with momentous impressiveness in every lithe movement, then sitting down on the beach, waiting to be interviewed. Karol Manoi, the Patusi immigrant, the present councillor-elect of the village, had been told nothing about the whole episode until the morning, and it was he who interviewed the three boys.

They were delighted to talk. They had run away, indeed, to make the matter come up. As Karol then outlined it to me, the whole trouble had started because Karol Manoi had sent a note to Peranis asking for some schoolboys to go on an errand to Bunai. Peranis had spoken to the schoolboys, and all of them had refused to go. Karol was angry and that evening, when the schoolboys were all perched idly on the railing by the sea, singing to the accompaniment of a ukulele, Karol had gone out and impounded the ukulele. The next morning, when the boys had gone to ask it back, he had scolded them and said that he had taken the ukulele because of their bad behaviour, and that now they would have to "buy" it back. His brother's son, a Patusi boy, had answered him back impudently, the old confusion between the former rule by senior kin and the present rule by elected officials reasserting itself. Karol, himself a Patusi immigrant, whose hold over Peri was fast loosening because of his failure to act as he ordered others to act, burst into a rage, berated the group of schoolboys from immigrant Patusi families, told them they were receiving for nothing an education for which they should have paid. In this accusation of receiving something for nothing, he invoked the same deep shame that had formerly held the young men enslaved by the old who had paid for their marriages—and added that if he, Karol, hadn't asked me to get them their present English readers, there would have been no books for them to read.*

The insulted and infuriated schoolboys reported this to their teacher, Peranis, a man whose own childhood rage was always close to the surface, despite his strong conscience and vigorous attempts at self-discipline. For four years he had kept school, devotedly, in return for one half-day's work from the schoolboys, every other week. And this the parents had grudged. They had said to a son who asked for the loan of a knife for this work, "Why don't you get it from Peranis?" To the son who asked to use a canoe, the father would say, "Hasn't Peranis a canoe for you to use?" or to the boys who returned for a meal after helping Peranis, "What! Hasn't Peranis given you any food!" The unsuccessful, mean-souled men of the village had jeered at Peranis' fine

---

*The Papuan readers which were especially sent to me from Port Moresby, through the courtesy of the Administration.

new house, "I would have one like that if I were able to have all the schoolboys work for me!" All this had smouldered in Peranis' mind for many months, and now, now Karol had claimed that he had got the school books, when *he*, Peranis, had written me within an hour of my arrival, asking me to help him with the school! And Karol had publicly shamed the Patusi boys, who would now probably go back to Patusi. Karol was trying to "break up the school." Furiously, Peranis demanded back the books, saying he would never teach them again, and the boys had responded by running away.

Bidden by Karol and Samol to return, they came back announcing that their one aim had been to bring the issue out into the open—the old Manus belief in truth, open threat, open resolution of every difficulty, coming to the fore. The schoolboys slept that night in Peranis' new house, and the young men, who usually gathered to gossip around Michael Nauna's front door, replaced them—in an odd symbolic move— on the moonlit beach which had been the schoolboys' play- ground. The girls and the little schoolboys who had been left behind joined the older ones in Peranis' house at the far end of the village.

Then the newly attained democratic procedures were invoked. Old Pokanau refused to discuss the matter privately with Peranis and would talk only in an open meeting. At "the line" the next morning, Pokanau called for a meeting and announced that all parents of school children must attend to thresh out their differences with the teacher. A parents and teach- ers meeting was born, before my eyes, out of the need for a better understanding between parents and teacher. The meeting began slowly, the parents drifting in, the school children marching in self-consciously in a body. Peranis, in a glow of righteous indig- nation, threw down the gauntlet:

I want every father and every mother here.
Let not one of them wander away.

All your wives are to stay too.
September 23, 1949 [this in a voice of thunder],
You all sent your children to me,
You said: 'All our children are to stay with you.
If they are insubordinate, you, you are to discipline them,
You teach them all.
As to their food, we will look after that,
But as regards school, this is your affair.'
I stand here. I want to hear what every man and woman
has to say!

Peranis then repeated the accusations that the parents had talked *bilas*, mocked him and shamed their children, and then he called on the parents to speak out. Had they, or had they not, said these things: "You want a canoe, has this man Peranis. Then, no canoe?" Had they, or had they not, instead of coming straight to him, talked behind his back? Then he turned to the schoolboys: "You boys, if you heard these things, stand up."

The group of schoolboys stood up solidly together. Then Peranis went on:

Very well, these remarks have been made over and over
  again.
You all know about it.
I've been accustomed to hear them.
However, I am a man of Peri,
I belong to this place.
I have heard all this
But I may not be angry.
If you had done this to a stranger who came from far away,
This school would have ended long ago.
I have had a little endurance, because of the children,
But this gossip and slander did not occur on one day,
On just the day I brought them to school.
It came up every day.

I think the meaning of it all is—*bilas*.
Gibes that I have no canoe, I have no knife, I have no
  tomahawk,
I have nothing.
The meaning of all this was to shame me
So that I would give up altogether
And find some other work.
I have understood this and it is wrong.
Now, part of the discussion which refers to you all is
  finished.
Now, let the children stand up
And bring it all to light.

Then each father, led off by Kutan, stood up and challenged
his son. Had his son heard him say such things? True, in the past
he might have, but he couldn't remember. Let the boy speak.
Each son stood up and answered, insisting that there had been
such talk. The fathers then, still protesting, called on the mothers
to say if it hadn't been they who talked. The mothers, embarrassed, reluctantly admitted that, yes, they had said these things.
To a burst of laughter, tall, impudent young Raphael told his
bad-tempered little father that he was a liar. Stephan Kaloi made
a humourous speech about no longer having a son in school—he
had got married. But when Josef was in school, he had said, "You
don't go to school to learn to work. You can learn this from your
parents. You go to school to learn to write." Back and forth the
controversy went, the children standing up straight and tall, the
fathers taking the initiative, on the whole affirming their own
goodness and blaming the tongues of the mothers.

The whole affair was remarkably gay and good-humoured,
although, if a boy giggled as he spoke, he was rebuked by the
humourless Petrus. The children and the issue were taken seriously, but the ritual of democracy, the way in which the women
were made to speak up, the shyness, embarrassment, the occasional wisecracks, were matters for laughter. Finally, when each
schoolboy had been challenged and had answered, Petrus Pomat

made a long summary speech. He went over the case again, scolded the parents:

> If you want to scold your child,
> If you want to discipline your child,
> You must do this straight without calling Peranis' name.
> It is wrong that Peranis' name should be brought into it.
> It is wrong to bring the names of others into such a quarrel,
> This is talking *bilas*. . . .
> You and I realize that we are not yet equal to our new tasks,
> It is as if we were paddling very hard [i.e., no wind to sail with]
> In making our village straight. . . .
> This talk among you, fathers and mothers and children, has been wrong.
> Now this talk must stop. It is wrong.
> As to the way to run a school, you do not understand well.
> But there are schools in the place of the white man.
> I have been there and I have seen them.
> They have schools for learning to write on paper, and work is done there, too.
> What do you think anyway! If in such a school only writing were taught, and nothing else is taught,
> By and by they would be ignorant.
> They teach how to make canoes, how to make boats, how to catch fish,
> They teach them about everything. This is the way of a real school.
> As for this talk you have been making, it must stop!
> This kind of talk results in quarrelling and in every kind of evil fashion.
> Quarrelling and fighting will result, and will not be finished.

But there are some other wrongs, too. . . .
This hasn't been a meeting, this has been like a court.
Like a court in which you boys and girls made court
    against your parents,
This isn't right either.
If your parents have said things, you should say them.
If not, be careful. And don't lie. And remember,
One week you go to school to Peranis and do what he
    says.
And one week you work for your parents.
Listen to them and do not provoke them.

The parents were asked if they wished Peranis to teach their children, and a unanimous *U*, yes, was given. Democracy had won, and won in the atmosphere out of which democracy has grown, an atmosphere in which each individual is treated seriously, and in which the exchange of roles can be greeted with good-humoured laughter. For laughter, that distinctively human emotion, laughter which springs from trust in the other, from willingness to put oneself momentarily in the other's place, even at one's own expense, is the special emotional basis of democratic procedures, just as pride is the emotion of an aristocracy, shame of a crowd that rules, and fear of a police state. Like the delicious ripple of laughter which runs over a meeting, weary with seconding motions, weary of "ayes" and amendments, when the chairman gives up the chair so that someone else can nominate him to reoccupy it, laughter is necessary to make the quiet, often boring ritual of democracy tolerable and cherished.

The schoolboys' runaway had begun out of the old angry, self-destructiveness of the Manus, out of the use of old patterns of insult and covert verbal attack, of shaming and hostility, and of running away in anger as the only way out. But the runaway had ended in orderly discussion, with problems aired and children allowed to speak on an equal footing with their elders—with good-humoured quip and laughter.

A month later, it was time for me to leave Manus. The two

nights and days of farewell feasting and dancing were in full swing; my house was decorated with flowers and leaves; people had come from afar. I had put on a bright cotton frock, stylishly cut to gladden the eyes of people who cared about style. Ted and Lenore Schwartz were dressed in white for the party. Then a message came that a child of Lukas Banyalo had cut his leg and that the bleeding wouldn't stop. There was still no trained medical assistant in the village. It would be months before the Peri boy who had been sent to Rabaul would return from his training. As we knelt on the slatted floor of Lukas' house, taking turns in pressing the boy's leg, for the bleeding would not stop, the whole incredible poignancy of this tiny island drama rose before my eyes. The ukuleles and the drums were playing, the people were dancing to honour my departure, "Before we had many of our old things and you took them with you. You did not take them without payment, you paid well for them. Now we no longer have the old things. It is now you who have the things that we want. You have brought them, and we, too, have not got them for nothing. We have paid for them. Now, like an old turtle, you are going out into the sea to die and we will never see you again."

And I had answered, "Years ago when I left, you beat the death drums. Neither you nor I ever expected to hear of each other again. There was no road. But now there is a road. Now you belong to the world. Now you can write to me and I can write to you. You can tell me how the council goes, what you do with your money in the bank, who dies and who is born, and how you rebuild the village." These were the things we had said the night before in a formal farewell meeting.

I lifted my fingers for a moment, but the arterial blood began to spurt again as the small boy tensed his body, proudly silent. The light in the hurricane lantern flickered. "You can tell me how the council goes," I had said. But would the council ever come? Would it come in time, or would their enthusiasm flicker out, apathy and suspicion and disgruntlement win, the faith in medicine and freedom from anger be replaced by a fear of the sorcery of strangers? Would the school, which the schoolboys and the villagers together had saved, survive? Was it all too ephemeral, too improbable, too unlike anything which the Western world was equipped to deal with? Were, in fact, these people doomed to lose their new-won identity, doomed just because they had risked everything on a sense of new-found community, on a willingness to move together into the modern world?

It was hours before we were able to stop the bleeding, and there were little spatters of blood on the pink tulips of my gay dress when we started back to the festivities. "When you leave, we will have no one to give us medicine." Chilled, stiff, saddened amidst the gaiety and music, we reached my house on the central square, so festive—and there, standing stiffly at attention, was a police boy with a letter from my friend the District Commissioner, sent specially over land so that I would get the news before I left the village. "The council has gone through."

18

implications for the world

"I have come back," I told the people of Peri, "because of the great speed with which you have changed, and in order to find out more about how people change so that this knowledge can be used all over the world."

What, in fact, can we learn from this single historical experiment, this one detailed account of how a handful of people on an isolated South Sea island entered our modern world, and about their efforts to stay here, and our efforts to keep them here? Before attempting to draw conclusions, it is perhaps necessary again to consider to what extent their experience can be generalized—for the people of undeveloped countries, for the mid-twentieth century, for mankind. Because one people have shown, under very special and quite unrepeatable conditions, a capacity to learn very rapidly, does this change our ideas about how rapidly other peoples can learn? To what extent can this unique little experiment be shrugged off with the statement that one swallow doesn't make a summer, that all over the world we have examples of people who are less developed technically, politically, socially, religiously, than others, and who, far from coming enthusiastically into the modern world, resist and withdraw, sabotage machinery, boycott schools, or put coal in bathtubs, retreat into ideas of witchcraft or sorcery or cheaply attained apocalyptic solutions, or deteriorate into forms of criminal sub-humanity?

Isn't "civilization" something that it takes a very, very long time to learn, perhaps many, many generations?

This question is the more important because of the special contribution that anthropological studies have made to our conceptions of change, of how fast and in what ways people could change, and what the ethic of changing people should be. For the first four decades of the twentieth century, anthropologists were concerned with demonstrating a series of propositions which had not, for some time, been regarded as to any degree self-evident: that the human race was one, and that the various "races" of mankind were specializations without any measurable differences in their capacity as groups of individuals to take on any civilization; that each people has a shared, learned way of life—a "culture"—of its own, within which dignity is accorded the individual, and continuity provided for the group, and that this "culture" should be respected in the same way that individual human beings should each be respected; and, finally, that, although the behaviour which differentiated an Eskimo from a Frenchman, a Hottentot from an Englishman, a Burmese from an American, was learned, the circumstance that it was learned and not inborn did not mean that, once learned, there were not very great differences between the members of these different societies. No human being once reared as an Eskimo would ever be able to be what he or she would have been if reared as a Burmese, a Frenchman, or a Hottentot. Even though, later, individuals migrated from one society to another, learned the language and beliefs and customs of another people, they would wear their second or third or fourth culture with a difference. Beneath the pattern of a second or third language would be found the "mother-tongue culture," first and differently learned.

If each man's culture was, then, an irrevocable part of his humanity, once learned never to be finally discarded, and essential to his dignity as a human being, the right of each culture to survive, its moral claim upon the protection and forbearance of those with power to alter, eradicate, or transform the culture of other human beings was very great.

Just as we based our whole political philosophy upon the dignity of each individual human being, and recognize with Dostoievsky that a state established by the intentional sacrifice of the life of a single child can be no state fit for man to live in, so we came to realize that a civilization which rode roughshod over the way of life of other peoples was incorporating evil in its own way of life. It was in terms like these that struggles were undertaken for the protection of native peoples, for the rights of minority groups to keep their own language and religion, and attacks were made on colonialism and imperialism, grave doubts were raised about the validity of foreign missions, and about head-taxes and indentured labour, and, after World War II, about the advisability of the export of either Coca-Cola, Hollywood films, or modern tractors. On one side of these discussions stood those who were convinced that our way of life, our culture, was the noblest and the best, so that all peoples everywhere should be exposed to it, and, if necessary, inducted into these higher values—whether these values were seen as the only true religion, public health, proper nutrition, or the secret ballot—by all means short of war and conquest. It was admitted that parts of Western countries were filled with backward peoples who failed to exemplify one or another of our various cultural superiorities, but this was no reason for not prosecuting vigorously our efforts to get other peoples, sunk in various forms of blindness and ignorance and stubborn resistance to the light, to see the error of their ways.

This battle had to be fought with vigour. We had to convince the skeptical that "savages" really had a "culture," that unwritten languages had forms comparable to and often more complex than written languages, that because people were undeveloped technologically or espoused a different religion it did not mean that they might not be very highly developed politically or artistically. Vigour was necessary because the ideas were new and strange, and because they conflicted with zeal and determination in those who took a different point of view. Why should anyone want to preserve the strange, barbaric, archaic, unenlightened practices that characterized every group except our own?

So, because the exponents of making men over in our own image had such a strong tendency to see other cultures in a dark light—where suttee and taboos on eating cows were combined with cow-dung floors and child marriage to create a nightmare for Western readers (as in the popular book *Mother India*[1] written in the twenties)—those of us who were charged, because of our special knowledge, to demonstrate that each culture had a dignity of its own and that its special practices of which we disapproved had to be seen in context, had to emphasize the positive values of each culture, however different from our own. The beauty of old harmonious ways of life, where hand and eye worked in perfect co-ordination, where each costume added to the lines of temple or courtyard, was stressed in counterpoint to those who spoke of unsanitary conditions, the infant death rate, or a failure to discriminate the value of always telling the truth.

Out of this controversy, there crystallized in the minds of those who were not immediately involved in it—the educated lay reader, the teacher of college freshmen, the contemporary social philosopher—an image of the anthropologist's view of the world, which, like all images, was culture-bound, time-bound, and peculiar to a particular climate of opinion. The anthropologist was said to believe that all people were just alike, that there were no differences among individuals—this was a transformation of the statement that within each race a comparable range of abilities could be found. The anthropologist, it was said, believed that all ethical values were relative, merely a matter of custom—this was a transformation of the anthropoligist's insistence that any given practice must be seen in context, as relative *to* a particular cultural, temporal situation. So, where the anthropologist pointed out that the Arapesh who married his brother's widow and thus acquired two wives who had to be cared for and supported was acting ethically and unselfishly, and that the Eskimo who walled up his aged parent to die was acting ethically, as an Eskimo, in the transformed statement the anthropologist was claimed to have argued in favour of polygamy, or killing one's grandmother.[2]

This incorrectly imputed advocacy meant, so the moral philosophers argued, the loss of all our dearly won ethics—as indeed it did, had it been so made. Finally, the anthropologist was believed to be in favour of leaving each people exactly as they were, of arguing that all purposive change, all attempts to change people's language, or beliefs, or living habits, were destructive and bad, and to be discouraged. The anthropologist, in a conference or on a team or as an expert, was opposed to the economist who wanted a viable new economic system which would raise the standard of living, the missionary who wished to convert the heathen, and the public health expert who wanted to eliminate the hookworm. This again was a distortion and an overstatement of a position in which the anthropologist insisted that changes must be congruent with the rest of the culture, that attempts at change which tore people from their moorings, separated children from parents, established sets of incompatible customs, put new wine in old bottles and old wine in new, and destroyed the wholeness of a culture, also destroyed the wholeness of each individual who had been made human through that culture.

These were transformations and distortions, bred of the newness of the ideas and the intensity of the battle, in which the anthropological position tended to be overstated and caricatured by its opponents. But the overstatement was also there in its advocates. Museum curators and artists both sigh when textiles woven on handlooms into lovely designs are replaced by machine-made textiles of poor quality and poorer design. When a member of an audience to which one had described the balance and beauty of life in a Balinese village asked worriedly, "But don't they want to progress?" the lecturing anthropologist, conscious of the many values in Balinese life—where every man was an artist of sorts—which were absent in that particular suburb, was a little inclined to answer impatiently, "Progress to what?" and to fail to accord the same dignity to the questioner's cultural belief that progress was a good in itself as had just been accorded to the Balinese belief that music was one of the great values in life. And, perhaps more seriously, there were even anthropologists who did statistics on the number of societies which approved pre-marital intercourse, and proved by the fact that more did than did not do so that somehow our present attitudes were therefore wrong—a sort of "Kinsey ethic," with cultures as units, which associated the frequent with the ethically desirable.

But, in addition to these sins of overemphasis, anthropological theory of change, developing as it has primarily out of a secularized Protestant, Western European ethic, however diverse its actual practitioners, was still compromised by the ethnocentrism, the cultural myopia against which it professionally fought.[3] In our theoretical discussions we opposed two concepts: that of *diffusion*—the fact that "traits," like trousers, or scissors, the potter's wheel, the alphabet, the story of Cinderella, were diffused around the world—and that of another process, *culture contact*—in which two peoples of different levels or types of culture met and one, against the will of the other, forced its way of life on the culture-contacted people, using persuasion, political power, bribery, economic sanctions, expressed contempt, and opprobrium to force the other people to accept a way of life essentially alien, incompatible, and unwelcome to them.

These donor people were seen as active, exerting power over the others, which, while it was done for the good of the recipients, was also, like all power over persons, somehow evil. The other side of this picture, the extent to which the recipient people wanted a more universal religion, a different kind of government, better food, literacy, or medical care, tended to be ignored. The implicit insistence that all attainment of "goodness" should be painful, whether the goodness consisted in telling the truth, eschewing idols, using a knife and fork, or eating the part of the potato which was next to the skin, subtly dominated theory as well as ethic. As a result, most anthropological theories of change up to and during World War II were permeated by attempts to protect the peoples whose cultures were threatened from the results of purposive attempts to change them, either by their own Western educated élite, or by the political, religious, or economic emissaries of foreign powers. Thus, in concentrating on the risks and dangers of purposefully induced change, we gave very scant attention to the other side of the coin, to what "Western" or "higher" or "more developed" peoples not only did not force on other peoples but actually *denied* them.

So the stock images of culture contact, perhaps inevitably, but certainly lamentably, emphasized negative coercion: the missionary who forced unwilling natives to wear clothes, which (getting wet in tropical rain) gave them pneumonia and decimated their numbers, traders who forced machine-made cotton on them and ruined their handicrafts, or who, after alienating their land and getting them into debt, forced alcohol and opium on a helpless and abstemious people, or government which, by insisting on a head-tax, forced villagers away from their wives and children to die in the killing work of the salt mines. These images, set against a background of palm trees under which the natives would otherwise have lived idyllic lives, blended with images of imperialism, colonialism, and later, with the spread of the uniformities of Soviet culture into Asia and Eastern Europe, of Communism.

Even after World War II, when the non-industrialized peoples of the world began clamouring for the blessings of the modern world, machine technology, universal literacy, medicine, the clamour was all too often seen as inevitable but regrettable, an unfortunate by-product of European contact which had made people discontented with their own way of life, or as the necessary answer to Communist propaganda. It was recognized that with the Communists promising the delights of progress to the villagers of the world, it would be necessary for us to accede to their requests for this same progress, to go in and help induce technological change.

And now our old sense that all change was one-sided and came from a misuse of power suffered a new transformation. All change was now seen as terribly difficult and against the real will of the people, who only thought they wanted tractors because these were symbols of Western superiority but who really hated regular hours, clocks, machines, hospitals, the dictates of nutritionists, sitting still in school, and learning to think in realistic Western terms. So the anthropologist on the technical assistance team tended to remain, in most cases, an expert on difficulties, who, while he insisted on taking the culture into account, also insisted that change must be slow, cautious, tentative, if resistances were not to be aroused, if social and personal disorganization was not to result. However much it had become clear that change was now inevitable,[4] that it would be rapid, that we must have as our goal the making of some members of every culture world-mobile in one generation, we were still trapped in a one-sided picture that something was being *done* to people, and that by insisting on working as slowly as possible, and through their own cultural values, we were protecting and cherishing them. So one important contribution of this record of change among the Manus is that it points up the completeness with which a people may want to change rather than merely submit to being changed; it shows culture contact as an active choice of the emigrants from the Stone Age as it is for the representatives of highly industrialized countries, and it points up the "resistance" to giving in the members of the more developed "culture" as well as the resistance to receiving in the members of the "under-developed culture."

How often has our Western attempt to preserve native dress, old customs, different styles of architecture, to respect native laws and customs, been only a thin disguise over an unwillingness to admit a people, newly entering into our way of life, to a full participation in the culture which we claim to value so highly? Yes, we want them to go to school; "education" is a beautiful thing, but we don't want too many of them to become aspiring white-collar workers; we want them—peoples of other countries, or of racial, ethnic, and economic enclaves in our own countries—to improve their standards of living, have better nutrition and running water, to be clean—but not to dance at our dances, join our clubs, or hold the offices of greatest sacredness and prestige in our societies.

Whether it is the white official saying of the New Guinea native, "It's such a mistake for them to wear trousers in this climate, you know; *laplaps* are much cooler, cheaper, and more practical, and so much healthier," or the commentator who regrets the disappearance of the lovely blue of traditional Chinese dress, or the old abbot, who after seventy-five years of presenting a picture of life of high religious abstinence into which no Australian native was considered fit to be inducted, remarks, "They just don't want to take our more complex ethical standards; their minds aren't able to deal with them. Generation after generation we educate the children here, the sisters teach them good habits, and then—they go back to the bush to live as their parents lived," each in his way is refusing to share the whole pattern, however willing he is to give bits of it. Resistance in the grudging and selective giver turns out to be as important as resistance in the grudging and selective receiver.

Once this is recognized, it is possible for us to scrutinize with newly opened eyes each situation in the world where people of a different sex or race, class or culture, seem to fail to accept or to use the opportunities which are offered them, to become "like" the members of another class or race or culture. Many different kinds of failure and refusal become intelligible—of girls to learn physics, of bright-eyed little African children who learn so quickly as small children and turn apathetic and disinterested at puberty, of new immigrants in model housing developments who don't "appreciate" their excellent plumbing, of the contrast between the adjustment of Negro boys and their sisters when both are asked to meet the middle-class standards of a high school in a small Pennsylvania town, of the restlessness and refusal of regular employment by modern Maoris in New Zealand or by American Indian immigrants to the big cities. The situation in which children are taught by individuals whose full status—as men, or members of an excluding race, or nuns, or persons free to travel—they cannot hope to attain makes the goals held up to the pupils seem not to be the wonderful opportunities which they are so often represented to be. So bright girls strangely have no "ambition," and children of discriminated-against minorities turn "dull" at adolescence, not because of intrinsic incapacity, but because the desire to learn is blocked by the knowledge that part of the pattern to which they aspire will be denied them.

Furthermore, as part of the pattern is denied—as for instance specially selected individuals from Africa or Asia are sent to school in the West and offered full "academic" but no "social" participation—the terms in which the participation is denied—physical self-identification as a "mere woman" or a Negro, or a national identification as an Indonesian or a Thai, religious identification as a member of a different creed, class identification as a member of "the masses" or "the proletariat"—are strengthened and intensified and become heavily loaded with both positive and negative feeling. The sex or race, nation, religious or class membership,

over-stressed and omnipresent, determines the lines of identification, so that individual children will quickly learn that they can
only do what a woman, a Negro, the foreign born, a Jew, or an
unskilled dock worker can do. Each such limitation of natural gift
and aspiration carries with it a kind of constriction, a denial of the
self, which, if once relaxed, provides channels through which great
energy can be mobilized and released. So we find that the first
groups of women who are admitted to some male occupation perform astonishingly well, learn faster than the norm for men, while
women who enter the same occupation after it has been defined as
something done by men and women will show no such conspicuous superiority. Throwing off colonial yokes which have included
definition as second-class human beings has the same releasing
effect on members of former colonial states.

The movements which spread from people to people—the
good news that is shouted from housetop to housetop until whole
sections of the people of the earth become one in religion or
political philosophy—characteristically rest on a reaffirmation
that all men are brothers. Those peoples who deny such brotherhood to men are characteristically members of shrinking, partial,
or limited societies, without the expansionist zeal of those who
greet all men everywhere as potential converts or comrades or
partners, because all are members of the human race.

"All men are brothers," say the Manus, "black, white, green, red
men, all are brothers." The addition of the words *red* and *green* adds
an extra blaze of glory to the statement which ennobles each of them
in his own eyes as he makes it. All men are brothers, not only the
black and white men who are known to exist, but the green and red
men who somewhere, somehow, might exist—all are brothers.

A second contribution of the Manus experience is the suggestion that rapid change is not only possible, but may actually be
very desirable, that instead of advocating slow partial changes, we
should advocate that a people who choose to practise a new technology or enter into drastically new kinds of economic relationships will do this more easily if they live in different houses, wear
different clothes, and eat different, or differently cooked, food.

Looked at from this point of view, the speed with which European immigrants adapted to American life will be seen not only as due to some problematical factor of selection through which all those who were willing to emigrate had more energy and flexibility and capacity to change than those who remained at home, but also to the transforming experience of entering a world where everything was different, to which one brought only the clothes in which one stood and which were easy to discard.

There was no old house style to remind one that the old social relationships no longer held. Instead, a different kind of house, lived in by those who practised the different kind of relationship, was ready to support the change. Children who came home from school to insist that a good American breakfast contained orange juice and cereal stormed up American steps and banged American doors; children, become far more active and free in the American environment, jumped on American sofas—if the springs were damaged there was at least no physical reminder of three generations of ancestors who had never jumped on any sofa as children. Unfamiliar foods were cooked on a new kind of stove, and served in a new kind of dish, whose pattern and design evoked no nostalgia for the old. Each detail of the new life supported each other detail. Peasants, unused to depending on an urban money economy, did not have to learn new attitudes toward the old money, which had been buried in the garden or hidden in old socks, but instead were confronted with a quite different-looking money, with different values, different shapes and sizes and names.

Partial change—installing new kinds of office furniture in out-of-date reconverted dwelling houses, turning carriage houses into garages, putting the engine of a car where the horse once stood—can be seen not as a bridge between old and new, something that permits men, slow to learn and fumbling at the unfamiliar, some respite from the unbearableness of change, but rather as the condition within which discordant and discrepant institutions and practices develop and proliferate—with corresponding discrepancies and discordancies in the lives of those who live within them. The alternative to the culture which has existed so long and changed so slowly that every item of behaviour is part of a pattern so perfect that it seems as if it must have sprung complete from the head of Jove, is seen to be not the culture in which necessary and wanted change is artificially slowed down and retarded but rather the culture in which—if there is to be purposeful change, by an Ataturk, an enterprising Maharajah, or the agricultural extension department—the whole pattern is transformed at once, with as little reminder of the past as possible to slow down the new learning, or make that learning incomplete and maladaptive.

As an analogy we may consider the new trends in the treatment of certain kinds of sprains, in which the injured foot is injected with a local anesthetic so that the foot may be walked on at once, because walking is actually beneficial, and the adjustments to the pain were the elements which did the harm and the reason why rest had to be prescribed, or the new post-operative treatment which gets the patient out of bed at once before crippling maladaptation can develop in response to the operation. In the same way, attempts to deal with some drastic alteration in a culture—to substitute wage labour for subsistence farming, assembly-line manufacture for handicrafts, democratic voting procedures for feudal rule, land ownership for sharecropping, non-segregation among castes for a former rigid segregation—may well work best if they are accompanied by as many other congruent changes as possible.

Just as the survival of some parts of an old pattern tends to reinstate the rest, and so continually acts as a drag on the establishment of new habits, so also the establishment of part of a new pattern calls for other congruent elements, facilitates their establishment, and each element supports the other. Let us consider the simple matter of the introduction of cloth or clothing among a people who have only worn G-strings or grass skirts. If cloth is introduced without soap, habits of handling cloth may grow up in which the infection rate rises; but if cloth and soap are introduced together, a pattern of sanitation can be established immediately. If fashioned clothes are introduced without sewing skills, a style of rags and tatters may be set up, simply because the art of mending, a by-product of sewing, is missing. But if the sewing machine is introduced with the ready-made clothes, this can be avoided.

Similarly, the acquisition of tailored cotton clothes without starch or irons means the possible establishment of a style of dress which will be greeted with ridicule by the very people—the Western model-setters—from whom it was meant to elicit a recognition of the human worth of the new wearers. The possession of clothing means that the clothing has to be stored. If new styles of housing with closet space are designed, it is possible to prevent the stage in which pieces of string are stretched across a house, once kept neat as a pin when only grass skirts had to be hung, tightly rolled, from the rafters. And with the introduction of clothing comes the need for some kind of handkerchief, for the old methods of nose-wiping accord poorly with the use of cotton cloth. If cloth is worn, and washed and starched, then sitting on a floor on which people walk either barefoot or in shoes becomes unfeasible, and it is necessary either to have furniture on which people can sit or new habits of taking off or changing footwear at the door.

The regularity with which most peoples who change from a life of highly traditional handicraft to one of dependence on the purchase of European-manufactured objects become slum dwellers, can be explained by the number and type of discrepant changes—in which wash basins become food bowls, objects are introduced without appropriate ways of keeping them clean or storing them. Each misuse breeds a new form of abuse, the whole in turn evoking contempt and ridicule as the aspirants toward a new way of life sink instead into one far lower than that from which they came.

As in such a concrete matter as clothing, so at every level the same rule may be seen to apply. Practically, this means that whenever a people wish to take over some invention or discovery or practice of another people, the real alternatives should be seen as between taking over the new idea in the most abstract form possible, so that it may be incorporated within the old pattern with a minimum of change, or else taking over as much of the culture in which the new idea is imbedded as possible.

The spread of the science of nutrition provides a good example here. If another people whose food and ways of cooking and eating are completely different from our own, who through all the centuries of their ancestral life have only considered food within a set of ideas quite different from those of the modern nutritionists, wish to learn to feed themselves and their children better, how can they most painlessly learn what we know? There seem to be two answers. Some of them can become highly trained nutritionists, able to analyze foods into their constituents and able to study the human beings who eat those foods and show where their diet is deficient. Then, armed with the knowledge of how to measure calories, assay proteins, spot vitamin deficiencies, they can readjust the food patterns of their traditional culture so that the missing ingredients are preserved or provided. This may have to be done by making a brew of pine needles or a jam of rose haws to provide the vitamin C which we get from oranges, or by feeding infants the water in which vegetables are cooked in place of cow's milk. If the adjustment is to be done this way, it is very important that none of our special ideas about kinds of food, number of meals, ways of cooking, should be included in the training which the young nutritionist from Japan or Burma receives. Otherwise, the mixture of a few of our ideas—like the sacredness of three meals a day, or the insistence that cow's milk is the only proper substitute for mother's milk—may confuse the picture disastrously. They should have only the purest and most abstract principles, and then be left alone to work within their own way of life with their own kinds of food, their charcoal fires or clay pots, to develop their own way of meeting the aspiration which they now share with us—a self-conscious application of the principles of nutrition to the problem of keeping their people alive and well.

The other alternative is to share as much of our pattern as possible. While the old civilizations of the world with ancient institutions of markets, banking, credit, and government may have a pattern within which they can take the new principles of medicine and nutrition and industrialism and develop their own style within the coherencies of their old way of life—as, for example, Japan did to a great extent—the more primitive peoples do not have such a pattern on which to build. Once the buffalo is destroyed, the once open plains enclosed, the spear and bow and arrow rendered useless, or, on the other hand, any need for real relationship with high civilizations develops, the very primitive and some of the simplest peasant peoples of the world have to change. Neither their clothes nor their manners, their economic ideas nor their political habits, fit them to live in the modern world as they are. It is then up to those societies which already have invented ways of living with these modern inventions to share their patterns in entirety with the peoples who wish to have them.

So this study of the Manus suggests the great importance of whole patterns, that it is easier to shift from being a South Sea Islander to being a New Yorker—as I have seen Samoans do—than to shift from being a perfectly adjusted traditional South Sea Islander to a partly civilized, partly acculturated South Sea Islander, who has been given antiquated versions of our philosophy and politics, a few odds and ends of clothing and furniture, and bits and pieces of our economics. I used to marvel at an individual Samoan, accustomed at home to go about barefoot, clad (except on Sundays) in a loincloth, and to sit cross-legged on the ground, who would turn up in my office perfectly and comfortably dressed in Western clothes, speaking English with grace and style. Without fully realizing the importance of changing from one whole pattern to another whole pattern, I used to attribute the Samoans' successful adjustment to their sense of style and to some particular security in their character. These are indeed there, but making a total shift seems now to be even more important. The same sense of style, which made Samoans refuse to speak *broken* English and insist on their few words of English being perfect, guided them in handling this total shift.

If we realize that each human culture, like each language, is a whole, capable of accommodating within it the wide varieties of human temperament, and that learning another culture is like learning a second language, or a third or fourth, then we can see that if individuals or groups of people have to change—leave their island homes, their mountain valleys, their remote fishermen's coves, give up their shell money, their old joint families, their hand nets, change because they wish to share, and to have their children share, in the benefits the great civilizations have made possible for mankind—then it is most important that they should change from one whole pattern to another, not merely patch and botch the old way of life with corrugated iron or discarded tin cans, in political peonage in the great cities of the world. While it is dreadfully difficult to graft one foreign habit on a set of old habits, it is much easier and highly exhilarating to learn a whole new set of habits, each reinforcing the other as one moves—like a practised dancer learning a completely new dance—more human even than one was before, because one has been able to do one more complicated human thing, learn something completely new.

But is this not a lonely path, to leave behind all the familiar sounds and sights, tastes and smells of childhood, and proceed alone, however perfectly, into another world? Here another suggestion from the Manus experiment comes in. For the people of Peri are not lonely or disoriented. They remember their past; their only poor memory is for the period of the nineteen-thirties when they were sharing little bits and pieces of Christian civilization, a little literacy with nothing to read, reading without arithmetic, Christian valuation of human life but no modern preventive medicine to keep it going, money but no economic system which made money function efficiently, aspirations to be civilized which were blocked by the fact that the white men they met both refused to treat them as equals, as human beings, and to believe that they were capable of sharing in the white men's superior civilization. But once they had taken their own modernization in their own hands, redesigned their culture from top to bottom, asserted their full dignity as modern Manus, the continuity with their personalities as they had been developed in the past was not broken, nor were their relationships with each other destroyed.

For here we have another part of the secret of felicitous change. The people of Peri all changed together as a unit—parents, grandparents, and children—so that the old mesh of human relations could be rewoven into a new pattern from which no thread was missing. As living individuals remembering their old ways and their old relationships, they could move into a new kind of village, live in new kinds of houses, participate in a new form of democracy, with no man's hand against another, no child alienated from the self or from the others.

It cannot be claimed, of course, that because a whole village or many villages act together as a group they will necessarily act progressively. We have ample evidence from groups like the Amish and the Hutterites that a radically new system, once established as a whole pattern among a group of related families who then live as closed communities, may be able to resist any further changes, may become indeed a fortress against the introduction of any new ideas, far more strongly walled than communities which have undergone no such radical initial alteration. The effort that goes into consolidating a change, combined with the reinforcement which is given each member of the family, each family in the group, because all move in step, may prove to be such a powerful condition of conservatism that it may well be too expensive for the world to encourage. While the mental health of each member of an isolated Hutterite community or a kibbutz may be protected by this strong group spirit, the protection given may be of a sort that disqualifies the individual member of the community from participating in other societies, and that keeps the group in an archaic or static state of adjustment, dependent after a time on a defensive attitude toward the rest of the world.

Here we may consider for a moment the significance of the apocalyptic cult aspect of the Paliau movement. The Noise—the mystical cult which ran through Manus with its full and familiar paraphernalia of prophecy and fulfillment, apocalyptic hopes, a utopia to be immediately established on earth, accompanied by seizures and quakings—was only a familiar example of what has happened many times in the world, as religious or political cults have swept through a population, bringing people into step with one another. Sometimes these cults die out altogether, sometimes they survive as small isolated sects, and sometimes they become part of a movement that sweeps the world, as a great religion or a great new political ideal.

One of the conditions which seem to determine which direction will be taken is the extent to which the leadership of what is potentially a cult or a movement with potentiality for change is able to modify the new system to allow for change—change of all sorts, flexibility in ritual practice and dogma, flexibility in personality type or in membership requirements. When the pattern is too narrow in relation to the possibilities of human variations and existing patterns in the outside world, then the system can only maintain itself by extruding or executing those who rebel against the narrowness, and those who remain become narrower still. Those who do not believe are "read out of meeting," exiled, liquidated. If this means that there are practically no children born into the group who follow the dictates of their parents, then we may have the strange phenomenon of a new membership recruited in every generation. This is a practice not likely to make a small group of enthusiasts into a world-wide movement, although it may be accompanied by extraordinary missionary zeal as the elders, seeing their children defect, search actively for those who will "see the light" and replace the lost children. In other cases, birthright membership becomes virtually the only way in which new members are recruited and rules against marrying outside develop so that those who marry outsiders are expelled from the group, disgraced in the party, or mourned as dead by their orthodox relatives.

When the new religion or new political ideal or new technology retains its original zeal and enthusiasm, born of a sense of a completely new pattern, counterpointed to the rejected old pattern and reinforced by the individual rejection of the imperfections in the old pattern, and when it is also able to accommodate change without branding it as heresy or deviation or subversion, it does not have to perpetuate itself by excommunication, exile, or liquidation, or to protect itself by theories of racial superiority which prevent intermarriage with, or absorption of, other groups not yet fully integrated into the great embracing new dream. Then we have what we may call world religions, world political and technological systems, systems with potential universality, whose claim to universality is based on a willingness to share what they have obtained by revelation or invention with all members of the human race.

If there were a reliable and irreversible dichotomy between closed and open systems so that it would be possible to say this movement has settled down to being a cult and lost its appeal for all of mankind, or conversely, this great new religious or political ideal is now safely launched on a path of potential universality, the problems which confront mankind today would be considerably simplified. But we have seen over and over again in history, a religion or a political system grown great by its openness, its receptivity to all men, contract and perish in a new throe of cultish exclusiveness. Often, perhaps always, it is reactive to some rival universalistic movement which threatens its hoary institutions, corrupted, as all old working institutions must be, by compromise and survivals. So the Spanish Inquisition can be seen as such a reactive phase in a church which had based its original appeal on inclusiveness. The cultivation of many religious orders with very different emphases, combined with self-conscious adaptation of canon law to different national cultures of present-day Catholicism, can be seen as re-institutions of universality, while the skirmishes fought by the Church over such issues as birth control fall again into the pattern of reactivity—the choice of battle lines which separate the sheep from the goats, on earth as they will in Heaven, and tend to preclude any accommodation which will include new groups of people within changing and flexible forms.

In similar fashion Western democracy has gone through various phases of being a pattern of life which could be shared with all mankind, and one which becomes narrowly exclusive and reactive. The France that gave the world the ideals of the French Revolution settled back into a society which equated "civilization" with "French culture," willing to welcome cultural immigrants and even to send out cultural missionaries, but whose members find it so difficult to think of any other culture as comparable with the French that the very phrase "patters of culture" has proved untranslatable. The British, who invented the British

Commonwealth and have been the great modern exponents of a reign of Law and Justice, permitted the universality of their system to be corrupted and jettisoned in many parts of the world by the snobberies, racial exclusiveness, and social, political, and economic exclusiveness of a colonial policy which treated "natives" as worthy of the Law but not of social equality. The fresh bright dreams of a system based on a belief that all men are created equal and with an equal right to life, liberty, and the pursuit of happiness, on which our American democracy is based, have been compromised from the beginning by the existence of slavery, by a general political and snobbish intolerance toward people of other races, creeds, and nations, and by a tendency to interpret our political forms and the accidents of our particular way of life as having wider and more universal qualities than they have, so that we often insist, with the voice of the cultist preacher, that the rest of the world, if they would share our wealth and our know-how, must adopt our manners and morals.

This brief set of references to events in the West can, of course, be paralleled in the East in terms which Western readers find it harder to grasp in outline because of the unfamiliarity of the details. These ups and downs in universality of those religious and political movements which had already established their ability to appeal to the whole of mankind, have been matched in modern history against the particular problems presented by Communism, with its high potential for universality, grown cultist, doctrinaire, and essentially closed within the circumstances of the political developments in the Soviet Union, and the various forms of Fascism, all essentially reactions against universality.

In the last twenty-five years, we have had to contend with new and horrible inventions, in which the pretence of universality is used as a missionary device for a system which has developed a control based on liquidation of all dissent. So we are now faced with the aftermath of the Communist invention of the United Front, and with the implications of a policy of recruitment in which Communist parties in countries not yet within the system are permitted a wide range of variation, such as expressed beliefs in a supreme being, harmony of human relations, local nationalism, etc. In turn, there is a persistent danger that, spawned from the experiences of psychological warfare which were developed during World War II, tutored and advised in method by deserters from the Communist apparatus, we may adopt some of the same devices—for example, offering the world health, education, and technical assistance, not because we have a system within which there are universal goods which we are willing to share with all members of the human race, but because it will provide us with allies against our enemies.

Thus it can be seen that throughout human history there has been a struggle between the proponents of closed and open systems, systems that could change their forms, accommodate to new ideas, retain the allegiance of new generations within them rather than goad them into rebellion or desertion, systems that welcomed the ideas, the questions, and the members of other systems, and those contrasting systems which hardened into exclusiveness and conservatism, so that wars of conquest, the rack, the ritual trial, the war on unbelievers in which one attained merit by killing them, became their destructive methods of self-perpetuation.

All this was true before the world was one, in the days when even the great civilizations of the world had no contact with each other—as in pre-Columbian days of the great civilizations of the Old and New World and in the days when the techniques which made rapid sharing of ideas and invention impossible. Today, when the fall of a cabinet, the death of a political leader, the capture of a spy, the echo of a prophecy of war, goes around the world in a matter of hours, this issue of universality has become a paramount one. Can we develop a system in which the belief in mankind is combined with a willingness to share, actively, all that we know about ways of increasing human dignity, in which we can offer to all those peoples of the world who want the dignity of law, the rewards of scientific research, the methods of government by consent of the governed, *all* of what we have, holding back nothing they wish to take, forcing nothing upon them which they do not wish to have, exacting no promises of aid or partisanship, binding them into no closed system, however closed the system against which we ourselves are battling? This is one of the great ethical issues of our age—perhaps *the* great ethical issue of our age. To meet this challenge, we need to know the conditions under which other men, if they wish to use our patterns, can use them safely and wisely.

This Manus record is presented as the material which the research scientist owes to the society of which he is a member, carefully collected materials on which considered decisions may be based. It is the story of a handful of men who twenty-five years ago seemed destined to live a life of anonymity and lack of wide significance. A series of historical accidents has transformed them—as any small group of individuals may be transformed within our reverberating world—into a group with world significance.

The Manus experiment itself is unique. The Manus were a people most favourably inclined toward change, conscious that cultural forms differed and could be changed, infused by their upbringing with an aspiration congruent with the more universal and humane forms of Western democracy, with the rare accident of a very gifted leader, and the unique experience of having a million men, members of a modern society intent on their own affairs, enact a large part of the pattern of Western democracy before their eyes. Whether they themselves survive, whether this unique attempt to move thousands of years in twenty-five is doomed, depends on that fantastic interlocking of world events which makes the existence of any individual who ties his identity to anything less than the human race hazardous in the extreme. The Manus have entered the modern world. As Manus, their humanity is dependent upon the preservation, in some sort of identified wholeness, of the small communities which they have built. In this respect they do not yet belong, nor do most members of the human race, to the age of the air, when the world becomes one great highway, and in any inn along the way there must be room and welcome for each and every guest.

notes to chapters

Chapter 1 Introduction

1. It was to answer such questions that the manual on *Cultural Patterns and Technical Change* was prepared by the World Federation for Mental Health for Unesco. *Cultural Patterns and Technical Change*, Unesco, Paris (1953) (reprinted as a Mentor Book by The New American Library, New York, 1955).
2. Huxley, Aldous, *Doors of Perception*, Harper & Bros., New York (1954).
3. Calas, Nicolas, "The Rose and the Revolver," *Yale French Studies*, Vol. 1 (1948), pp. 1–6.
4. Freud, Sigmund, *Collected Papers*, Vols. I–IV, Hogarth Press, London (1924–1925); Haddon, Alfred C. (ed.), *Reports of the Cambridge Anthropological Expedition to Torres Straits*, Vols. II, Pts. 1 and 2 and III–IV, University Press, Cambridge (1903–12); Piaget, Jean, *Language and Thought of the Child*, Harcourt, Brace, New York (1926); Mead, Margaret, "An Investigation of the Thought of Children, with Special Reference to Animism," *Journal of the Royal Anthropological Institute*, Vol. 62 (1932), pp. 173–90.
5. For records of this work see: Mead, Margaret, "The Application of Anthropological Techniques to Cross-National Communication," *Transactions of the New York Academy of Sciences*, Ser. 2, Vol. 10 (1947), pp. 133–52; Mead, Mar-

garet, "A Case History in Cross-National Communications," in *The Communication of Ideas*, Lyman Bryson (ed.), Harper & Bros., New York (1948), pp. 209–29; Mead, Margaret, *Soviet Attitudes Toward Authority*, McGraw-Hill, New York (1951), and William Morrow, New York (1955); Mead, Margaret, and Métraux, Rhoda (eds.), *The Study of Culture at a Distance*, University of Chicago Press, Chicago (1953); Mead, Margaret (ed.), *Cultural Patterns and Technical Change*, Unesco, Paris (1953) (reprinted as a Mentor Book by The New American Library, New York, 1955); Report of the Committee on Food Habits, 1941–1943, *The Problem of Changing Food Habits*, "National Research Council Bulletin," No. 108, Washington, D. C. (1943); Report of the Committee on Food Habits, *Manual for the Study of Food Habits*, "National Research Council Bulletin," No. 111, Washington, D. C. (1945).

6. Rabi, Isidor I., "Physicist Returns from the War," *Atlantic Monthly*, Vol. 176 (October, 1945), pp. 107–14.

7. Mead, Margaret, *Coming of Age in Samoa*, William Morrow, New York (1928) (reprinted in *From the South Seas*, William Morrow, New York, 1939).

8. For a summary of this problem, see Mead, Margaret, "Unique Possibilities of the Melting Pot," *Social Welfare Forum* (Official Proceedings of the 76th Annual Meeting of the National Conference of Social Work, June 12–17, 1949), Columbia University Press, New York (1950).

9. My interim field experience between 1929 and 1953 has been: Omaha Indian, 1930; Arapesh, 1931–32; Mundugumor, 1932; Tchambuli, 1933; Bali, 1936–38, and 1939; Iatmul, 1938; plus war and post-war research on contemporary cultures.

10. See Keesing, Felix, *Culture Change*, Stanford University Press, Stanford, Calif. (1953), for a general discussion and bibliography on the subject.

Chapter 2   Arrival in Peri, 1953

1. Soddy, Kenneth (ed.), *Mental Health and Infant Development*, 2 vols., Basic Books, New York (1955).
2. *Fundamental Education. Report of a Special Committee to the Preparatory Commission of Unesco*, Macmillan, New York (1947).
3. For a description of these *other* people see Fortune, Reo F., *Manus Religion*, American Philosophical Society, Philadelphia (1935), "Who's Who in Chapter Four," pp. 358–72.
4. In 1929, Pomat was described by everyone as the "adopted son" of Isali. In 1953, he spoke of her as his mother, and renewed investigation disclosed that he was indeed her son who had been adopted by other relatives and then re-adopted by Isali, his own mother.
5. There was a brief revival of mysticism in late 1953 and early 1954 which the leaders of the New Way succeeded in suppressing. This will be described by Theodore Schwartz in a forthcoming monograph on "The Paliau Movement in the Admiralties, 1946–54."

Chapter 3   Old Peri: An Economic Treadmill

1. Bateson, Gregory, and Mead, Margaret, *Balinese Character: A Photographic Analysis*, New York Academy of Science, Special Publication II, New York (1942).
2. *New Testament*, Matthew, Chap. 25, Verse 29.
3. For discussion of calendrical and non-calendrical cultures, see Mead, Margaret, "More Comprehensive Field Methods," *American Anthropologist*, Vol. 35 (1933), pp. 1–15.

Chapter 4   The Wider Context in 1928

1. For an account of Manus previous to World War I, see Nevermann, H., *Admiralitäts-Inseln*, Vol. III of "Ergebnisse der

Südsee—Expedition 1908–1910," G. Thilenius (ed.), Friederichsen & Co., Hamburg (1934), and Parkinson, R., *Dreissig Jahre in der Südsee*, Strecker und Schröeder, Stuttgart
(1907). For a general account of New Guinea between wars,
see Reed, S. W., *The Making of Modern New Guinea*, American Philosophical Society, Philadelphia (1943).

2. Bateson, Gregory, "Pidgin English and Cross-Cultural Communication," *Transactions of the New York Academy of Sciences*, Vol. 6, Ser. 2, No. 4 (1944), pp. 137–41, but see also
Hall, Robert, Jr., *Hands Off Pidgin English!* Pacific Publications Pty. Ltd., Sydney (1955), with its stress on the usefulness
of Pidgin English—Neo-Melanesian—for two-way communication today.

3. Fortune, Reo F., *Manus Religion*, American Philosophical
Society, Philadelphia (1935), p. 246.

4. Fortune, Reo F., *Manus Religion* (1935), pp. 28–29.

5. Fortune, Reo F., *Manus Religion* (1935), pp. 358–72.

6. Mead, Margaret, "Americanization in Samoa," *American Mer-
cury*, Vol. 16 (1929), pp. 264–70.

7. Mead, Margaret, *The Changing Culture of an Indian Tribe*,
Columbia University Press, New York (1932).

8. Williams, Frances E., *Orokaiva Magic*, Oxford University
Press, London (1928).

9. Mooney, James, *The Ghost-Dance Religion and the Sioux Out-
break of 1890*, "Fourteenth Annual Report of the Bureau of
American Ethnology," Pt. 2, pp. 641–1136, Washington,
D. C. (1896).

Chapter 5   Yesterday's Children Seen Today

1. Weber, Max, *The Protestant Ethic and the Spirit of Capitalism*,
Allen and Unwin, London (1930).

2. Frank, Lawrence K., *Society as the Patient*, Rutgers University
Press, New Brunswick (1948).

3. Malinowski, Bronislaw, *Sex and Repression in Savage Society*,
Harcourt, Brace, New York (1927).

4. Mead, Margaret, *Coming of Age in Samoa*, William Morrow, New York (1928).

5. Lasswell, Harold D., *Psychopathology and Politics*, University of Chicago Press, Chicago (1930).

6. Benedict, Ruth, "Psychological Types in the Cultures of the Southwest," in *Proceedings, Twenty-third International Congress of Americanists* (1928), pp. 572–81 (reprinted in *Readings in Social Psychology*, Theodore M. Newcomb, Eugene L. Hartley, *et al.*, eds., Henry Holt, New York, 1947, pp. 14–23).

7. Mead, Margaret, *Social Organization of Manua*, Bernice P. Bishop Museum, Honolulu (1930), pp. 80–86.

8. I owe my own increasing clarity in the field to contact, particularly with John Dollard, at the Harvard Seminar, 1934, directed by Lawrence K. Frank, to conversations with Erich Fromm, and to the work on *Co-operation and Competition* done under the auspices of the Social Science Research Council, 1934–35. Dollard, John, *Caste and Class in a Southern Town*, Yale University Press, New Haven (1937); Dollard, John, *Criteria for the Life History*, Yale University Press, New Haven (1935) (reprinted by Peter Smith, New York, 1949); Fromm, Erich, *Escape from Freedom*, Farrar and Rinehart, New York (1941); Mead, Margaret (ed.), *Co-operation and Competition*, McGraw-Hill, New York (1937).

9. Junod, Henri A., *Life of a South African Tribe*, 2 vols., Imprimerie Attinger Frères, Neuchatel, Switzerland (1912–1913); Macmillan and Co., Ltd., London (1927).

Chapter 6   Roots of Change in Old Peri

1. Benedict, Ruth, "Psychological Types in the Cultures of the Southwest," in *Proceedings, Twenty-third International Congress of Americanists* (1928), pp. 572–81 (reprinted in *Readings in Social Psychology*, Theodore M. Newcomb, Eugene L. Hartley, *et al.*, eds., Henry Holt, New York, 1947, pp. 14–23).

2. Sachs, Hans, "The Delay of the Machine Age," *Psychoanalytic Quarterly*, Vol. 2 (1933), pp. 404–24.

3. Gorer, Geoffrey, *Himalayan Village*, Michael Joseph, London (1938); and Mead, Margaret, "Review of *Himalayan Village*, by Geoffrey Gorer," *Oceania*, Vol. 9 (1939), pp. 344–53.

4. Myrdal, Gunnar, *et al.*, *The American Dilemma*, Harper & Bros., New York (1944).

5. Mead, Margaret, *Male and Female*, William Morrow, New York (1949), Appendix II, "The Ethics of Insight-Giving," pp. 431–50.

6. Preface to my *Growing Up in New Guinea*, Mentor edition, published by The New American Library, New York (1953).

Chapter 7   The Unforeseeable: The Coming of the American Army

1. Berndt, Ronald M., "A Cargo Cult Movement in the Eastern Central Highlands of New Guinea," *Oceania*, Vol. 26 (1952–1953), pp. 40–65, 137–58, 202–34.

2. Gorer, Geoffrey, *The American People*, Norton, New York (1948), Chapter IX, "Lesser Breeds," pp. 220–46.

3. Feldt, Eric A., *Coastwatchers*, Oxford University Press, Toronto (1946).

4. Mair, Lucy P., *Australia and New Guinea*, Christophers, London (1948).

5. Read, K. E., "Effects of the Pacific War in the Markham Valley, New Guinea," *Oceania*, Vol. 18 (1947), pp. 95–116.

6. Mead, Margaret, *Growing Up in New Guinea*, William Morrow, New York (1930) (reprinted in *From the South Seas*, William Morrow, New York, 1939; Mentor edition, The New American Library, New York, 1953; English editions: George Routledge, London, 1931, and Penguin Books, London, 1954). Mead, Margaret (ed.), *Co-operation and Competition among Primitive Peoples*, McGraw-Hill, New York (1937); Mead, Margaret, "Primitive Society," in *Planned Society, Yesterday, Today and Tomorrow*, T. F. Mackenzie (ed.), Prentice-Hall, New York (1937); *Natural History*, Cover of March-April issue, 1930; Parker, Lockie (ed.), *Story Parade*, Vol. 1,

No. 8 (1936), Cover, "The Children of New Guinea"; Benedict, Ruth, "Continuities and Discontinuities in Cultural Conditioning," *Psychiatry*, Vol. 1 (1938), pp. 161–67 (reprinted in *Personality in Nature, Society and Culture*, Clyde Kluckhohn and Henry A. Murray, eds., Knopf, New York, 1948, pp. 414–23, and in *Childhood in Contemporary Cultures*, Margaret Mead and Martha Wolfenstein, eds., University of Chicago Press, Chicago, 1955, pp. 21–30); Spitz, René A., "Frühkindliches Erleben und der Erwachsenenkultur bei dem Primitiven; Bemerkungen zu Margaret Mead *Growing Up in New Guinea*," *Imago*, Vol. 21 (1935), pp. 367–87.

7. Macaulay, Rose, *Orphan Island* (pocket ed.), Collins, London (1929).

8. O'Brien, Frederick, *White Shadows in the South Seas*, Garden City Publishers, New York (1928).

Chapter 8   Paliau: The Man Who Met the Hour

1. Schwartz, Theodore, "The Paliau Movement of the Admiralty Islands, 1946–54" (in press).

2. Mead, Margaret, "Living with the Natives of Melanesia," *Natural History*, Vol. 31 (1931), pp. 62–74.

Chapter 9   What Happened, 1946–1953

1. See Bibliography on "cargo cults" (issued by the South Pacific Commission) and the journal *Oceania* (passim, since the war).

2. Based on personal communications from those involved.

3. Father Lamers very kindly permitted me to consult these records, which are now in his keeping.

4. The origin of the word *besman* for the first village leader of the Paliau movement is obscure. Ted Schwartz thinks the best reconstruction is from *face-man*.

Chapter 11   The New Way

1. After I left it was finally decided to give up the communal ownership of sago palms, as people were showing too much irresponsibility by cutting young trees.
2. Compare Henry, Jules, "The Economics of Pilagá Food Distribution," *American Anthropologist*, Vol. 53 (1951), pp. 187–219; and Leet, Glenn, "They Did Not Wait for a Tractor," *The Survey*, New York (March, 1951).
3. Berndt, Ronald M., ms. in preparation which contains materials on the native courts in New Guinea.

Chapter 12   "And unto God the Things That Are God's"

1. *New Testament*, Matthew, Chap. 22, Verse 21.
2. For an extensive discussion of the old religious system, see Fortune, Reo F., *Manus Religion*, American Philosophical Society, Philadelphia (1935).
3. In Dr. Fortune's discussion of the working of the Manus Sir Ghost system, he laid great stress upon the way in which the cult maintained the strength of the patrilineal line against the matrilineal, and the accompanying checks and balances which were provided by such mechanics as the cursing power of the "son of the women's side." But any system which separates out one line at the expense of another, however this is then balanced by possible retribution if the power is exercised too hard, is also a system which interferes with the growth of community. Different New Guinea societies have attempted to overcome the divisiveness of clans in various ways, with dual organization, village men's houses in which each clan is subordinated to a whole, etc. The Manus situation is conspicuous because of the high moral tone of the household cult, the more humanly retributive tone of the clan cults, and the lack of any morality in the occasional affairs which involved all the men of the village.
4. Mead, Margaret, "Some Anthropological Considerations Con-

cerning Guilt," in *Feelings and Emotions, The Mooseheart Symposium*, Martin L. Reymert (ed.), McGraw-Hill, New York (1950), pp. 362–73.

5. *New Testament*, Matthew, Chap. 22, Verse 21.

Chapter 13   Rage, Rhythm, and Autonomy

1. See my *Growing Up in New Guinea*, William Morrow, New York (1930), p. 322, for the description of birth practices in 1928. See Appendix I in this book (pp. 479–81) for Lenora Schwartz' contrasting description of a Bunai Usiai delivery. There are very great differences in methods of delivery in different villages. In Bunai, interaction between the extreme passivity of the Usiai and the vigorousness of the Manus, combined with some of the pseudo-medical practices of the New Way, have produced a style which is more contrary to tradition and hard on the mother. Cf. Lenora Schwartz' forthcoming "Comparative Study of Manus and Usiai Infancy and Childhood."

2. Erikson, Erik H., "Growth and Crises of the 'Healthy Personality,' " in *Symposium on the Healthy Personality*, Milton J. E. Senn (ed.), Josiah Macy, Jr. Foundation, New York (1950), pp. 91–146.

3. Mead, Margaret, "Social Change and Cultural Surrogates," *Journal of Educational Sociology*, Vol. 14 (1940), pp. 92–110; Mead, Margaret, "Administrative Contributions to Democratic Character Formation at the Adolescent Level," *Journal of the National Association of Deans of Women*, Vol. 4 (1941), pp. 51–57; Mead, Margaret, "Collective Guilt," in *Proceedings of the International Conference in Medical Psychotherapy*, Vol. III, International Congress on Mental Health, London, and Columbia University Press, New York (1948), pp. 57–66.

4. Mead, Margaret, *And Keep Your Powder Dry*, William Morrow, New York (1942).

Chapter 14   New Working of Old Themes

1. For discussions see: Bateson, Gregory, "Morale and National Character," in *Civilian Morale*, Goodwin Watson (ed.), Houghton Mifflin (1942), pp. 71–91; Farber, Maurice L. (issue ed.), *New Directions in the Study of National Character*, Journal of Social Issues, Vol. 11 (1955); Gorer, Geoffrey, "The Concept of National Character," *Science News*, No. 18 (1950), pp. 105–22, Penguin Books, Harmondsworth, Middlesex; Mead, Margaret (ed.), *Co-operation and Competition Among Primitive Peoples*, McGraw-Hill, New York (1937); Mead, Margaret, "National Character," in *Anthropology Today*, Alfred L. Kroeber (ed.), University of Chicago Press, Chicago (1953), pp. 642–67; Mead, Margaret, and Métraux, Rhoda (eds.), *The Study of Culture at a Distance*, University of Chicago Press, Chicago (1953).
2. Grinnell, George B., *The Cheyenne Indians*, Vol. I, Yale University Press, New Haven (1923).
3. For contrast between Manus children's drawings and those of other primitive children, see Mead, Margaret, "Research on Primitive Children," in *Manual of Child Psychology*, Leonard Carmichael (ed.), John Wiley and Sons, New York (1954), pp. 735–80.
4. For the relationship between generation and social change, see Abel, Theodore F., *Why Hitler Came to Power*, Prentice-Hall, New York (1938); Mead, Margaret, *Soviet Attitudes Toward Authority*, McGraw-Hill, New York (1951), and William Morrow, New York (1955); and Tannenbaum, Frank, *Mexico, The Struggle for Peace and Bread*, Knopf, New York (1950).

Chapter 15   The Sunday That Was Straight

1. Mead, Margaret, "More Comprehensive Field Methods," *American Anthropologist*, Vol. 35 (1933), pp. 1–15.

Chapter 17 Reprieve—in Twentieth-Century Terms

1. I am indebted to Mr. Gaywood for lending me a paper on the early period after The Noise.
2. For a discussion of morale problems in a democracy, see Bateson, Gregory, and Mead, Margaret, "Principles of Morale Building," *Journal of Educational Sociology*, Vol. 15 (1941), pp. 206–20.
3. Savarese, Julia, "The Outing," in *Discovery*, No. 1 (1953), pp. 37–51, John W. Aldridge, Vance Bourjaily, and Chandler Brossard (eds.), Pocket Books, Inc., New York (1953).
4. Savarese, Julia, "The Outing," in *Discovery*, No. 1, p. 51.

Chapter 18 Implications for the World

1. Mayo, Katherine, *Mother India*, Blue Ribbon Books, New York (1930).
2. Mead, Margaret, "The Comparative Study of Culture and the Purposive Cultivation of Democratic Values," in *Science, Philosophy and Religion*, Second Symposium, Lyman Bryson and Louis Finkelstein (eds.), Conference on Science, Philosophy and Religion, New York (1942), pp. 56–69.
3. Mead, Margaret, "Applied Anthropology, 1955," in *Some Uses of Anthropology*, J. Casagrande and T. Gladwin (eds.), Washington Society of Anthropology Symposium (in press).
4. Mead, Margaret (ed.), *Cultural Patterns and Technical Change*, Unesco, Paris (1953) (reprinted as a Mentor Book by The New American Library, New York, 1955); Mead, Margaret, "Professional Problems of Education in Dependent Countries," *Journal of Negro Education*, Vol. 15 (1946), pp. 346–57.

notes to plates

Since photography has become a serious adjunct of ethnological recording, we have realized that a photograph is of very little value unless it is firmly placed within the matrix of recorded material so that we know when it was taken, by whom, of whom, and under what circumstances.* In *Balinese Character*† and in *Growth and Culture*,‡ we followed the procedure of giving complete details in the captions of each picture. However, these were both photographic studies.

In the present volume, designed for the non-technical reader, I have given only the details which are immediately relevant, the names of identified individuals, place, date, and photographer. It cannot be too strongly stressed that the identification of the photographer is as essential as identification of the subject. For these plates the choice was first made by subject matter—pictures of such and such an activity—and then from among the assembled

---

*Tax, Sol, *et al.* (eds.), *An Appraisal of Anthropology Today*, University of Chicago Press, Chicago (1953), pp. 191–217.

†Bateson, Gregory, and Mead, Margaret, *Balinese Character: A Photographic Analysis*, Special Publication of the New York Academy of Sciences, II, New York (1942).

‡Mead, Margaret, and Macgregor, Frances C., *Growth and Culture*, A Photographic Study of Balinese Children, Putnam, New York (1951).

prints. Not until the prints for each plate had been made did I know whose photographs had been selected. The scarcity of Lenora Schwartz' photographs is due primarily to my desire to keep her large collection of pictures of infants and children for her publication on the subject. One of the pictures, Fig. 2 on Plate IV, was taken as one of a series of repeats of pictures published in *Growing Up in New Guinea*, the subject gaily posing in the posture of the old picture. I have also noted whether the subject was posing, conscious of being photographed but not fully posing, or not consciously relating to the photograph. In Manus it can be said that no one was ever totally unconscious of the photographer.

Abbreviations for photographers are given by initials: RF (Reo Fortune), MM (Margaret Mead), TS (Theodore Schwartz), LS (Lenora Schwartz). Plate XI, Fig. 6, was taken by an unidentified American in 1946, and reproduced from a print.

PLATE	FIGURE	TITLE AND DATE	BY	POSED OR NOT
I	1	Stephan Posanget (Kapeli), Dec. 14, 1953	TS	posed
	2	Petrus Pomat, Dec. 15, 1953	TS	part of larger group
	3	John Kilipak, Dec. 14, 1953	TS	posed
	4	Johanis Lokus (Loponiu), Dec. 14, 1953	TS	posed
	5	Kapeli, Pomat, Yesa, Kilipak, Loponiu, 1928	MM	posed
II	1	New Peri, July 6, 1953	TS	
	2	Site of Old Peri, Sept. 10, 1953	TS	
	3	Old Peri, 1928	RF	
III	1	The volcano, from Patusi patrol post, June 30, 1953	MM	

PLATE	FIGURE	TITLE AND DATE	BY	POSED OR NOT
	2	Petrus Pomat speaking to the meeting about returning to Peri, June 30, 1953	MM	not consciously related
	3	People listening to Petrus, June 30, 1953	MM	not consciously related
IV	1	Josef Bopau, Dec. 14, 1953	TS	posed by old photograph
	2	Teresa Ngalowen and the "starved baby," Dec. 14, 1953	TS	conscious
	3	Bopau, 1929	MM	conscious
	4	Ngalowen, daughter of Ngamel, adopted by Pwisio; and Ponkob, her brother, 1929	RF	conscious
V	1	Canoes raising sails for canoe race off Peri, Dec. 13, 1953	TS	unrelated
	2	Canoe at sea, burial of Popwitch, Feb. 3, 1929	RF	unrelated
VI	1	Peri men, fishing with the *kau* [two-man net] off New Peri, Aug. 29, 1953	TS	unrelated
	2	Peri men, fishing with the *kau* [two-man net] between Peri and Patusi, 1929	MM	unrelated
VII	1	Feast given for Paliau by Kompo of Lahan, Bunai, European style.	TS	conscious, taken with strobe at night

PLATE	FIGURE	TITLE AND DATE	BY	POSED OR NOT
		Paliau's son John is seated beside him, Aug. 1, 1953		
	2	Peri child floating a model of a European pinnace, 1929	MM	unrelated
	3	Ponkob, son of Ngamel, seated on one of our chairs, playing at being a European, 1929	MM	conscious
VIII	1	Paliau shaking hands in Peri, July 27, 1953	TS	unrelated
	2	Pokanau and his son, Matawai. Pokanau is playing panpipes, 1929	MM	posed
	3	Pokanau making a speech at the farewell feast given me by the village, Dec. 12, 1953	TS	unrelated
IX	1	Stefan with his child, 3-year-old Selan, whom he carried a great deal, Oct. 16, 1953	MM	conscious
	2	Piwen bullying her mother, Molung, insisting on betel nut, 1929	RF	unrelated
X	1	Raphael Manuwai and his daughter, Nyawaseu, 5 months old, Sept. 10, 1953	TS	posed
	2	Raphael Manuwai dancing his daughter, Nyawaseu, 3 months old, July 24, 1953	LS	conscious

PLATE	FIGURE	TITLE AND DATE	BY	POSED OR NOT
	3	Own mother, Marketa Itong (rt.), and adopting mother, Tonia Ipau, walking Changkal, 8 months old, Sept. 5, 1953	LS	conscious
	4	Luwil Bomboi, with his daughter, Piwen (see Plate IX, Fig. 2), 1929	MM	posed
	5	Nyatchunu, wife of Manuwai's father (mother of Teresa Ngalowen, Plate IV, Fig. 2) and her child, Imol, 1929	MM	unrelated
XI	1	Marriage in Church, Peri, of Martin Paliau of Bunai and Ludwika Molung of Peri, June 13, 1954	TS	conscious, taken with strobe
	2	Karol Matawai, son of Talikai, and his wife, Aloisa Ngakakes, daughter of Poli, new form of marital behaviour, Dec. 14, 1953	TS	posed
	3	Pokanau and his fourth wife, Benedikta Nyaulu, new form of marital behaviour, Sept. 10, 1953	LS	posed
	4	Nyalen, bride of Malean (Alois Manoi as of 1953), bridal canoe, bride dressed in currency, Jan. 26, 1929	MM	unrelated

PLATE	FIGURE	TITLE AND DATE	BY	POSED OR NOT
	5	Betrothed Peri girl, Taliye, wearing calico head covering so she could hide from her in-laws, 1929	MM	conscious
	6	Wedding of Johanis Lokus (Loponiu) and Pipiana Lomot in Peri, 1946.* Catholic ceremony illustrating how the old economic forms of marriage, represented by the bride's finery, remained, 1946		
XII	1	Peri children in their play canoes. Topal I, Topal II, Bopau, son of Sori, and Kalowin, son of Luwil, 1929	RF	unrelated
	2	Children playing on land in New Peri, Oct. 10, 1953	MM	unrelated
	3	Temper tantrum in the lagoon at low tide, Old Peri, 1929	MM	unrelated
	4	Temper tantrum on land, child covered with sand, New Peri, Oct. 16, 1953	MM	unrelated
XIII	1	Nauna and his father, Ngamel, 1929	MM	posed

*Taken by an American soldier.

PLATE	FIGURE	TITLE AND DATE	BY	POSED OR NOT
	2	Michael Nauna and his youngest son, Pwochelau, aged 16 months, Nov. 10, 1953	MM	conscious
	3–4	Rosina Pwailep, wife of Michael Nauna, and their son, Pwochelau, aged 16 months, Nov. 10, 1953	MM	conscious
XIV	1	Karol Manoi, gaily relaxed, with "ukulele," Oct. 23, 1953	TS	conscious
	2	Karol Manoi, withdrawn to the sidelines with his youngest child, Talawan, aged two, July 12, 1953	MM	unrelated
	3	Karol Manoi, speaking at a meeting, rigid and anxious, Sept. 17, 1953	TS	unrelated
	4	Karol Manoi, watching a disapproved activity balefully, Aug. 1953	MM	unrelated
	5	Karol Manoi in set of farewell pictures, Dec. 14, 1953	TS	posed
XV	1	An economic exchange in Old Peri, Pomasa's *metcha* ["silver wedding" exchange] Left, Lomot, wife of Talikai, with Matawai (see Plate XI, Fig. 3) on her back;	RF	unrelated

PLATE	FIGURE	TITLE AND DATE	BY	POSED OR NOT
		Pomasa (standing); Isopwai, his wife, with basket on head; parents of Simeon Kutan; Molung Talawan, a daughter of Isali; Isali and Kemwai (standing), Feb. 26, 1929		
	2	Tax collection, Peri, with James Landman standing behind table, Paliau keeping records, May 11, 1954	TS	unrelated
	3	Women ready to pay taxes, May 11, 1954	TS	unrelated
XVI	1	Small girls making fishing baskets, New Peri, July 27, 1953	MM	unrelated
	2	Ponowan and his son's son, Kanawi, working together, he with the new axe, the child with the old adze, July 27, 1953	MM	unrelated
	3	Peranis Cholai, the schoolteacher of New Peri, clerk of the South Coast council, gay at the meeting over the runaway schoolboys, Nov. 5, 1953	MM	unrelated
	4	Peranis Cholai, stiff in the New Way, Dec. 14, 1955	TS	posing

PLATE	FIGURE	TITLE AND DATE	BY	POSED OR NOT
	5	Peranis Cholai in the empty church, preparing an arithmetic lesson on a homemade blackboard, Sept. 18, 1953	TS	unrelated
	6	The school children of New Peri with their teacher, Peranis Cholai, and myself, in a picture which they had asked to have taken to send to my daughter with whom the older ones had corresponded, Dec. 13, 1953	TS	posed

FIGURES

Map.	This map is based on, and is a modification of, the map published on p. 193 of *Kinship in the Admiralties*, by Margaret Mead, American Museum of Natural History, *Anthropological Papers*, Vol. 34, Pt. II (1934).
Village Diagram of New Peri.	This is a tracing of the map which I made on the spot, after making measurements of houses and distances. Drawn July, 1953.
Figures in Appendix I.	Tracings of the sketches that Lenora Schwartz drew on the spot to illustrate posture. August 28, 1953.

appendix i

Methods Used in This Study

I have experimented in the past with various sorts of complicated methodological appendices in books designed for the general reader and found them unsatisfactory; the technical reader ignores them and the general reader does not need them. Publishing a schematic census of an entire tribe, as I did in *The Changing Culture of an Indian Tribe*, using up fourteen good pages, or the replies of my entire sample to a long check list, as I did in *Coming of Age in Samoa*, or a list of householders, as I did in *Growing Up in New Guinea*, seems always to fall between two stools.

For this book, I am simply going to summarize the methods used and give such illustrations as seem necessary for clarity, without providing long lists which no one will read or use.

This was a restudy, and a restudy as such is only as good as the original work. This point is often obscured when the restudy is made by someone else, as the new field worker luxuriates in commenting on his predecessor's deficiencies. When one has done both the jobs oneself, one has no such temptation, and it is easier to measure improvements in technique and theory and at the same time stay within the limits of the earlier material. In the body of this book, I have discussed in Chapters V and VI how our advances in theory have made it possible to reinterpret the materials of twenty-five years ago. It remains to describe those materials and the methods by which they were collected.

When we went to the Admiralties in 1928, I had already studied the adolescent girl in Samoa and Dr. Fortune had made his study of sorcery in Dobu. I had worked on a limited series of individuals of one age and sex placed within their household structure in three adjacent villages. He had had to derive the whole pattern of a complicated culture from a tiny little population on the island of Tewara. He had learned to get maximum information from events, past and present; I had learned to work by case studies and the use of home-made projective tests. (The phrase *projective tests* did not yet exist.) The methods used in Manus grew out of these two field-work techniques, plus the standard methods then in use.

The old study was firmly based in the village of Peri, and other villages were visited only to attend feasts or to collect specimens for the museum. The single community, with its relationship to the other ten Manus villages known, was studied as a microcosm of the whole Manus group. The village was mapped, a household census was taken, and the genealogical ties of each person were worked out. All events in the village—illnesses, quarrels which had extra-household repercussions, séances, payments of expiation to ghosts, exorcisms, *rite de passage* ceremonies— were carefully observed and recorded. In *Manus Religion* by Dr. Fortune, there is a detailed record of the interlocking of these events over a period of four months, each actor in each event placed in time and space in relation to each other.

The texts of speeches and séances were not taken down at the time but were redictated by Pokanau (the intellectual of the village, still alive in 1953; see Plate VIII) and analyzed for esoteric terms, obscure linguistic forms, and sociological and cultural allusions. Descriptions of behaviour at a séance, or during a feast, ran to half a dozen pages. In the samples that follow, words in brackets are explanations added in 1955:

FROM MY NOTES

Page 1   January 31, 1929

Lulu Visit (for collecting mortuary contributions)

Woman wore a headdress made of slender coconut leaf-
lets, forming a screen and bordered with beads, a band of black
stuff, probably *arit* [paraminium nut], around her breast,
above the breasts, supported by two straps of the same material
running over shoulders, all possible armlets and leglets and
earrings, a rope belt, two *wanke* [chains] done up in undyed
kop [native string], and two baskets beaded in front, pandanus
sacks containing human hair, one basket on each hip. No
painting. Simply entered a house and sat down. The woman of
the house went through a box; meanwhile the *lulu* visitor sat,
played with the children, the people of the house took her children and played with them. Her little boy tied a string to the
fetish (a desultory unrebuked play), a baby had fallen off the
back of the *paleal* [house platform], and everyone, including
the visitor and her children, but not the rummaging mistress
of the house, rushed back to look at the baby which was being
cuddled by the fire. Nothing formal occurred. The wife of
Manawei's father [Plate X, Fig. 5] gave the woman a piece of
cloth and an incomplete belt of beads. Paleao [Josef Lalinge]
threw at her a stick of tobacco and an equally incomplete belt
of beads which Sain had dug out of a chest. Bonyalo and Manawei poled her to Manawei's* house and Bonyalo joined her in
Lalinge's house. To these she gave the tobacco she received
(first Bonyalo refused, second took it, and Bomboi took a stick
[see Plate X], although he hadn't poled her). They said she
gave them because they went with her. Nothing formal was

---

*Raphael Manuwai of New Peri.

said on either visit except Manawei's mother said: "I have nothing good. This is but little."

(Then followed the Manus terminology for the mourning costume)

She is making this visit for her father's sake, because her father is dead, helping to collect death duties. Next day to house belong
Tchauwan—got cloth and tobacco
Tcholai—got cloth and tobacco

Polau, *papun* [father]
Paleao, *nat* [son]
Tchauwan, *pision* [brother]
Tcholai, *nat* [son]

There are several striking omissions in this account from one which I would do today. There is no indication whether this took an hour or a half-day. The emphasis is on the cultural structure of the event. Small details of behaviour on the part of the children or the other participants, if noted, were noted on separate slips, under headings like "fathers and children," detached from the context. The notes are brief and summary, and *lend themselves to no more analysis* than is given here in the actual write-up. There is no reference to photography, although actually we took two pictures of this woman [now called Teresa, mother of Michael Nauna] in her *lulu* attire. The event is placed in time, by date, and the relationships of every person are known, in terms of household, genealogy, and terminology used, but there is no fine detail out of which a subsequent study of rhythm, pace, sequence, or theme could be made.

If we look at Dr. Fortune's report of a séance in the same household, we find the same focus on the essence of the event.

FROM *MANUS RELIGION*, PP. 143–45

On January 21st, just as Ngamel [Plate XIII] was preparing to go to a ceremony at another village, his small son, Pakob [Ponkob, Plate VII], fell into a fit of unconsciousness from a malarial attack. There was general assembly, crying and stirring of bowls. I brought Pakob back to consciousness with spirits of ammonia. This relieved the tension. Ngamel's action was credited with the recovery, not the spirits of ammonia, of course. Ngamel had immediately made divination to see if the trouble was not that his Sir Ghost was angry at his having taken out hoarded tobacco to give away at the ceremony he was setting out for, instead of keeping it to pay for communal work towards building a house for his titular son, a project he had also in mind. The answer to his divination had been affirmative so he had hastily locked up the tobacco again.

On January 23rd the child fainted again, and the performance was repeated. The divination this time was that Ngamel had not made sufficient speed towards that house building. That evening a séance was held, Isole being the medium. The séance followed the lines that had preceded it earlier in the day.

"Go ask *nyame* (Ngamel's Sir Ghost) why he strikes this child."

"He strikes Pakob (the child) because you, Ngamel, and Pwiseu with you are slow in building a house for Manuai."*

(Manuai was the son of Pwiseu, brother of Ngamel. He had just returned from working for the white man. He had brought home two pigs as part of his wages. It was felt that a bachelor's house should be built for him, as he was too old to stay in his father's house. Prudery entered into this consideration, but the general euphemism was that some return should be made for his two pigs by his fathers, Pwiseu and Ngamel.)

---

*Raphael Manuwai of New Peri

Ngamel replied:

"So: *nyame*, restore my son to health and I can make it. Already today the posts were cut. Tomorrow I shall set up the marking sticks" (used to chart out the positions where the posts will be placed later).

nyame said:

"So. All the constabulary, and all the assistant constabulary and all the common people may buy rice from the traders, and you two may build a house and pay for the rice for all (feast to validate house building). And this coral rubble territory the house of the white man has gone by it (a new house of mine being built near Ngamel's house); it has gone to all the whites. One (coral rubble) up at Pontchal (the barrack) has gone also to all the whites. One (coral rubble territory) in Peri belongs to Korotan. And this new house let it go in the sea between the two coral rubble territories."

The fact that Ngamel had already attended to post cutting did not relieve him from the weight of the system. If a man is not working his child may fall ill because he is idle. If he is working his child may fall ill because he is not working at unprecedented speed. There is no real loop hole for a man harassed by ghosts or by Sir Ghosts. But he strains every effort towards speed in work after such a séance as the above.

The striking thing about this account is the starkness of the conversation, the absence of repetition. This was a function of: the way the material was collected, by dictation from a trained informant; the fact that it was taken down as a text in longhand; and, in general, as in the description of the *lulu* visit above, the lesser interest in fine-grain note-taking. Pokanau told me in 1953, when I had perfected a Neo-Melanesian typed shorthand which could go as fast as he could talk, that now he would always tell me everything that was said but that dictating in Manus, as he had done in 1928, when it was being taken down in longhand with pen or pencil, was too boring.

We may now compare a piece of my note-taking in 1953.

Page 1   AIE-mm August 11, 1953
(date typed)   (date it occurred)

Elisabeta Makes Court Against Raphael Manuwai's
Daughters, p. 1

[If there had been photographs they would be noted here.]

Raphael Manuwai (42) [Plate X]	cf. "RM's account of his marriages"	Segmentation of quarrelling
Elisabeta his young 4th wife	"Anna's wedding"	struggle
Monica (41) his step-daughter by former marriage and Litau, her three-year-old son	"Straightening ceremonies for Peranis Cholai and his Patusi relatives-in-law."	between old and new kin lines, court procedure, lack of warning, reconciliation style.
Anna Nyawaseu (43) his own daughter by former marriage		
Ngaoli (43) mother of the husband of Anna, Josef Tapas		

1   This morning at "the line" the announcement was made that
2 RM had a "court case" coming up. The present method of mak-
3 ing court means simply that someone goes to the "council" and
4 announces that they have a case to bring, or a letter is sent to the
5 "council" from some other place and the defendant's name is
6 called out at "the line." RM had no previous intimation that this
7 case was coming up and knew nothing of Canoe Episode no. 1.
8 He did know about a Canoe Episode no. 2. In between Quarrel
9 no. 1, which he knew nothing about because he was away in
10 Lorengau and Quarrel no. 2 which took place when he got back,
11 there had been a temporary *rapprochement* between Monica and

12 Elisabeta—they had all gone and worked sago together and then
13 Monica's request to borrow the small canoe reactivated Elisabeta's
14 anger over the previous misunderstanding with Anna.

15   General outline of plot. KM [Karol Manoi, Plate XIV] had
16 borrowed RM's canoe, and RM had told Anna and Josef to go
17 and get it when KM came back, but he had not told Elisabeta he
18 had told them this. Anna and Josef got the canoe, and the next
19 day went fishing in it. Elisabeta protested to Anna and Ngaoli
20 took Anna's part. (This is forbidden under NFF [The New
21 Way] law.) The second quarrel took place when Monica asked
22 to borrow the canoe and RM consented but Elisabeta refused.

23   The real theme is Elisabeta's relationship to her husband's
24 daughters, Monica and Lusia (Lusia is the adopted daughter of
25 his former wife) and his real daughters, Monica Nauna and
26 Anna. RM fancies himself in relation to all these dependent
27 women; he looks after them all with a fine expansiveness and
28 Elisabeta is jealous of them.

29   8:30   (A.M.)   RM and Anna on rail. KM writing on
30   table at his verandah (Karol Manoi, the
31   councillor Ngaoli, Monica, and Litau
32   on the verandah bed).
33   8:35   KM still writing. Monica going over
34   child [for scabs, etc.]. Elis. chewing
35   *mbue* [betel] comes and stands against
36   room wall.
37   8:36   KM asks me the date.
38   8:38   KM gives his interrogative: HMMM!
39   KM and Anna sym[metrically seated]
40   on rail. Ngaoli and Monica sym. on
41   ver[andah] bed. Elis. squats by room
42   wall.
43 KM: *Now today me like t liklik*
44   [Today I want to make a short speech]
45   *Today i no ll bl no 9 moon, 1953*
46   [Today is the 11th day of the 9th
47   month, 1953]

48	*All i got talk, all i come al court*
49	[Those who have problems, come to
50	court]
51	*Now alt. fashion you me sv finish*
52	[Now the whole (New) Way, you and I,
53	we understand]
54	*Now alt fashion you me hear em finish*
55	[Now the whole (New) Way, you and I
56	have heard]
57	*Now pt i nother f something*
58	[Before it was different]
59	*You f makim finish*
60	[The things which you did]
61	*Today i nother f something*
62	[Today it is something else]
63	*You me like work em*
64	[Which we want to deal with]
65	*i no bl cross*
66	[This is not a matter for quarrelling]
67	*i no bl talk more*
68	[This is not a matter for further argu-
69	ment]
70	*i no bl talktalkplenty more*
71	[This is not a matter for a lot of further
72	argument]
73	*i bl straightin*
74	[This is a matter of setting it straight]
75	*Time bl t true i court*
76	[The court is the time for telling the
77	truth]
78	*Court i no bl gammon*
79	[The court is not a place for lying]
80	*Court i no bl cross*
81	[The court is not a place for quarrelling]
82	*Court i no bl fight*
83	[The court is not a place for fighting]

84		*Court i bl straight em something wrong*
85		[The court is a place for straightening
86		out that which is wrong]
87		*Court i bl straight em something i wrong*
88		*finish*
89		[The court is a place for straightening
90		things that have really gone wrong]
91		*Ta bl straight em, no bl cross more*
92		[This talk now is to straighten things, to
93		have no more qaurrelling]
94		*No bl fight more, no bl gammon*
95		[It is not to be a matter of more fight-
96		ing, nor of lying]
97		*All road bl NFF you me sv finish*
98		[The New Way you and I understand]
99		*Alt man, mari, now pick you me alt sv finish*
100		[Men, women, and children, all of us
101		altogether understand it]
102		*Em liklik t bl me i go st ats*
103		[This little speech of mine goes like this]
104		*Now whosat i got t no l i talk em.*
105		[Now whoever has the first talk to make,
106		let him make it.]
107	8:41	Pause 15 sec. (Note. These pauses seem
108		very long. I would subjectively say at
109		least a minute. *Why* do they seem so
110		long, is it because everyone is *pausing*, or
111		because the rapid rate of speech in the
112		intervals when there is talk?)*
113		Elis. stands up.
114		KM: Cut your talk into bits so that MM
115		can write it down.

---

*By placing questions of this sort in the account, it is possible to trace back
to the source of a theoretical point.

116		Elis. smiles at me as I write.
117		KM repeats (M) [in Manus] that court
118		belongs to
119		straighten, talk true, *tundrun, pwen* [no
120		quarrelling]
121		*no paun pwen* [no fighting], etc.
122	8:44	Elis. stands.
123		Pause 35 sec. Elis. hangs by her arms
124		from the side of the room wall. Grins.
125		KM (M) repeats about *kaiye apalan* [the
126		ways of before]
127		At 55 sec. Anna interjects.
128		KM repeats (M) about not fighting and
129		telling the truth
130		At 1 min. 15 sec.
131		KM: *Whosat i got talk?*
132		[Well, who has something to say?]
133	8:46	Pause 35 sec.

In 1928, observations on behaviour were separated out from the main matrix of large events and written on small slips, with an immediate classificatory heading. Example (originally in ink):

older and younger children Mar 5 (1929)

Kilipak [Plate I] holds Ponkob [Plate VII], plays with him with rough, gusty affection, asks him the following questions and threatens him that he will throw him in the sea if he doesn't answer.
"What's your name?"
"What's this?" (his hand)
"Who sleeps in your house?"
ans.: "Nauna" [Plate XIII]
"Who else?"
ans.: "Lots of people."

"Call their names."

He names them all and under threats does it over to prove
to Pomat [Plate I] that he said them right.

I collected hundreds of these small behaviour sequences,
partly verbatim, partly translated, partly summarized, and without records of time beyond the date.

Then there was another level of analysis in which I used protective tests of a variety of improvised sorts, Japanese paper flowers
which opened in water, a mechanical mouse, an artificial snake, ink
blots, form recognition, adaptation to form of drawing paper, etc.
In most of these I kept the protocols, but in one case I ranked the
children, recorded their rank order, and threw the details away,
which makes them quite useless today. The ink blots, suggested by
the remark in the work of an English psychologist that "ink blots are
a good test of the imagination," provide us today with a base line
of popular responses to compare with Rorschach responses. But I
was content with one answer, so that when a child said, "cloud,"
I accepted it. It took the meticulous insistence on timing of
Rorschach administration to reveal that if given enough time—perhaps as long as twenty minutes—a Manus adult will identify other
forms within a form designated as a "cloud." So where for 1928, I
had a small copy book, with a home-made ink blot on each page,
containing a single response by each one of my central experimental
group of children and young adolescents, in 1953, Ted Schwartz'
Rorschach protocols averaged four typewritten ten pages and took
an average of one hour and twenty-four minutes for the free association, with the inquiry of about the same duration or slightly less.

The children's drawing collections are comparable, except that, as
I found I was demonstrating a negative point in 1928—the absence
of animistic drawing—I had to collect an enormous number. So I
have today 32,000 drawings, identified and dated, with lists of which
children were drawing together, and each child's explanation of what
he had drawn. These can be compared for style with drawings collected in 1953, and also provide material on the personality of individuals who are adults today. Additionally, we have today moving

pictures and stills of children drawing, photographs of their clay modelling and mobiles, the latter which they made today spontaneously but did not make in 1928. We have also added a whole battery of modern tests: TATs, Mosaics, Bender-Gestalts, Stewart Ring Puzzles, Gesell Infant Development Tests, Caligor Eight-card Redrawing Test, Minnesota Paper Form Board, which provide material on the psychology of the Manus today. In 1928, we took about 300 still photographs, all of which had to be developed the same day, and no movies or tape recordings. In 1953, we took approximately 20,000 still photographs, 299 100-foot rolls of cine film, and 100 half-hour tape recordings. We also made, with a specially constructed turntable, somatotype photographs of a sample of the Peri population.

When it comes to verbatim life-history materials, or interviews with informants about the culture, the convention twenty-five years ago was to take texts in the native language in a way which slowed up the informant so much that both spontaneity and linguistic accuracy on matters such as repetition and tempo were sacrificed. In contrast to such texts—of myths, speeches, séances—of which Dr. Fortune recorded a considerable body, discussions with informants were recorded in *ratio obliqua*.

So on a page headed, "*palits* [ghosts] and Social organization," February 17, 1929, my notes run:

Interview with Bonyalo
Anent spirits and the bringing up of children

Children are not gammoned about child birth or death, but they are not told things young if they do not know them. If Ponkob asked what a *garamut* [slit gong beating] made for a *bris* [sex offence], he would be told, oh nothing, by and by Kiap [government officer] he come . . . Children are lied to to keep them at home and to make them go to bed. Popoli is scared about the belly of the Lotja man and about Ngaleap's leg . . . Children are not very afraid of *palits* in the daytime, only at night . . . As they grow older they learn to be afraid of *palits* . . . Furthermore, *mary palits* [female ghosts] do not go about in the *namel* [middle] and as a result women are less subject to attack than men. The reason that the *pinpalit* [female ghost], the daughter of Tunu, fought Alupwai was because she could go safely to the house of her father and the *moen palit* [Sir Ghost] could not be angry because she was the true daughter of the house . . .

With Tultul *of Pontjal*

Some people keep the skulls and the bones of women if they wish to, particularly if the woman has young children the *pinpalit* is called upon to look after them. A *pinpalit* can't stay in the house of her widower but she can marry his *moen palit* [Sir Ghost] and stay. The skull, if kept, is always kept by the kin group.

With Kukerai *[Pokanau]*

A woman can only harm her very young child. If a *pinpalit* leaves a young child she will often have compassion on its motherless state and call it to her. A man however does not fear his mother, but his father's sister can punish him.

This type of interview, in which gist and sequence are preserved but the actual complete text is never given, can be compared with the type of recording that is made possible by a typewritten shorthand of Neo-Melanesian.

From JK's *[John Kilipak] life history.*
Sept. 17, 1953
lines 41–46

(first memory) me liklik yet now me th long liklik canoe i small f me liklik yet me no can th al nother f sthg. Now me can

thinkim ta me like go shoot im fish some f time, alt me sleep
morning time me no can thinkin nother f something me must
th hurry up long kis em liklik spear bl me, liklik spear bl
punaro—you sv long punaro (bow and arrow) me f wk em
somf now Pomat i wk em f now master i kis em, etc.

Ted Schwartz has prepared the following samples of the
methods that made it possible for him to take down verbatim
interviews at high speed.

These abbreviations have no consistent linguistic relationship
to Pidgin English; some are derived from phonetic spelling of
the Neo-Melanesian form, some from orthographic form of
the English cognate. The purpose was to find abbreviated
forms of the most frequently recurrent Neo-Melanesian words,
such that each abbreviation was often the first and most rapidly occurring association to the Neo-Melanesian for each
transcriber. Thus we each had our own "most natural" abbreviations and used them rather inconsistently, sometimes writing the word in full, sometimes shortening, depending on the
informants' rate of verbal output. Most informants fell instantly
into step with our typing speeds, dictating as one experienced
in working with a stenographer.

TS *Talk with JK about Lokus' version of Noise,* Feb. 2, 1954.

JK's account of Noise supplementing other interviews with
him. This was offered when I asked him about details in Lokus'
version.

Alr, na lg time mf st lg Ndropwa, Noise i kisim fin Peri.
Alr, na all Peri i thth als, all i harim disf t bl Noise
em i km up als lg hap bl Tawi. na i ks (kisim) Patusi,
na hap bl em i km ks Peri now. Now all Peri i harim als.

ym mas tromwe algt smtg ins lg h bl ym, rausim now. i no gt l f ll smtg i k i st ins lg h. sp i gt smf ll smtg i st ins

lg l f h, bb i no gt smtg i km up lg em. sp ym tromwe algt smtg i st ins lg h bl ym i fin, alr na all papa bl ym, als all tumbuna lg ym all smf brata too, na all mama bl ym, all i k bringim all smtg bl ym i km now, ins lg h bl ym bl alipim disf smtg ym tromwe fin. na sp hs i no tromwe smtg lg h bl em, bb all disf man i lkm i gt smtg ins lg h, all i no k bringim all smtg ins lg h bl disf man. alr na all i tt fin lg em lg all arakeo, bh all i go lg h now, all i tromwe algt smtg ins lg h. algt suspan, algt spear, alg t pul, mojtjel, algt gdf laplap, algt bigf sheets bed. na I t algt smtg bl kamenda, na algt smtg bl gdf bilas bl all. all i tromwe algt.

The same passage written in a tentative phonemic spelling of the Manus Neo-Melanesian dialect. (This is not a suggested spelling for general use.)

orajt, na lonk tajm mipela stap lonk Nropwa, Najs i kisim pinis Peri. orajt, na ol Peri i tinktink olosejm, ol i arim tispela tok pilonk Najs em i kamap olosejm lonk ap pilonk Tawi. na i kisim Patusi, na ap pilonk em i kam kisim Peri naw. naw ol Peri i arim olosejm. jumi mas trowmwej oloketa samtink insajt lonk aws pilonk jumi, rawsim naw. i now kat wanpela liklik samtink i ken i stap insajt lonk aws. spows i kat sampela liklik samtink i stap insajt lonk wanpela aws, pajmpaj i now kat samtink i kamap lonk em. spows jumi trowmwej oloketa samtink i stap insajt lonk aws pilonk jumi i pinis, orajt, na ol papa pilonk jumi, olosejm ol tumpuna lonk jumi, ol sampela prata tu, na ol mama pilonk jumi, ol i ken prinkim ol samtink pilonk jumi i kam naw, insajt lonk aws pilonk jumi pilonk alipim tispela samtink jumi trowmwej pinis. na spos usat i now trowmwej samtink lonk aws pilonk em, pajmpaj ol tispela man i lukim i kat samtink insajt lonk aws, ol i now ken prinkim ol samtink insajt lonk aws pilonk tispela man. orajt, na ol i toktok pinis lonk em lonk ol arakeow, piajn ol i ko lonk aws naw, ol i trowmwej oloketa samtink insajt lonk aws. ologeta suspen, oloketa sipia, oloketa pul, mojtjel, oloketa kutpela laplap, oloketa pikpela sispet. na wantajm oloketa samtink pilonk kamenra, na oloketa samtink pilonk kutpela pilas pilonk ol. ol i trowmwej oloketa.

Translation of the same passage into English:

All right, when we were on Ndropwa, The Noise came to Peri. All right, the people of Peri thought like this, they had heard the talk about The Noise having arrived over at Tawi, then it took Patusi, then part of it came taking Peri now. Then the people of Peri heard as follows: "We must throw away everything inside our houses, throw it all out now. Nothing at all may remain within a house. If there is anything at all left inside a single house, later nothing will appear there. If we throw away everything that is inside our houses completely, all right. Then our fathers as well as our grandparents, some of our brothers also, and our mothers, they can bring everything that is ours to come now into our houses to replace all that we have thrown away. But whoever does not throw away something that is in his house, when these people [the returning dead] see that there is something inside the house, they will not bring everything into the house of this man." All right, when they had finished speaking of this on the *arakeo*, they went to their houses then. They threw away everything inside the houses— all of the basins, all spears, all canoe paddles, all *mojtjel* [mats or mat curtains], all good cloth, all of the large bed sheets, along with everything for carpentry and everything used for dressing up. All of this they threw away.

Lenora Schwartz concentrated on studies of infants, projective tests (Mosaics, TATs for adolescents and women, Gesells, Bender-Gestalts), child and adult art and fantasy, and studies of posture and gesture, movement and organization in groups. When photography was impracticable, she used rapid sketches to illustrate her notes. She worked in the mixed village of Bunai, with two Manus hamlets and four Usiai hamlets, and used the

Manus-Usiai contrast as a background for many of her insights. The following description of the first Usiai childbirth she witnessed illustrates some of the facets of her recording.

The First Usiai Birth
(Nahalengan's child)

Aug. 28, 1953

8:30   PHYSICAL ARRANGEMENT OF THE ROOM

The woman and the women helping her were in the house cook. The first contrast to that of the Manus birth was that the woman was placed more to the centre of the room rather than near some wall. The same casual air that was in the Manus births was here also. The pregnant woman leaned against her sister, who had her legs outstretched and kept the pregnant woman between her legs. Sitting in front of her with her legs up and hands around the pregnant woman was the sister of the father of the pregnant woman's husband. Around the room in sort of a circular order sat Njausai, a neighbour, Nabuway and her child, a *lapoon mary* [old woman] who was no relation to the pregnant woman, and two other neighbours, one with a child. Everyone was sitting quietly except for the pregnant

woman and the two women working on her. From the pregnant woman there was a soft moaning sound. And the only action was the seesaw movement back and forth of the woman in back of her and the one in front of her.

DESCRIPTION OF MOVEMENT

The pregnant woman had her feet outstretched and her head on her sister's shoulder. Her hands lay perfectly relaxed at her side. The woman in front kept her hands around the pregnant woman's sides, and would rub from the back to the side. The pregnant woman is sweating profusely, but does not twist her body or grab the woman in front of her, as the Manus women did. Instead, when she moans she just turns her head and puts her hands on the woman in front of her. Her hand is actually just placed and not clutching. The sound she makes is A—o,

 A—o, A—na na. Fingers and toes are relaxed, the fingers lay slightly spread over the arm of the first woman, and her legs lay quietly without any movement of the toes, no contracting and expanding. (Another element that existed in the Manus women.)

8:37   PAIN

As soon as the pain begins the woman in front pulls forward and the woman in back pulls backward. (Group is too interested in what I am doing to watch the woman in pain.) As the pain gets more intense the pregnant woman sits up straight.

HEAD MOVEMENTS

At the peak of the pain period the pregnant woman presses her head against the woman in front of her, and they both look down with their arms around each other. The back woman puts her head on the pregnant woman's shoulder and keeps her arms around the pregnant woman's abdomen. A third woman, nearing the age of a *lapoon*, comes over and starts to shake the pregnant woman's abdomen up and down. She uses an arbitrary time. She shakes it, then looks at the pregnant woman, then shakes it again. The woman in the back does not do as much work as the woman in the front.

In our use of both still and cine cameras, we worked as a team, one person taking notes, one using the Leica, and one using the cine, when all three of us worked in Peri. When Ted and Lenore Schwartz worked in Bunai, one recorded and the other photographed. Alone, I used the Leica and took notes. I only attempted using all three methods by myself once, and then decided that it was better to keep our major photographic work for a team approach, as each of the media requires full attention. For several major meetings we have moving pictures, stills, written notes, and tape recordings of the speeches.

In exploiting the early materials, I first checked my old household census and determined the whereabouts of each individual (if dead, when they had died) and each individual's marriage or subsequent marriages in the interval of my absence, etc. The diagram of New Peri gives an approximate picture of the relationship between the old and new material. Everyone under twenty-five had, of course, been born since I left. Two of the boys and five of the girls in my original group of children were dead. One of the adolescent boys whom I had known best had moved to another village, and one of the adolescent girls was dead. One of our chief male informants, Pokanau, and two of my female informants, Sain and Molong, were still alive, all old people.

On the 1928–29 trip I had not made any attempt to reconstruct the past since there was no way of checking whether it would be accurate or not. However, when I found out in 1953 with what extraordinary accuracy people could reproduce the events of twenty-five years ago (I had diagrams, photographs, and lists against which I could check their accounts), I decided to use Pokanau to build up earlier history of Peri and found that the village had been abandoned and rebuilt due to internal friction and warfare far more often than we would have thought from the synchronic materials collected in 1928. I made up a photographic memory test using pictures of persons still alive and of some who were dead and gave it to a selected sample of old and young. The extremely high accuracy of recall was the principal point demonstrated. In discussing the past, people were able to "set" themselves, so if I showed a photograph of a woman and child they would recognize the woman but would be likely to call the baby

by the name of the one of her five children which they remembered best as a baby; but if I then reminded them, "But *I* took
that picture," they would say without a moment's hesitation,
"Oh, then that baby is Sofia." They could tell me where everyone
lived during my stay in the village and which of the some five or
six of their names I had used or other people had commonly used
during my stay. I discuss in the body of the book their confusion
about how I had hurt my foot and how much poorer people's
memories were for the period of early Christianization than for
the pre-Christian and post-Paliau periods. They also forgot things
like the tests I had given them, intrusive meaningless details
from another world, while they could describe minutely every
time I had ever left the village, who went with me, what we said
on the way.

     It has become fashionable in recent years to use the term
"participant observer" rather loosely to mean either an observer
who becomes part of something, or someone who hides under a
formal participant role his real function as an investigator—
something which anthropologists have fortunately been debarred
from doing by their high visibility. On this field trip both households, mine in Peri and the Schwartz' in Bunai, became integral
parts of the community with definite functions assigned to us.
The most important of these were medical. We were furnished by
the Territory medical authorities with a stock of standard supplies in addition to the medicines we had brought, and, after it
became clear that the amount of illness among the seven hundred
people of Bunai was far more than the Schwartz' could cope
with—and get any work done—the medical authorities stationed
a trained medical assistant in Bunai, who would also come to Peri
to give NAB injections and for emergencies. One of my former
small boys, Kapeli, now Stefan, had worked as a medical orderly
on a plantation, and he acted as my clinic assistant in Peri so that
for the first time in my New Guinea field work I was freed from
the actual daily hours of dressing and bandaging sores. I had a
clinic every morning where diagnoses were made and where I
kept a record and supervised what was done. In addition, at any

hour of the day or night, emergency calls would come in, and there was seldom a day when I did not have at least one desperately sick patient to keep track of—to be sure that he took the next dose of sulfa and drank the necessary amount of soda bicarbonate, etc.

In Old Peri there were only two deaths during our entire six months; a visitor from Loitja died of an obscure blood infection, and little Popwitch, the son of Nane. In those days, children were spaced more widely apart and there were fewer older people alive, for there had been an epidemic of influenza a few years before. Furthermore, today, with the new drugs, many children are saved—to die later—who would once have perished as infants. During the six months in 1953, there were nine deaths in Peri. Even so the new antibiotics, the new antimalarial drugs, and sulfa made it possible to save many people, especially adults with pneumonia and children with malaria, who would certainly have died without them. All of this meant that I carried the continuous awareness of illness and death, not only as an anthropologist recording the events, but as the only therapist available.

The new drugs also made a great difference in our own efficiency. In 1928–29, I had malaria a third of the time, and, although I worked on the off days and often used even the bad days as an excuse for private conferences, still my efficiency was very much impaired. On this trip the three of us lost only about half a dozen days from very mild fever, as the suppressant drugs, aralen and primoquine, were effective throughout our stay. The interruptions of illness are somewhat balanced today by the interruptions from improved communications; air mail means more frequent mail with its reminders and demands from the outside world—in one case this involved a loss of five days when I was sent for to come into Lorengau for an important telephone call from New York, in which a market research company wanted to ask me if I smoked! (Air transport makes one other important difference; when something broke in the past we simply wrote it off. There was never time to get a camera or a typewriter repaired. Today, the possibility of getting spare parts or replacements, or sending things out for repairs at enormous cost, while it increases efficiency, also greatly increases the cost of an expedition.)

I had other roles besides that of medical therapist: informal retail storekeeper for people who needed a lamp mantle or batteries for a torch; banker for those who wanted to put their money where no one else would know how much was there; letter writer and recorder for a variety of purposes; and curriculum adviser for the school. The people treated me as a resource, using competently every skill or supply that I possessed. I was part of the web of borrowing, gift-giving, advice-asking in the village. I was called upon to intervene in circumstances like attempted suicide, or the birth of the anacephalic baby, or an individual's request to explain what had happened at a meeting, or requests for suggestions for some new course of civic action. I was regarded as completely "inside" as far as their identification of themselves with the social and political aspects of Paliau's movement was concerned. When it came to the mystical elements, my position was more mixed.

Two of the religious functionaries, Tomas and Lukas Banyalo, were Patusi men with no childhood memories to bind them to me. Furthermore, I felt that the mystical aspects of the "cargo cult" were still sufficiently alive—especially in the unstable—so that it was unwise to stir them up. It was not until the very end of my stay that I began collecting dreams and was told also the dreams which different people had had at the time of The Noise. This meant that the fringe effects of the small abortive revival of the "cargo cult" which started on Johnston Island, and which was to burst out strongly among the Usiai of Bunai after I left, were not confided to me. The constituted authorities in each village set themselves against the mystical revival, and it was the constituted authorities—themselves subversive from the point of view of the Administration—whose trust I had initially won. This illustrates some of the problems which beset the modern field worker who must operate in a series of cross-currents of great urgency which did not worry us twenty-five years ago. The factionalism in the village, between Patusi immigrants and Old Peri, between John Kilipak and Karol Manoi, etc., would have proved a much greater handicap to a new field worker than they did to me, returning to an old field.

We used diaries as a permanent index to events, for the placing of photographic sequences, etc., and I will give here two sample days.

Tuesday, September 1, 1953

Waked at 12:30 A.M. Paulus' baby in convulsions. Worked over it. (See "Two instances of intervention" and "Birth and death of Paulus' baby.") Went back to bed at 1:10 in my clothes. Called again at 2:40. Baby had died. Stayed at house all night. Burial completed at 5:30 A.M. Went to sleep. Mentun (Rene's) baby born sometime between 5 and 6, Stefan didn't call and (afraid I'd be cross because I wasn't called) claimed he couldn't wake me. Kutan away in Ndropwa, no court, Katerina [Kutan's estranged wife] still at John Kilipak's. JK went to the bush to get wood for the outrigger float. Cool and rainy. Straightened out the anacephalic case and suggested ways of dealing with birth, "Two instances of intervention."

Visited Ngalowen. 2 L(eicas) Fr(ame) 1 and 2 of Ilan (h. 26) with a headdress made of a garland of seeds and a cooky. Rainy all day, day divided between brief sick calls, writing up, visitors. See pages of Running Account.

Ponowan's anger against his sons.

RM's comment on *lapan* and *lau*.

Details of JK's relations to BK and SK to R.

Calabus [our medical assistant from Lorengau] brought over Cine and Cine film, Diary (LS) and notes, and returned to Bunai.

Mouk canoe left with Letter to J. Landman. Landmans on Rambutjon this week.

Rumours keep coming of when T. Edgell is expected.

Poor bung [market] because people were waiting for Kiap [patrol officer] and because of death of KM's maternal relative in Londru. KM goes.

Monica is now able to get about a little, moved to house of Alois Paneu (15).

Joanna's quarrel with BK.
Meeting to start the week. (Stefan's account.)
Paulus' wife's fainting spells, 2 visits. 1 tin apricots, small.
Bed with little expectation of staying there.

Saturday, September 5, 1953

Very small line. Lukas Banyalo came in with his child and I
started to explain to him about heredity and Ngalowen's baby.
Interrupted by arrival of T and L [Schwartz].

Loaded cameras and planned day's photography.

L's of Tjankal in Sepa's arms.

5 scenarios done in course of day.

Morning. "Francisca's Nyapin's house of four" L and C(ine).
"Children's parallel play in the sand" (at the beach) L and C.

Lunch

Discussion of state of the Brush [tape recorder], decision to
let it wait, new exposure metre not working, planned for testing,
time budgeting, discussed implications of this kind of apprentice-
ship versus individual first field trip, plans for subsequent study of
the Usiai, LS's observing Manus as background for writing up
Usiai in detail, keeping record of her major Manus observations.

TS plan to do linguistics of thought and feeling, then trial
tests. MM to try some trial TAT's, both types, LS to do some
trial Mosaics, TS to write Rawdon Smith [our adviser on tape
recorders] giving him all details, but telling him to do nothing
unless we cable, which we won't if the Magnecorder holds out.

Afternoon scenarios: "The two youngest children of Stefan
Posangat" L and C, "Adults and children on canoe by the sea" L
and C. Very pleased with the total of the day's work, and general
set of plans. TS and LS sailed away into the sunset—not literally,
no sail yet—with JK and Stefan going around the point with
them and I wished I had a canoe for the sheer beauty of it. Then
JK and Kutan came back and sat—looking empty-handed. [They
had been the carpenters of the new canoe.] Pokanau told TS that
if Kisekup wanted to get up a race it was his business and came
back later to arrange to borrow ten of the thirty sacsac [sago pack-

ets] I had ordered through him to go to Mouk. Put up the chimes today to test memory. Mouk canoe here, fastened by the wind. Ngalowen had a headache. Popwitch [whose mother Katerina had run away from her husband] cried all night. Gave LS seven pounds [petty cash]. Raphael Manuwai arrived back with small order I put in last night. Very very sleepy.

The great bulk of our fine-grain material, which is so incomparably more detailed than the material of twenty-five years ago, has not been used in this study except as a rough cross-check against generalizations based on the old methods. So, for example, my discussions of the characters of the various Peri people have, as background, the first inspection of their records on projective tests, but the test results have not yet been analyzed. Detailed analysis of the cine film, detailed analysis of the linguistic texts, the tape recording, the Gesell tests, etc., for all of which we do not have comparative material, will be analyzed and published in separate studies. The study has benefitted by the twenty-five years of experience in between and by the new skills and new theoretical orientations which could be brought to it by Ted and Lenore Schwartz.

appendix ii

A LOOK AT MANUS FROM THE LEVEL OF THE UNITED NATIONS

This book has been concerned with the fate of a few thousand people on one remote island in one trust territory. Each year the Trusteeship Council of the United Nations is concerned with matters of policy which are ultimately reflected in the lives of members of villages like Peri. As the members of the Trusteeship Council deliberate, they have before them massive reports compiled as schedules, schedules which make it possible for them to make comparative statements about different trust territories. In these reports, under such headings as "Status of the Territory and its Inhabitants," "International Peace and Security, and Maintenance of Law and Order," "Economic Advancement," "Political Advancement," and "Social Advancement," the district of Manus and the Baluan Council Area appear. By contrasting the level of analysis in this book and the level possible in the report, we may arrive at one aspect of the problem of comprehension and responsible action which faces the modern world.

DISTRICT ADMINISTRATION AS AT 30TH JUNE, 1954

District.	Land Area.	Head-quarters.
	Square Miles.	
North-East New Guinea (called the Mainland)—		
Eastern Highlands	6,900	Goroka
Western Highlands	9,600	Mount Hagen
Sepik	30,200	Wewak
Madang	10,800	Madang
Morobe	12,700	Lae
	70,200	
Bismarck Archipelago—		
New Britain	14,100	Rabaul
New Ireland	3,800	Kavieng
Manus	800	Lorengau
	18,700	
Solomon Islands—		
Bougainville	4,100	Sohano
	4,100	
Total Area of the Territory	93,000	

Report to the General Assembly of the United Nations on the Administration of the Territory of New Guinea from 1st July, 1953, to 30th June, 1954. Government Printing Office, Canberra, 1954, p. 21.

ENUMERATED AND ESTIMATED INDIGENOUS POPULATION AS AT 30TH JUNE, 1954.

District.	Enumerated.									Estimated.	Grand Total.
	Children.			Adults.			Persons.				
	Male.	Female.	Total.	Male.	Female.	Total.	Male.	Female.	Total.		
Manus	3,494	2,981	6,475	4,444	4,095	8,539	7,938	7,076	15,014	—	15,014

Ibid., p. 105.

NUMBER OF VILLAGE OFFICIALS AND COUNCILLORS AS AT 30TH JUNE, 1954.

District.	Luluais.	Tultuls.	Medical Tultuls.	Total Village Officials.	Councillors.		Total Village Officials and Councillors.
					Official.	Unofficial.	
Manus	93	71	54	218	23	—	241

Ibid., p. 128.

NATIVE WAR DAMAGE COMPENSATION: CLAIMS AND PAYMENTS DURING 1953–54 AND TOTAL AS AT 30TH JUNE, 1954.

District.	1953–54.		Total as at 30th June, 1954.	
	No. of Claims.	Amount Paid.	No. of Claims.	Amount Paid.
Manus	—	£ —	2,443	£ 43,210

Ibid., p. 128.

HOLDINGS OF 1 ACRE OR MORE USED FOR AGRICULTURAL OR PASTORAL PURPOSES, BY DISTRICT, AS AT 31ST MARCH, 1954.

District.	Area of District.	Holdings Being Used.	Land in Holdings Being Worked.						
			Land Tenure.		Total Area of Plantation or Farm.	Land Under Crops excluding Retired Crops.	Established Pastures.	Cleared Areas Not Under Crops or Established Pastures.	Balance of Holding.
			Owned by Administration.*	Alienated in Fee Simple.					
Manus	Acres. 512,000	No. 14	Acres. 2,727	Acres. 9,179	Acres. 11,906	Acres. 7,077	Acres. 64	Acres. 224	Acres. 4,541

Ibid., p. 172.
* Mainly land leased to operators engaged in rural activity.
Note.—Holdings unoccupied or unused are excluded Where two or more holdings are operated conjointly, they are enumerated as a single holding.

AREAS UNDER ADMINISTRATION CONTROL OR INFLUENCE AS AT 30TH JUNE, 1953 AND 1954.

(Area in Square Miles)

District.	Total Area.		Area Under Control.		Area Under Influence.		Area Under Partial Influence.		Area Penetrated by Patrols (Restricted Area).	
	1952–53.	1953–54.	1952–53.	1953–54.	1952–53.	1953–54.	1952–53.	1953–54.	1952–53.	1953–54.
Manus	800	800	800	800	—	—	—	—	—	—

Ibid., p. 127.

LIVESTOCK ON HOLDINGS BY DISTRICT AT 31ST MARCH, 1954.

District.	Cattle.	Sheep.	Horses.	Donkeys.	Mules.	Pigs.	Goats.	Poultry (Commercial Flocks).*	Beehives.
Manus	28	—	—	—	—	15	—	120	—

Ibid., p. 174.
* Excludes non-commercial flocks.
No information is available of the livestock owned by indigenous inhabitants which mainly comprises pigs and fowls.

The Baluan Council Area appears in tables like this:

COUNCILS PROCLAIMED PRIOR TO 30TH JUNE, 1954.

	No. of Villages in Council Area.	Population.	No. of Members.
New Britain District—			
Rabaul	18	6,000	19
Reimber	24	4,500	20
Livuan	18	3,400	20
Vunamami	27	4,200	24
Vunadadir-Toma-Nanga Nanga	27	6,013	24
Manus District—			
Baluan	30	4,487	32

Ibid., p. 23.

ANALYSIS OF ACTUAL EXPENDITURE ON PUBLIC SERVICES FOR FINANCIAL YEAR
ENDING 31ST DECEMBER, 1953.

Council.	Council Administration.	Medical and Sanitation.	Education.	Agriculture.	Forestry.	Roads and Bridges.	Water Supply.	Total Expenditure.
Baluan	£ 1,073	£ 175	£ 307	£ 2	£ —	£ —	£ —	£ 1,557
Total Expenditure for All Councils	18,328	3,413	3,122	1,896	3	279	480	27,521

Ibid., p. 25.

WORKS COMPLETED, IN HAND AND ESTIMATED FOR DURING PERIOD 1ST JULY, 1953–
30TH JUNE, 1954, BY NATIVE VILLAGE COUNCILS.

Council.	Completed during Period.	In Hand.	Estimated for.
Baluan	Teacher's house Purchase of saw bench and engine Purchase of radios Water pumps	School furniture Tools and workshop equipment Purchase of water pumps Well sinking	Purchase of Council site Council house furniture Tools and workshop equipment One aid post Well construction Boat purchase fund One native medical assistant's quarters Water supply School furniture Sawmill equipment Purchase of livestock Navigation markers Purchase of radio set

Ibid., p. 28.

DETAILS OF ESTIMATED TOTAL REVENUE AND EXPENDITURE FOR FINANCIAL YEAR
ENDING 31ST DECEMBER, 1954.

Details of Revenue and Expenditure.	Baluan.	Total for All Councils.
Revenue.	£	£
1. Council Tax	1,900	27,267
2. Fees and Rates—		
(a) Dog Licenses	5	5
3. General—		
(a) Vehicle hire	—	2,900
(b) Rice huller hire	—	30
(c) Interest on Reserve	—	50
(d) Contributions towards Market by other Councils	—	96
Total Recurrent Revenue	1,905	30,348
4. Non-Recurrent Revenue, Miscellaneous	—	15
Total Revenue	1,905	30,363 *
Capital Expenditure.		
Purchase of Council site	175	175
Council house furniture and fittings	50	459
Tools and workshop equipment	30	471
Underground tanks	—	730
School buildings	—	4,862
Tanks for schools	—	174
Aid post buildings	100	1,050
Motor vehicles	—	3,200
Land Settlement scheme	—	500
Well construction	10	710
Bicycles	—	160
Local Government Bulk Store	—	660
Boat purchase funds	250	350
Boat house	—	200
Council Rest houses	—	350
Women's Education Fund	—	200
Clerks' houses	—	500
Native medical assistants' quarters	55	805
Water supply	125	1,730
Tanks for clerks' houses	—	45
Rice huller	—	225
Tanks for aid posts	—	180
Wash basins for aid posts	—	80
Teachers' houses	—	250
School furniture and fittings	10	463

* Does not include balance from 1953 or cash equivalent of building stocks on
hand as at 1st January, 1954.

DETAILS OF ESTIMATED TOTAL REVENUE AND EXPENDITURE FOR FINANCIAL YEAR
ENDING 31ST DECEMBER, 1954 (*continued*).

Details of Revenue and Expenditure.	Baluan.	Total for All Councils.
Capital Expenditure.	£	£
Sawmill equipment	85	85
Purchase of livestock	74	154
Navigation markers	20	20
Radio sets	20	20
Capital Works Price Variation Funds	—	500
(1) Total Capital Works Expenditure	1,004	19,308
(2) Total Salaries, Wages and other Emoluments	942	9,754
(3) Total General Expenditure	685	10,157
Total Expenditure	2,631	39,219

Ibid., p. 27.

COUNCIL TAX RATES FOR 1954.

Council.	Adult Males over 21 Yrs.	Males, 17–21 years.	Females over 21 years.
	£	£	£
Reimber-Livuan	4	1	1
Vunamami	4	1	1
Vunadadir-Toma-Nanga Nanga	4	1	1
Rabaul	4	1	1
Baluan	4	4	1

Ibid., p. 25.

CO-OPERATIVE SOCIETIES FORMED DURING YEAR 7/1/53–6/30/54.

District.	No.
Madang	1
New Britain	20
Bougainville	1
Manus *	2

* Co-operative organization in this District is still in the early stages of development. Two societies are in existence marketing produce and operating retail stores. *Ibid., pp. 40–41.*

VEHICULAR ROADS AND BRIDLE PATHS.

District.	Bridle Paths.		Vehicular Roads.				
	Mileage at—		Mileage at—		Heavy Traffic.	Up to One Ton.	Jeep Traffic.
	6/30/53.	6/30/54.	6/30/53.	6/30/54.			
Manus	200	200	90	50	15	—	35

Ibid., p. 177.

ADMINISTRATION SCHOOLS BY DISTRICT AS AT 30TH JUNE, 1954.

| District. | Schools. | | | | Pupils. | | | | | Teachers. | | | |
	Euro-pean.	Asian.	Mixed Race.	Na-tive.	Total.	Euro-pean.	Asian.	Mixed Race.	Na-tive.	Total.	Euro-pean.	Asian.	Mixed Race.	Na-tive.	Total.
Manus	1	—	1	5	7	44	—	20	278	342	4	—	—	11	15

Ibid., p. 205.

ADMINISTRATION SCHOOLS — NATIVE — AS AT 30TH JUNE, 1954.

District.	Place.	Type of School.	Pupils.				Teachers.							Teachers Total.
			Males.	Females.	Total.	Age Groups.	European.			Native and Mixed Race.			
							Males.	Females.	Total.	Males.	Females.	Total.	
Manus	Lorengau	Central	56	—	56	14–18	1	—	1	4	—	4	5
	Liap	Area	36	18	54	8–14	—	—	—	2	—	2	2
	Baluan	Village Higher	71	46	117	8–14	—	—	—	3	—	3	3
	Aua	Village Higher	14	12	26	8–14	—	—	—	1	—	1	1
	Bipi	Village Higher	15	10	25	8–15	—	—	—	1	—	1	1

Ibid., p. 208.

RELIGIOUS MISSIONS OPERATING IN THE TERRITORY AS AT 30TH JUNE, 1954.*

Name of Mission.	Index Letter.	Head- quarters.	Districts of Operation.	No. of Non- Indigenous Mission- aries.	Estimated No. of Adherents.
Bismarck Archipel- ago Mission of Seventh Day Adventists	D	Rabaul	New Britain, New Ireland, Manus, Bougainville	59	6,700
Catholic Mission of the Most Sacred Heart of Jesus	G	Kokopo	New Britain, New Ireland, Manus	210	69,764
Evangelical Lu- theran Mission	J	Lorengau	Manus	5	3,000

Ibid., p. 222.
* We have included only those with missions on Manus.

ABSTRACTS OF ESTIMATES FOR FINANCIAL YEAR ENDING 31ST DECEMBER, 1954.

Baluan Council.

Revenue, 1954	£	£	*Expenditure*, 1954	£	£
Council Tax	1,900		Personal Emoluments	942	
Other Revenue	5		Other Charges	685	
			Capital Expenditure	1,004	
Total Revenue, 1954		1,905			
Balance from 1953		1,557	Total Expenditure		2,631
			Balance to 1955 (Reserve)		831
Total		3,462			
			Total		3,462

Ibid., p. 26.

NUMBER OF POSTS, ATTENDANTS AND KNOWN TREATMENTS FOR YEAR ENDING
30TH JUNE, 1954.

District.	No. of Aid Posts.		No. of Medical Attendants.		No. of Known Treatments.	
	1952–53.	1953–54.	1952–53.	1953–54.	1952–53.	1953–54.
Manus	16	16	22	23	4,594	8,213

Ibid., p. 74.

COMPARATIVE PERCENTAGES OF NATIVE HOSPITAL IN-PATIENTS TREATED FOR
DISTRICTS.

District.	1951–52.	1952–53.	1953–54.
Manus	1.9	1.9	1.9

Ibid., p. 77.

EUROPEAN STAFF: NUMBER BY DEPARTMENT AND DISTRICT OF EMPLOYMENT AS AT
30TH JUNE, 1954.

Department or Branch.	*Manus.*
Administrator	—
Government Secretary—	
Central Administration	—
Police and Prisons	1
Works	3
Public Service Commissioner	—
Health	4
District Services and Native Affairs	10
Treasury—	—
Stores and Transport	1
Government Printer	—
Postal Services	1
Telecommunications	1
Law—	—
Registrar-General	—
Supreme Court	—
Public Curator	—
Land Titles Commissioner	—
Education	6
Agriculture, Stock and Fisheries	1
Lands, Surveys and Mines	—
Forests	—
Customs and Marine	2
Total	30

Ibid., p. 125.

TOTAL ACREAGE OF PRINCIPAL CROPS BY DISTRICT, YEAR ENDED 31ST MARCH, 1954.

District.	Permanent Plantation Crops.*					Other.			
	Cacao.	Coffee.	Coconuts.	Rubber.	Tea.	Peanuts.	Rice.	Vegetables for Human Consumption †	Other Seasonal Field Crops.
Manus	—	—	7,080	—	—	1	2	9	9

Ibid., p. 173.
* Excludes retired crops.
† Includes tuber and root crops.

ANALYSIS OF ESTIMATED EXPENDITURE ON PUBLIC SERVICES FOR FINANCIAL YEAR
ENDING 31ST DECEMBER, 1954.

Council.	Council Administration.	Medical and Sanitation.	Education.	Agriculture.	Forestry.	Roads and Bridges.	Water Supply.	Total Expenditure.
	£	£	£	£	£	£	£	£
Baluan	(a) 1,860	349	183	84	—	20	135	(b) 2,631

Ibid., p. 26.

(a) Includes all expenditure not strictly chargeable to other services, e.g., Council Houses, furniture and fittings, Clerks' Houses, stationery, vehicles, transport running costs, cartage of materials, insurance and maintenance of buildings, celebrations, competitions, Councillors' allowances, wages of clerks, constables, carpenters, drivers, etc.

(b) Does not include Building Materials Price Variation Fund.

PATIENT DISTRIBUTION.

District.	Population.	Daily Average In-patients.	Total In-patients.	Total Out-patients.	Total Deaths.
Manus	15,014	76.50	1,539	1,049	20

Ibid., p. 78.

appendix iii ·

Technical Assistance Policies and the Future of the Manus

In the body of the book I have stressed the importance of a people choosing their own way, choosing it together, and making the transformation a rapid and total one. I have indicated and also I have stressed the enormous release of energy that occurs in the members of a less complex culture when the members of the more complex culture admit them into a shared world as full participants. I have not dealt with the tremendously complicated problem of how these changes are to be underwritten economically. The future of the Manus people depends not only upon events in the outer world far beyond their control but also on whether the course of economic development in New Guinea—itself dependent on trends in the modern world—is toward keeping people in their traditional village communities or developing a territory-wide labour force.

During the years between World Wars I and II, there was a strong emphasis in responsible circles on preserving village ties, protecting the native's land rights, and not permitting a system of indentured labour which would keep natives away from their villages for too many years or alienate them altogether. Within this protective system, a native of New Guinea had a future only as a member of his own small tribal community, sometimes a group only a few hundred strong. Despite the many attempts made to educate individuals, as teachers, medical assistants, clerks, there was actually no urban, mobile, New Guinea citizenship to which an educated man or woman could belong. Away from home, they were essentially in the custody of some individual enterprise, as an employee of a firm, a mission, or of a government department. Their social or ethnic identity, as opposed to their work identity, remained tied firmly to their own village, the village book. Of course, the responsible efforts to protect the native's rights in his own land intensified this parochialism. Much of the post-war labour legislation, with its cutting down on indenture time and requiring employers to support wives and children if the employee wishes to bring them, has had the same effect, because employers cannot afford to import or support such workers from a distance. Transportation costs by air have contributed to this effect, so that there is an increasing tendency in Manus for men to seek employment near home, with or without their wives, and for wages, or on commission. But this means that each island group is more dependent on its local resources for any sort of economic development.

The present situation in Manus highlights the problem as it exists in most of New Guinea and in many underdeveloped areas of the world. If a people are to accept the values of the modern world, modern education, health, political democracy, respect for the dignity of each individual, they can do this very well, if not best, by moving as communities—a whole village, or cluster of villages, or a district moving together to inaugurate, learn, and support a different way of life. But such a forward thrust means a new sense of ethnic identity, a new pride in the village or district or newly established little nation, and a corresponding willingness or even pressure for the whole population to remain *within these limits*. Emigration is frowned upon; immigration is likely to be suspect, and there may be strong pressures to increase the birth rate.

Development on a much wider scale, with a free labour supply, easy trade conditions, ease of travel, means the development of an urban population without village roots, a population which is much harder to motivate in the direction of any kind of culture change except economic change, with the principal incentive a monetary one. From the standpoint of dignified participation in a world culture, the picture of independent, self-supporting, responsible local communities is much more promising. But, without economic resources, the legitimate demands for a rising standard of living cannot be met, and costs of education, medical care, and government soar well beyond the resources of the group.

Stated concretely, in Manus at present the Australian air and naval bases supply a market for employment which may or may not continue, and which is grounded in the strategic, not the economic, position of Manus. The economic resource which the military bases have established is comparable to the resources of a backward country with large tourist appeal—an economic relationship which retains something limited and parasitical unless military service can be introduced. And what of other resources? The soil of Manus is exceedingly poor; the present plantations are old, dating from German times, and new plantations take many years to mature. There is fishing, and it would be possible with a good deal of expert help and encouragement to develop a fishing industry which might supply smoked fish for the district market and frozen fish which could be exported by air. To pay, this would mean organizing the community to service a fishing industry as they might supply labour to a factory, but with the easier involvement of women and children. If such an industry were built up with very careful government financing and expertise, there would still be the question of supporting population increase, on the one hand, and on the other, of how the people of Manus were to be further integrated in a wider system.

The sharpest alternative is provided by the development of a New-Guinea-wide economy, where even such an enterprise as fishing in Manus for commercial exploitation would be more centrally managed, whether the commercial initiative came from

Europeans, Asians, or New Guinea people with greater commercial experience. At the same time the school system would be organized so that the brighter children went on to higher schools and jobs elsewhere. The labour system would be such that, at least for all skilled jobs and possibly for the first period of contact between a tribal people and modern society, work at a distance would be the rule.

The development of city life, with freedom to travel, hotels, dwellings that can be rented, a press, a territory-wide language—which Neo-Melanesian makes possible—for press and radio, and territory-wide forms of political government, would begin to provide a framework for a New-Guinea-wide way of life. One would not lament the fact that many young people would leave Peri any more than one need lament the fact that young Americans or Englishmen leave their villages to play an economic and social role on the national or international scene. It would be possible for government to utilize the special energies of an awakened community like Manus, to give the necessary educational background for the development of a group of practical skilled engineers, for example. Just as today many of the most highly gifted of our scientists, technicians, statesmen, etc., come out of the most educated homes in small communities, especially the homes of ministers, so we might expect that small communities with especially high social motivation might become the breeding ground for the new intelligentsia which any new country needs.

Every institution which ties a small group like the members of the Baluan council into the wider world increases this possibility, while the break with world churches and preservation of local unaffiliated sects and the tendency to develop local, but extra-legal, adaptive institutions—like informal courts, banks, stores, etc.—decreases it. It seems very possible that economic moves which are strictly local in character, such as village councils and co-operatives, however sound they may be in underwriting immediate subsistence needs and immediate social improvements, may, unless rapidly supplemented with institutions capable of expanding, become in their turn narrowing and preventive of progress. New-found ethnic identity—as a member of the New Way South Coast Manus—is a step which, unless it is rapidly followed by other steps, means the establishment of groups which have to be bolstered up, as groups, by a larger economy, rather than giving to individuals a wider mobility.

The enthusiasm which is often displayed for the revolutionary economic change—a tractor, a new form of boat, a truck—and the equally ubiquitous pessimistic accounts of how the whole fabric of a traditional culture was torn to shreds by such an economic change,* both miss the possibility of a non-economic pattern transformation of the social attitudes of a people, coupled with their integration as individuals into the pattern of world economy, which is not a pattern of self-sufficient villages. The question should be shifted from "What can the Manus people do, as Manus?" to the question of "What can the Manus people do as *people?*" which leads inevitably to a question—not of better fish-drying methods in Peri and Bunai or a little coco planting on Baluan—of economic plans for New Guinea and of educational plans for Manus.

And here we see the peculiar importance of what the Manus did. They did not start with an economic change; they kept their basic subsistence institutions intact while they changed everything else first. The women who have a vote in the council fish exactly as their forebears did; canoes differ only in having canvas sails. Paliau's plan provided for saving what money they had and getting ready, in social terms, to live in a different world.

This is, in fact, what so many of the cults attempt to do. They offer a model, too static and unrealistically conceived, of a whole way of life in terms of which individual behaviour can be transformed into a new pattern. If they are seen as a measure of extreme aspiration which can be rapidly harnessed in the interest of transforming socio-political movements, each one may become the centre of a new source of social energy. For people to

---

*Berreman, Gerald D., "Effects of Technological Change in an Aleutian Village," *Arctic*, Vol. 7 (1954), pp. 102–07.

want change, to work for change, to cherish change, these community involvements are essential. But to have the kind of economy in which these aspirations can be realized, wide-scale economic planning and modern economic methods are equally essential. Meanwhile, the most astute methods are necessary in those communities where people once hoped to make a rapid change and were disappointed, and became stabilized at a level which is very low economically, much lower than their previous cultural state. They no longer possess the old and are hopeless of ever attaining the new. But among those newly entering the modern world we may, with wise planning, quickly implemented, skip this stage of disappointment and apathy and take advantage of the wholeness of the old pattern, now to be abandoned, by giving people a knowledge of the new, just to be learned.

appendix iv

IMPLICATIONS FOR MENTAL HEALTH AND EDUCATION

I have oriented the body of this study about the capacities of *groups* of individuals, whether they be primitive peoples, peasants, submerged classes, or submerged racial or religious groups, to change their cultural behaviour when certain conditions are met by those who embody the different culture. But inevitably many readers will ask—as have those to whom I have presented this material orally—what implications does this material have for the understanding of *individual* mental health, for educational policy?

I dealt with some aspects of this problem in the 1954 Kurt Lewin Memorial Lecture,* but I should like to recapitulate briefly here. What the members of the village of Peri did as a group is analogous to what is done by an individual who migrates from one culture to another after maturity, and takes on the new culture. This material suggests that we should distinguish sharply between the process of learning a culture and the process of learning *another* culture, just as we should distinguish between the process

*Mead, Margaret, *Cultural Discontinuities and Personality Transformation*, Journal of Social Issues (Kurt Lewin Memorial Award Issue, Supplementary Series No. 8), 1954.

of learning to speak and the process of learning to speak another language. It furthermore suggests that all cultures may be seen as patterns which are of the same order in that they all provide a set of forms which can be learned by all human beings who fall within the normal range of human variation. (This does not preclude the gradual extension of these forms to include more deviant and more handicapped individuals—as the blind who can use Braille in literate cultures, and the deaf who have modern hearing aids.* The inclusion of the necessary redundancy to accommodate human variation is one of the attributes of a "natural language" as contrasted with an artificial or logical language, and one of the attributes of a culture that has survived through time and bears the imprint of the teaching-learning experience of generations. All the great religions and all the great cultures of the world share this characteristic of having accommodated the known variations of human capacities to learn and to function culturally. On the other hand, new ideologies, cults, and sects may be seen as selections from a much narrower range of human experience, and traditionally perpetuate themselves by the rejection, ejection, liquidation of those who do not fit, and the recruitment of individuals whose temperament and character make them suitable converts.

Conversion experiences in the individual may be seen as analogous to the role of revitalization movements—the felicitous phrase which Anthony Wallace has given to the sort of movements which we may classify as nativistic cults, messianic cults, apocalyptic cults, "cargo cults"—in their potentiality for facilitating a complete transformation of the personality. These experiences in themselves may take the form of a momentary complete conviction of sin and/or redemption evoked by evangelistic revivalist religions, of mystical experiences mediated by no formal religious system, of the sort of experience on which Alcoholics Anonymous relies, of the

---

*Tanner, J. M., and Inhelder, B., *The Psychobiological Development of the Child*, Vol. I, "Conferences of the World Health Organization," Tavistock Press, London (1955), and International Universities Press, New York (1955).

brain-washing techniques associated with Communist conversions and Communists' repudiations, or of the experiences evoked by drugs, electric shock, or lobotomy. In all of these cases, the individual's previous total pattern of response is shaken—literally in the trance and trembling of religious sects—and the shift to another pattern is facilitated. Organizations like Alcoholics Anonymous and Christian evangelical movements, certain aspects of Zionism, the extreme proliferation of zealous extremism in disturbed areas like Indo-China—all are examples of movements which work for the sudden, absolute, and complete repudiation of a previous pattern and the equally complete acceptance of a different pattern.

Extreme forms of repudiation of the past—which in the individual means cutting all ties with former associates and friends, selling all your goods or giving them to the poor, entering a monastic order, fleeing to or from a Communist country, etc.— are the analogues of the complete destruction of the past—which in its cult form is expressed by destroying property, giving away money, etc., such as occurs in "cargo cults" and in small apocalyptic cults in our own society. Only by the destruction of every vestige of the past can the new order be ushered in. It seems likely that this will turn out to be one of the universal characteristics of human psychology. The denunciation of one's associates in one's former but now repudiated state—demanded of those who recanted under torture during the Inquisition and demanded today by Communists and ex-Communists—is another illustration of the same psychological principle. Everything that one had, property, friends, position, must go to make way for the new. The more complete the repression of all doubt or criticism, or the more unconscious—in the sense of unexamined and inarticulate—the commitment to the old has been, the more it is necessary to make a holocaust of the old in order to make way for the new. The rituals of many societies provide for such clean sweeps,*

---

*Cf. van Gennep, A., *Rites de Passage*, to be issued in English with an introduction by Solon T. Kimball, University of Chicago Press, Chicago (in press).

at initiation, marriage, divorce, and death, to eradicate the old commitment to a status or to an individual. The destruction of the clothes which belong to the former state and the donning of a complete new costume are widespread accompaniments of the acceptance of a new state.

It is possible to distinguish between these violent complete personality transformations, in individuals and in society, which are the route by which individuals or groups *enter another state*— itself already a full and inclusive pattern—and those instances in which the individual or group attempts to *maintain* the transformation state itself. Sometimes we find social expressions of both of these mechanisms in the same institution; so, for example, an individual may be converted at a Salvation Army meeting, and then stop drinking, marry, settle down as a good husband and father, a supporter of the local church and community, without any need ever again to participate in a revival meeting. Other individuals may experience the conversion itself as the state which they wish to live in forever and become continuous enthusiastic proselytes, giving up all worldly goods and devoting their entire energy to repetition of the original ecstatic experience. So alcoholics may give up drinking and simply become responsible members of their communities, or they may have to reinforce their original transformation by attending biweekly meetings in which they nominally relive the conversion experience in the form of converting others.

We are also familiar with the phase in any sort of psychotherapy in which the patient wishes to become a therapist, however inappropriate this aspiration may be. Drugs, such as mescal or LSD, may be taken as aids to therapy, which make it possible for the patient to reorganize himself and accept an existing new pattern, or they may, as among American Indians, become a way of life in themselves. This distinction was acutely recognized by our American Indian informant, who remarked: "We who cannot read must take peyote, but the next generation will not need peyote." Here again we find the distinction between those individuals who gave up Communism without any fanfare of public confession and denunciation, and those to whom these exercises, and the eliciting of such exercises from others, become a way of life. The rituals of self-criticism within the party, the attempt to maintain each member at a high pitch of dedication, are aspects of Communism which make it still more of a cult than a complete political system.

In a slowly changing society, it is possible to rely on various mechanisms of personality transformation to cure the sick, stabilize the emotionally disturbed, reorient the ethically deviant, discipline the morally intransigent, etc. The transformation experience will facilitate the change toward a stable, existent pattern, and only a certain number of individuals will be so caught by the experience itself that they will wish to give their lives to it, becoming the saints and physicians, prophets, teachers, and evangelists. But in a rapidly changing society, the patterns to which the transformation experience can admit one are much less reliable, and the possibility of building little closed social systems to perpetuate the transformation experience institutionally would seem to be greater. The experience of leaving Communism for Catholicism belongs to the first type in which one total commitment is substituted for another, made the easier because Catholicism, as contrasted with Communism, has room for every human type. But those individuals who leave Catholicism, or Orthodox Judaism, or Fundamentalist Protestantism, or the secular commitments of Zionism, Marxism, Communism, Nazism—in those forms which commanded real dedication—and leave them in moments of sudden revulsion against the old, unaccompanied by any sense of direction toward the new, are exceedingly vulnerable, as are the individuals who enter psychotherapy out of dissatisfaction with their relationships to family, spouse, profession, etc. However much psychotherapists may affirm their intention to return their patients to the bosom of their families, there is a great tendency to connive at the simpler devices of encouraging the patient to leave home, get a divorce, and change his job or profession. The new integration seems easier to maintain when there is destruction of the old and a complete change of setting.

There is one big distinction between the sort of change which modern psychotherapy promotes and such changes as conversion to a great religion or even acceptance of the conventions of the responsibilities of one's own society, as when an alcoholic begins to get to work on time and to support his family. In the cases of religious or social conversion, the pattern is already there, built through generations of human experience, ready with its feasts and fasts and ritual licence to receive and provide for the special gifts and weaknesses of the new member. Where the individual goes through a transformation, not overtly oriented toward such an organized system—as a result of shock therapy, drugs, lobotomy, psychotherapy in general—and finds no pattern within which to fit lovingly, but is left instead to work out an entirely new way of life, the burden on the individual may be almost intolerable.

It is here that the support of the group—the other alcoholics, the other reformed convicts, the other graduated patients—is sought, and that the likeness to a cult is often greater than the likeness to a whole pattern.

The Manus instance provides a useful commentary on such situations. The Mission failed to utilize the device of total destruction of the old and failed to offer the people complete participation in the new, but left them stranded between old and new patterns, similar to the position of so many urban secularized people of the modern world. Paliau offered a new pattern which would revise and integrate what they had learned from Christianity with their wartime experience of a complete new pattern, but it was The Noise, the extreme convulsion in which individuals were swept beyond themselves and the old was thrown into the sea, which made it possible for the whole group to accept a new pattern. The steps by which their old way of life was to be replaced by the new remained laborious, unexplored, and the temptation to preserve the transformation experience by rigid, repetitive ritual was very strong. Paliau demonstrated his genius by working steadily to minimize the cult aspects and to stress the importance of moving toward a new, changing, flexible emerging pattern.

Against this background we may ask how we can educate children to deal with problems of change of pattern as they grow up in a world in which different solutions and violent personality transformations, as a prelude to new commitments, may become more rather than less frequent. If the future is to remain open and free, we have to rear individuals who can tolerate the unknown, who will not need the support of completely worked-out systems, whether they be traditional ones from the past or blueprints of the future.

Our existing cultural materials suggest that vulnerability to conversion may result from a complete, uncritical commitment to one pattern, which therefore must be destroyed if any change is to occur, and that this is particularly so if the commitment has involved the suppression or repression of all ambivalence in people who have felt their way of life was absolute and best. If they are to change, they must change entirely. Also the period of confusion during the change of patterns may make individuals vulnerable to transformation experiences. On the other hand, a recognition that the pattern of life of one's society, class, region, religious and political group is one among many great human patterns is in itself a protection against the type of fanatical, drastic conversion which treats the new as as much of an absolute as the old. It is hard to treat all great human experiments with respect and affection unless one is firmly rooted in one which is known to be one among many. It was the Manus understanding that cultures did differ and could change that made it possible for them to redesign their own culture. Their articulate ambivalence toward the old culture provided the leverage for change which could be articulate and conscious, just as the experience of growing up in a democratic society, where opposing viewpoints may always be expressed, equips the individual to treat his own society with voiced rather than repressed ambivalence. An expressed continuing mild dissatisfaction with one's culture or the functioning of a religious or political movement is essential if there is to be continuous and orderly adjustment and innovation in a changing society.

Without repudiating the techniques of the therapist and the evangelist, we may well work toward a type of education which makes changing patterns less traumatic, and produces individuals who, because they know that each great culture and each great religious or political system is one among many great human experiments, are able to espouse, to love, to cherish one of these solutions, without the need for fanaticism and without vulnerability to sudden conversion. Conversion and therapeutic methods would then become what wise religions and wise systems of medicine have traditionally wanted them to be—ways of treating those unfortunates for whom the ordinary processes of education, socialization, and enculturation have been inadequate.

references

SELECTED READING LIST

In publishing this bibliography it seems worthwhile to comment on the present very confused state of mind which exists about bibliographies, and the space and expense that should be devoted to them. A reviewer in a professional journal commented on my *Male and Female* by saying, "Unlike her earlier works, this is well documented," utterly ignoring the fact that "the documentation" was to those "earlier works," which, as they were original reports, obviously could not be so documented.

This bibliography is designed to answer the questions about bibliography which may be asked by various types of readers. I am assuming that the reader who wishes to follow up obscure German references on the Admiralties will have access to the sort of library where he can get a copy of Nevermann or Taylor—the two publications with the fullest bibliographies—and I do not propose to publish here a long bibliography on the Admiralties, on culture and personality, or on cultural change, but merely to refer the reader to the appropriate sources. Full references to specific works quoted in the text are given in the Notes to Chapters (pp. 441–51).

Publications on the 1928–29 Expedition

Fortune, Reo F., "A Note on Some Forms of Kinship Structure," *Oceania*, Vol. 4 (1933), pp. 1–9.

————, "Manus Religion," *Oceania*, Vol. 2 (1931), pp. 74–108.

————, *Manus Religion*, American Philosophical Society, Philadelphia (1935).

Mead, Margaret, *Growing Up in New Guinea*, William Morrow, New York (1930) (reprinted in *From the South Seas*, William Morrow, New York, 1939; Mentor edition, The New American Library, New York, 1953; English editions: George Routledge, London, 1931; Penguin Books, London, 1942 and 1954).

————, "Melanesian Middleman," *Natural History*, Vol. 30 (1930), pp. 115–30.

Mead, Margaret, "Living with the Natives of Melanesia," *Natural History*, Vol. 31 (1931), pp. 62–74.

————, "An Investigation of the Thought of Primitive Children with Special Reference to Animism," *Journal of the Royal Anthropological Institute*, Vol. 62 (1932), pp. 173-90.

————, *Kinship in the Admiralty Islands*, American Museum of Natural History *Anthropological Papers*, Vol. 34, Pt. II (1934).

————, "The Manus of the Admiralty Islands," in *Co-operation and Competition among Primitive Peoples*, Margaret Mead (ed.), McGraw-Hill, New York (1937), Chap. VII, pp. 210–39.

————, *Male and Female*, William Morrow, New York (1949) (Mentor edition, The New American Library, New York, 1955; English edition: Victor Goliancz, London, 1950; German edition: *Mann und Weib*, Diana Verlag, Zurich and Stuttgart, 1955; Netherlands edition: *Man en Vrouw*, Erven J. Biljleveld, Utrecht, 1953).

Spitz, René, "Frühkindliches Erleben und der Erwachsenenkultur bei dem Primitiven; Bemerkungen zu Margaret Mead *Growing Up in New Guinea,"Imago*, Vol. 21 (1935), pp. 367–87.

Publications on the 1953–54 Expedition

Foerstel, Lenora, "Cultural Influences on Perception: A Comparative Study of the Development of Manus and Usiai Children, Using Gesell Diagnostic Techniques and Projective Materials," M.A. Dissertation, Temple University, in press.

Lowenfeld, Margaret, *The Lowenfeld Mosaic Test*, Newman Neame, London (1954) (which includes a brief report on one Manus Mosaic protocol with the methodological additions in test administration made by Lenora Schwartz).

Mead, Margaret, "Manus Revisited," *Annual Report of the Papua and New Guinea Scientific Society* (1953), in press.

————, "Manus Restudied: An Interim Report," *Transactions of the New York Academy of Sciences*, Ser. 2, No. 8 (1954), pp. 426–32.

————, *Cultural Discontinuities and Personality Transformation*, Journal of Social Issues (Kurt Lewin Memorial Award Issue, Supplementary Series No. 8), 1954.

————, "Applied Anthropology, 1955," in *Some Uses of Anthropology*, J. Casagrande and T. Gladwin (eds.), Washington Society of Anthropology Symposium (in press).

————, "Twenty-five Year Reunion at Manus," *Natural History*, Vol. 63 (Feb., 1954), pp. 66–68.

Schwartz, Theodore, "The Paliau Movement in the Admiralty Islands, 1946–54," *Anthropological Papers of The American Museum of Natural History*, 49, Pt. 2 (1962), pp. 207–422.

General Works on the Pacific Islands

Elkin, A. P., *Social Anthropology in Melanesia—A Review of Research*, Oxford University Press, London (1953) (See especially bibliography).

Hall, Robert, Jr., *Hands Off Pidgin English!* Pacific Publication Pty. Ltd., Sydney (1955).

Hogbin, Ian, *Experiments in Civilization: The Effects of European Culture on a Native Community of the Solomon Islands*, Routledge, London (1939).

————, *Transformation Scene, The Changing Culture of a New Guinea Village*, Routledge, London (1951).

Keesing, Felix, *The South Seas in the Modern World*, John Day, New York (1941).

Mair, Lucy P., *Australia and New Guinea*, Christophers, London (1948).

Oliver, Douglas, *The Pacific Islands*, Harvard University Press, Cambridge (1951).

Reed, S. W., *The Making of Modern New Guinea*, American Philosophical Society, Philadelphia (1943).

Stanner, W. E. H., *The South Seas in Transition*, Austral-Asian Publishing Co., Sydney (1953).

Taylor, C. R. H., *A Pacific Bibliography*, The Polynesian Society, Inc., Wellington, New Zealand (1951) (includes references on Melanesia and New Guinea).

On the Admiralties

Meier, P. Josef, "Mythen und Sagen der Admiralitätinsulaner," *Anthropos*, Vol. 2 (1907), pp. 646–67, 933–41; Vol. 3 (1908), pp. 193–206, 651-71; Vol. 4 (1909), pp. 354–74.

Nevermann, Hans, *Admiralitäts—Inseln*, Vol. III of "Ergebnisse der Südsee—Expedition 1908–1910," G. Thilenius (ed.), Friederichsen & Co., Hamburg (1934).

Parkinson, Richard H., *Dreissig Jahre in der Südsee*, Strecker und Schröder, Stuttgart (1907).

On Culture Change

Keesing, Felix, *Culture Change, An Analysis and Bibliography of Anthropological Sources to 1952*, Stanford University Press, Stanford (1955).

On Culture and Personality

Honigmann, John J., *Culture and Personality*, Harper and Bros., New York (1954).

Mead, Margaret, and Métraux, Rhoda (eds.), *The Study of Culture at a Distance*, University of Chicago Press, Chicago (1953).

Mead, Margaret, "Research on Primitive Children," in *Manual of Child Psychology*, Leonard Carmichael (ed.), John Wiley & Sons, New York (1954), pp. 735–80.

On Revitalization Movements

"Bibliography on the cargo cults." Issued by the South Pacific Commission (mimeographed).

Knox, Ronald, *Enthusiasm: A Chapter in the History of Religion with Special Reference to the 17th and 18th Century*, Oxford University Press, New York (1950).

Wallace, Anthony, "Revitalization Movements," *American Anthropologist*, 58, No. 2 (1956), pp. 264–81.

Wallis, Wilson D., *Messiahs: Their Role in Civilization*, American Council on Public Affairs, Washington, D. C. (1943).

On Field Methods

Gottschalk, Louis, Kluckhohn, Clyde, and Angell, Robert, *The Use of Personal Documents in History, Anthropology and Sociol-*

ogy, Social Science Research Council Bulletin, No. 53, New York (1945).

Kroeber, A. L. (ed.), *Anthropology Today*, University of Chicago Press, Chicago (1953).

Mead, Margaret, *The Mountain Arapesh*, American Museum of Natural History, *Anthropological Papers*, New York: Methodological Prefaces to: II. "Supernaturalism," Vol. 37 (1940), Pt. 3, pp. 317–451; III. "Socio-Economic Life"; IV. "Diary of Events in Alitoa," Vol. 40 (1947); Pt. 3, pp. 163–419; V. "The Record of Unabelin, with Rorschach Analyses," Vol. 41 (1949); Pt. 3, pp. 289–390.

Tax, Sol, *et al.* (eds.), *An Appraisal of Anthropology Today*, University of Chicago Press, Chicago (1953).

*index**

* References to quoted text from *Growing Up in New Guinea* appear in italics. Proper names have been indexed under second names, and, in cases where the Christian name is used frequently or exclusively, under their Christian names also.

Child development, ref. to studies in, 103
Child nurture: bathing, 242–43; carrying, 355, 366; motor freedom, 335, 336; new baby, *114–15,* 333–34; nursing, 333–34, 337; parents, 337–42 *passim;* sphincter control, 245, 336, 337; talking, 124–25; tidying-up, 349; walking, *118,* 335, 336
Children in adult life, 257–58, 259, 261, 352; following behaviour, 243; "Knee," 334, 334 n.; in 1928, *106–10, 131, 143–44, 148, 149,* 349–50; reversal in treatment of, 256; shyness, 369; temper tantrums, 369; and work, 281, 350–51
China, 153, 359–60, 426
Chinese, 177; contrast to Americans, 154; troops, 21
Chinnery, E. P. W., 97, 218
Cholai (father of Peranis), 230
Cholai, Peranis, 37, 40, 183, 187, 224, 233, 271, 358, 361, 363–67, 368, 381, 402, 407, 408, 410–14 *passim,* 416, 460–70, (Plate XVI, Figs. 3–6)
Christian Science, *147*
Christianity, 380, 388, 434, 514; and marriage, 214, 387; movements in, 359, 511; New Way version of, 315, 328; and old religion, 171–72; preparation for, 229; teachings of, 391. *See also* Mission.
Christmas, 382, 383
Clan, 26, 57–58, 59, 68–69, *136,* 217, 271, 274, 278, 373, 448 n. 3
Cloth, bark, 53

Clothing, bridal, 45; children, 250; girls, 45, 250; men, 26, 45, 68, 250, 375; women, 26, 250. *See also Laplap.*
Coast Watchers, 161, 175; native loyalty to, 167
Coconut oil, 24, 48, 97
Colonial administration. *See* Australian administration; German administration; Japanese.
"Committee" (local usage for committeemen), 20, 20 n., 188, 268
Commonwealth, ref. to, 401
Commonwealth government, 297
Communal, methods, 232; responsibility, 45
Communism, 7, 424, 438, 511–12 *passim;* China, 359; propaganda, 425
Community, 292, 293–94, 306–7, 373. *See also* Village.
Competition, 348
Confession, 90, 311, 312–13, 326–27
Cook, Captain, ref. to, 210
Cooking, 46–47, *130*
Co-operation, *108, 143,* 267, 279–80
Co-operation and Competition, 179
Copra, 80, 96, 172, 173, 242, 285
Corvée, 293
"Council" (local usage for councilor), 20, 20 n., 188, 268; councilor-elect, 42, 340
Council, house, 199; and partition of Paliau's group, 187, 397–99; as self-government, 186, 198; South Coast, 233, 268, 339–40, 418; and taxation, 233; "wait-council," 202. *See also* Village.

Perennial

Books by Margaret Mead:

COMING OF AGE IN SAMOA

A Psychological Study of Primitive Youth for Western Civilisation
ISBN 0-688-05033-6 (A Perennial Classic)

Focusing on adolescent girls, this book, now a scientific classic, presented to the public for the first time the idea that the individual experience of developmental stages could be shaped by cultural demands and expectations.

GROWING UP IN NEW GUINEA

A Comparative Study of Primitive Education
ISBN 0-688-17811-1 (A Perennial Classic)

Mead's 1928 analysis of the play and imaginations of the younger children on Manus Island, and how they were shaped by adult society's attitudes toward sex, marriage, the rearing of children, and the supernatural.

SEX AND TEMPERAMENT

In Three Primitive Societies
ISBN 0-06-093495-6 (paperback)

Mead's study of gender in three New Guinea tribes advances the theory that many masculine and feminine characteristics are not based on fundamental sex differences, but reflect the cultural conditioning of different societies.

MALE & FEMALE

ISBN 0-06-093496-4 (paperback)

First published in 1949, Mead draws on her anthropological research of seven Pacific island cultures to explore the inherent meaning of "maleness" and "femaleness," then delves into the complex sexual patterns that drive American society.

LETTERS FROM THE FIELD, 1925-1975

ISBN 0-06-095804-9 (paperback)

This volume stands not only as a classic on the conducting of field work, but also a window on the life and thoughts of one of the most unique women and scholars of our century, from her beginning explorations through the peak of her career.

NEW LIVES FOR OLD

Cultural Transformation—Manus, 1928-1953
ISBN 0-06-095806-5 (paperback)

A record of a people's self-transformation, offering key insights about a society's capacity for change, with the daring thesis that many of the ills of the present world come not from too much change, but from change that is too little and too late.

Available wherever books are sold, or call 1-800-331-3761 to order.